The Mobile Workshop

The Mobile Workshop

The Tsetse Fly and African Knowledge Production

Clapperton Chakanetsa Mavhunga

The MIT Press
Cambridge, Massachusetts
London, England

This book was set in ITC Stone Sans Std and ITC Stone Serif Std by Toppan Best-set Premedia Limited. Printed and bound in the United States of America.

Library of Congress Cataloging-in-Publication Data is available.

ISBN: 978-0-262-53502-1

10 9 8 7 6 5 4 3 2 1

For Mildred Maidei

Contents

Preface: Before We Begin …

The Mobile Workshop: The Tsetse Fly and African Knowledge Production is a project about African understandings of their surroundings. The archive mobilized is composed of indigenous idioms that are often studied more for linguistic and literary value, rather than, as I see them, as philosophical modes representing knowledge of the majority cultural group in Zimbabwe today, who share the same language, commonly called "Shona." This is the language I think with.

Zimbabwe (from whence the country's name came) means "a big house of stone," and refers to the biggest and most extensive stone-built structures observers appropriately called "Great Zimbabwe." "Great" here means it is the biggest and most majestic among many other *dzimbahwe*. While *zimbabwe* (*zimbahwe*) refers to a large stone house, an ordinary house of stone is called *imba yemabwe*/*imba yemahwe* in the family of "Shona" languages that scholars have grouped together as *ChiKaNdaMaZeKo*. This acronym is drawn from *Chi-* (prefix, meaning "the language of") *-Karanga*, *-Ndau*, *-Manyika*, *-Zezuru*, and *-Korekore*.

Growing up, I used to hear elders greeting each other with the morning salutation *"E, mamukaseiko vedzimbahwe?"* (E, how did you sleep *vedzimbahwe*?). When I asked, my father told me that *vedzimbahwe* is a more culturally appropriate term to describe us instead of "Shona" or Zezuru. When being more specific, he said I could refer to us as *Mwendamberi* ("those who always go forward, backward never"), of the *Nondo* totem, descendents of Chirau, who hail from *Chirorodziva cheChinhoyi* (the shimmering blue pools of Chinhoyi), who settled in the lands of *vaBudya* (the Budya people) of Chihota. Then again, we were just a part of a bigger confederacy of same-language speakers, first under the Munhumutapa and Rozvi kings, to whom we paid tribute. Besides the language, what identified us as one people was the stone architecture.

The term "Zimbabwe culture" as deployed recently by Innocent Pikirayi (2001) is what I call *chidzimbahwe*, but with a few caveats. Pikirayi's focus was on the rise and fall of indigenous states distinguished from archaeological findings according to similarities in stone architectural design, ranging from mighty capitals of kings to humble homesteads. His book borrows its title from a term popularized by German writers of the late-nineteenth and early twentieth centuries. To these writers, "the Zimbabwe culture" referred to culture displayed in stone buildings and ceramic arts (Hall 1905).

Vedzimbahwe (singular *mudzimbahwe*) simply means dwellers (or those) of the houses of stone. A builder of houses of stone, or any stone structures, is called *ndongamahwe* (arranger of stones); and a *sarungano* (by virtue of building and telling a story) also becomes *ndongamahwe*. Building from the word *dzimbabwe* or *vedzimbahwe*, I invoke *chidzimbahwe* (the language of *vedzimbahwe*), also called *chivanhu* (ways of the people), or *chinyakare* (ways of the past), as a better term in place of the wieldy, meaningless lexicon of *ChiKaNdaMaZeKo*. This is a deliberate move to reintellectualize African terminology—to move away from 'Shona' (which means nothing to those of us arbitrarily called Shona by people other than us) and towards restoring the larger economy of knowledge and practices resident in our own terminologies, which during the European occupation reduced to a mere aesthetics of language or means of communication.

Dzimbahwe or *dzimbabwe* is many things. It is not just a house of stone, but the headquarters of the king, built in carefully chiseled and laid stones of the *maware* (granite) type. It is also the place where the king's deceased body is hidden *(kuvigwa)*—especially in a mummified state—and stood, sat, or rested in a cave, or buried in the ground. *Dzimbahwe* is also the grave, where the remains of the ancestor are laid to rest (*kuradzikwa*), flesh corrupting, but *mapfupa* (bones) incorruptible, and the spirit having left them to be *mhepo* (the air) or *vadzimu* (ancestral spirits), which, after proper ritual returns to possess a mortal kin (*wehama*), speaks through and turns them into a spirit medium (*svikiro*). Every religion has its own myths and legends that people canonize into truths; to be faithful means not questioning whether such spiritualization makes sense when subjected to physics, biology, or chemistry experiments, but to allow such spirituality to be the driving force of life itself.

I bring a different sensibility to matters of faith: I never question anyone's faith. I would rather deal with someone who has faith in something—in themselves, in a stone, a mushroom, their cat, ghosts, witches, ancestral spirits, Crucifixion and Resurrection, the whole lot. I therefore approach *dzimbahwe* not from my perspective, but from the perspectives of the people who made such values and live by them.

Dzimbahwe interests me for an entirely different reason: as a space from which I can initiate a *chimurenga* (rebellion) against always starting from concepts emanating from Europe or North America. Very few scholars from these two regions ever cite Africans in their writing, even when they are writing about Africans. Africans very rarely write about other people; we tend to write about ourselves in the world. I only worry that the way we order our thoughts exhibits a certain inferiority complex; we always want to cite the big names in the North, virtual strangers and at best fleeting visitors to our experience, as if our own ancestors did not think.

Everyday language, as an oral expression of the many transient workspaces that animate life in Africa, constitutes a rich philosophical space from which to initiate a view from the Global South, from Africa, and, here, from Zimbabwe, and to produce a knowledge meaningful and usable to Africa.

Everyday language expresses realities and imaginations at the intersection of African inventions and inbound idioms and thus testifies to the creativities of Africans who strategically deploy them. Language constitutes an archive from which people of Africa can come to knowledge without a sense of intellectual inferiority; the French, British, Germans, Japanese, and South Koreans made their languages the official languages of knowledge production. Language is where the mastery of knowledge is won or lost.

The Mobile Workshop is an experimental space to test new forms of writing the African experience. Specifically, it is an invitation to each of us to go back to our languages, to recover the meanings of things that have come to dominate contemporary life. The book's central analytic is *mobile*—that is, knowledge produced by the mobility of an insect carrying an invisible thing that kills people and their domestic animals. It is a *workshop* in the sense of an experimental mode of writing. At stake is an effort to take vernacular concepts as starting points in writing a narrative of *ruzivo* (what is commonly called *knowledge* or *science*), means and ways to things (*technology*), and the creation of things (*innovation*). Can we not find within our own African languages terms that describe our creativity, the means and ways to it, and how we know and what we know, without enslaving ourselves to other cultures' meanings? Can we take seriously the intellectual forces animating Africa's encounters with things inbound into Africa on one hand, and those animating Africa's intellectual endeavor in the world on the other?

What is required urgently is an intellectual program unlimited by European Enlightenment traditions of analysis and expression, one that takes seriously the concepts people of Africa invent and deploy to talk about their own experiences—in short, treating the realm of living and life as an experiential location from which to approach questions about and meanings of the scientific, technological, and innovative in our own time.

The Mobile Workshop is an attempt to encourage fellow Africa scholars to write a narrative the keywords of which are readily visible to and derived from African tongues, first and foremost, so that the rest of the world is able to learn about us through our own keywords. This is contrary to the prevailing norm, wherein the keywords are foreign (colonialism, capitalism, democracy, technology, science, innovation, entrepreneurship, etc.) and we are forced to understand Africa through them. That we have allowed ourselves to continue using such concepts uncritically constitutes the greatest intellectual laziness (or mischief) of our time. Africans meeting in Africa or overseas, cannot speak and socialize in Amharic, Wolof, kiSwahili, or *chidzimbahwe*; those from Senegal and Zimbabwe, colonized by the French and British, respectively, cannot even talk on the taxi, even while they share so much of everyday life in common. One speaks French, the other English. Why are we doing this to ourselves?[1]

I am yet to see any word in *chidzimbahwe* which translates to *colony*, *colonization*, *colonialism*, *colonial*, or *coloniality*, as experienced. It was the European imperialists—in this case, the British—who used the term *colony*, created a *colonial office*, and became the *colonial power*. Later, especially after 1945, African nationalists strategically deployed "colonialism" to analytically demarcate a system of European imperialistic domination that had to be confronted and dislodged. "Colonialism" was a category describing how the "colonial power" exercises agency over those lands and people it sees as having been "colonized." It is not a category emanating from those on the receiving end of aggression. The idea of "postcolonial" or "decolonial" is first and foremost an acknowledgment of the legitimacy and validity of a top-down category affirming the claims of Europe. Hence, this text moves to a noncolonial language that removes any reference to the *colon* root. Africans cannot continue giving life to a dragon that they have already slayed: Zimbabweans, in particular, defeated the European oppressor through the barrel of the gun. That physical and political independence frees us to focus on slaying even more dragons in our time. For that we need keywords derived from none but our own tongues.

Even the mere mention of the root word *colon*, including what Mignolo calls the *decolonial option*, still retains the Western as the central referent. Do not even *decolonialize* or *decolonize*. I am saying *re-Africanize*. Write other stories—our own stories, our own vocabularies: *chimurenga* vocabularies made possible through epistemologically chasing away the colonizer far into the sea. Even when we talk about the period in which the Europeans were here, let us re-Africanize as much as is possible to re-Africanize. The decolonial option is still a colonial option. It is a distant cousin of what we are trying to do.

In that language, those who appear in the *colon* narrative as "the colonized" have their own categories for who they are. In my *chidzimbahwe* language, they refer to themselves simply as *vanhu* (people; singular *munhu*). What distinguishes them from *mhuka* (animals) or *zvipukanana* (small animals, i.e., insects) is *hunhu* (behavior, being), which often entails respect for *tsika* or *chivanhu* (custom, culture, the last translating to personhood). When they first arrived, "white people" were called *vasinamabvi*—those without knees—because they wore trousers and their knees were invisible.

There is no word for "Europeans" in *chidzimbahwe*. People cared nothing about where these *vaeni* (aliens, strangers, foreigners, or unknowns) or *vasinamabvi* were coming from. What they cared about was *what they looked and behaved like, what they did, and how it felt*. Thus, they looked white (*vachena* or *varungu*; singular *muchena* or *murungu*), whereas *vanhu* (people, those who are familiar, kin, or known) looked black (*vatema*). Before *vachena* came, *vedzimbahwe* called themselves and all peaceful people (neighbors) simply *vanhu* (people); those who raided, who took away the women, children, and cattle, were called *madzviti* (pillagers), in specific reference to the Ndebele (west) and Gaza (southeast). The new strangers were very different; they did not look like *vanhu*. They were *vachena* in a very alien sense.

Vachena are not remembered as "colonizers," but for what they did to *vatema*. First, they are *vapambi*, those who abducted (our independence), forcibly and murderously seized, plundered, robbed, ravaged, savaged, pillaged, raided, ransacked, raped, looted, or sacked. The prefix *vapambi-* is applied to two things critical to the survival and prosperity of *vatema* that *vachena* pillaged. The first is *ivhu* (land, the soil)—hence *vapambevhu* (those who plundered and seized the land and turned *vanhu* into *nhapwa* [slaves]). *Vapambi* seized not only the agricultural soil, but the wealth of *vatema*: *zvicherwa* (minerals), *mombe* (cattle), and Africans' labor value. Hence the second reference, *vapambepfumi* (robbers of our wealth) also includes how *vachena* engaged in *chibharo* (forced labor) to turn the land into mineral and agricultural wealth. By pulling the labor and intellectual resources critical to *vatema*'s existence in the countryside to the mine, farm, and town, *vachena* became *vasvetasimba* (drainers of energy; singular *musvetasimba*). The white man in general was called *bhunu* (*Boer*, an Afrikaans word for farmer), the word denoting the cruelty and slave-driving attitude rather than the farming prowess. *Bhunu* was the *musvetasimba*-in-chief.

Vachena were remembered principally for what they did to *vatema* and how it felt to be treated that way. From this experiential realm, *vachena* became *vadzvanyiriri* (oppressors), the experiential meaning of which is best captured in the word *downpressor* (the one who presses someone down, especially with a boot to the face on hard, rocky ground).[2] The downpressor also deliberately reorganized human relations so that skin color became a marker of intellect, civilization, access to resources, and human-to-human relations. *Hudzvanyiriri* became *rusaruraganda* (racism; literally, "discrimination on the grounds of skin color"). *Vachena* became *vadzvanyiriri*. All of the looting, downpression, and racism led *vatema* to deploy an even harsher term; they saw their conditions as those of *nhapwa* (slaves) in their own country.

Hence we have the imperative of *kuzvisunungura* (self-liberation) from *hunhapwa* (slavery). *Kuzvisunungura* (untying oneself) refers to *hunhapwa* as bondage. *Kuzvisunungura* is also what this book is about—and *rusunungiko* is the state of feeling and being free. Whoever coined the phrase *lost in translation* certainly knew the limits of completely translating the languages of other people into English, French, or German. The erasure of those attributes of *vatema*'s naming (usually descriptive of a thing's properties and actions) as *vachena*'s names enveloped them constitutes a serious problem that we as Africa scholars have failed to correct.

In this book, readers are forewarned that they will see a lot of *chidzimbahwe* names; these are not just there for show or to annoy the reader, but are used because they restore the intellectual weight removed from them by translation into English. To have written this book otherwise was, quite simply, impossible.[3] I think, analyze, and write in *chidzimbahwe*; the hope is that others fluent in *isindebele*, Yoruba, *kiswahili*, and thousands other languages, and all other marginalized knowledges throughout the world, can do the same, so that one day I can also understand the world from their own categories. That way, we enrich the global store of knowledge through diversity rather than with a monoculture of concepts.

There will be moments when I get deeper into describing what the European was doing after turning *vatema* into *nhapwa*, and English became the official language of knowledge. When *vachena*'s keywords are used, they are actor categories (terms *vachena* themselves used); as the mobile workshop gathers momentum, even that imperfect language will eventually be Africanized. For now, it is sufficient to retain only a core set of vocabularies for the main actors—*mhesvi, hutachiwana, mhuka, vachena, vatema,* and so on.

Acknowledgments

My PhD adviser, Gabrielle Hecht, and my committee members, Mamadou Diouf, Arun Agrawal, Nancy Rose Hunt, and Rebecca Hardin, told me countless times that my dissertation, "The Mobile Workshop: Mobility, Technology, and Human-Animal Interaction in Gonarezhou (National Park), 1850–Present," was a career's worth of publications. I never agreed with them; I was sure they were limiting my ambitions.

My career is still ahead of me, but two books later, I realize they were probably right. And so having thanked them in *Transient Workspaces: Technologies of Everyday Innovation in Zimbabwe,* I only repeat: I am deeply honored to have been their student. Outside the committee, I cannot forget the ever-open doors of advice and support from David William Cohen (the "production of knowledge" influence came from him), Kathleen Canning, Michelle Mitchell, John Carson, Paul Edwards, Joshua Cole, Farina Mir, Jonathan Sheehan, and Marty Pernick, Kevin Gaines, and Sean Jacobs, among others.

To my dear friends in grad school, who stood beside me when times were hard and friends are usually few—Christian Williams, Sarah Hillewaert, Anneeth Kaur Hundle, Esabel Cordova, Ken Garner, Kaluke Mawila, Brian Maguranyanga, Menna Demessie, my big brothers and mentors Moses Ochonu and Vukile Khumalo—know that you were a solid rock of support and it still means a lot to me.

One-fifth of this dissertation benefited from the financial support of the following: Andrew W. Mellon Research Fellowship on Migration and Urbanization (Wits University, South Africa 2003–2006) as well as multiple internal awards from the University of Michigan: Global Ethnic Literatures Seminar (GELS) Fellowship, Department of Comparative Literature (2007–2008); Rackham Dissertation Fellowship, Rackham Graduate School (2007–2008); Rackham Humanities Research Candidacy Fellowship (2006–2007); Rackham International Research Award, International Institute (2006–2007); Melvin and Janey Lack Fellowship, Eisenberg Institute for Historical Studies (2006–2007); South Africa Initiative and African Initiative Travel Grants, Center for Afro-American and African Studies (CAAS) (2004, 2006); Pre-dissertation Research Award, International Institute (2004);

Rackham International Travel Grant (2004); and the Institute for Research in Women and Gender (IRWIG) Travel Grant (2004).

The bulk of the research expanding on the tsetse fly theme was undertaken during my tenure-track assistant professorship at MIT. I was privileged to have received the unfailing support of the then dean of the School of the Humanities Arts, Social Sciences (SHASS), Deborah Fitzgerald, and her staff, especially Magdalena Reib. When Deborah returned to the Program in Science, Technology, and Society, her successor Melissa Nobles carried on this support under my Black Bvekenyas and African Chemistry projects. I was honored to have received the SHASS Research Fund (2013; 2016; 2017) and numerous Dean's Fund for Professional Development grants for this research.

A significant chunk of writing happened while I was a fellow and senior fellow in two of Germany's intellectual paradises. I was privileged spend six months of fall 2011 at the Rachel Carson Center for Environment and Society in Munich, where I completed the polished draft of a book that would have combined *Transient Workspaces* and *The Mobile Workshop*. Working all night in the Rachel Carson Center, drowned by the music from the nearby Immobilien—the decibel increasing markedly when FC Bayern had won!—I emerged in the morning to go home to sleep. For the honor of being in this beautiful space in the heart of Munich, I owe a huge debt of gratitude to the center's two directors, Helmuth Trischler and Christof Mauch, and my colleagues, among them Cotton Seiler, Gijs Mom, Riin Magnus, Maohong Bao, and many others.

I returned to MIT in December that year, polished manuscript in hand, for submission for my promotion case. But there was one problem: I did not like what I had written. The guns I was writing about were still *vachena*'s guns, every trigger-pull, every shot—the language of *Glossina morsitans*, *Glossina palpalis*, and *Glossina pallidip*es and all these funny Latin and English names meant nothing to the ordinary people from whose world I write. What would a book that saw the gun and the tsetse from their perspective, expressed through their keywords, look like?

At that moment, I had a choice between submitting everything I had to get promoted to associate professor without tenure, the first of two milestones to tenure at MIT and starting afresh. I began unbundling the gun and the tsetse fly into two books. It was painful to commit treason to my own handiwork. *Transient Workspaces* was the foundation of the creative defiance that underlines my writing. I wanted technology and innovation to speak in my father and mother's tongue, so that *vatema* like me could be legible in it.

Meanwhile, I was working on *The Mobile Workshop*. I needed a space where I could sit down and write peacefully, somewhere quiet but scenic. If I went to Makuleke or my own village—called Mavhunga Village—I would get distracted by my passion for farming and taking photos and videos of everyday life. I was privileged to be invited as a senior fellow to the Internationales Kolleg für Kulturtechnikforschung und Medienphilosophie, shortened to IKKM at the Bauhaus, in beautiful, historic Weimar. I wish to extend my sincere thanks to its directors, Lorenz Engell and Bernhard Siegert. Thank you, Christina Terberl, for helping me settle down, open a bank account, and function in a town where everything is done in German. And to Katarzyna Wloszczynska, for your energy and hard work that inspired those of us around you to work even harder. And to all the fellows of the IKKM, whose arrival and welcoming ceremony coincided with my departure, I hope you enjoyed Weimar and IKKM the way I did, because how could you possibly be in the presence of Lorenz and Bernhard and have a bad day?

In my life I have been privileged to work with Solomon Bvekenya, son of the notorious white poacher, Cecil Bvekenya Barnard. Having now finished writing about the tsetse fly, I can now honor my promise to the black families the poacher left in southeastern Zimbabwe (then Rhodesia). In the northwest, I similarly thank Asa Mudzimu, who tirelessly conducted the interviews in Nembudziya, west of a place—Chiroti—where I lived intermittently from 1998 to early 1999.

Since 2001 when I began this research, and now as I publish the book, I have gone from being a father of one to a father of three. I started the project when my daughter, Cleopatra Nyasha, was a year old. When this book is published, she'll have less than two months before entering university. The most enduring memory of this project are the names that always draw a laugh from my children's mother, Mildred. *Glossina morsitans. Glossina palpalis. Glossina pallidipes. Glossina brevipalpis. Glossina* ...

This book is in honor of that laughter and the person who found the names so funny, so ridiculous as to compel me to see their role as technologies of erasure, completely covering the tracks of the knowledge so that its source could never be known, never be traced.

Introducing *Mhesvi* and *Ruzivo Rwemhesvi*

How did it happen that an insect that is not itself harmful, bar its painful bite, forced *vachena* to defer to *vatema*'s knowledge and skills to control it? How do we tell this story of knowledge mobility from indigenous knowledge to the very core of *vachena*'s project to research and control a pestiferous mobile insect deadly to people and their livestock—an insect that virtually turned the space it occupied as it moved into a transient workspace?

Mhesvi is the name that *vedzimbahwe*—the people associated with the *zimbabwe* (stone building) culture from which the name Zimbabwe came—gave to the insect that *setswana* speakers (*vedzimbahwe*'s neighbors to the west) called *tsetse*. After encountering the insect for the first time and being told its name, white travelers then publicized *tsetse fly* as the official name of the insect. Of course, *vedzimbahwe* are only the majority language group in Zimbabwe, not the only people who know *mhesvi*. *Vedzimbahwe*'s neighbors to the southeast—the Hlengwe, who spoke *xitsonga*—called it *ndedzi*. Their *xitshangana*-speaking conquerors, the Gaza, called it *inthesi*, clearly a corruption of *ndedzi*. The Ndebele—speakers of *isindebele* who subjugated the western *vedzimbahwe* and raided and exacted tribute from those in the east—called the insect *mpukane*.

These terms will be deployed as actor categories (in the way the people themselves used them) whenever these neighbors of *vedzimbahwe* are discussed in the book. However, the central focus remains *mhesvi* as known and experienced among *vedzimbahwe*; *mhesvi* is the preferred analytical name of choice throughout the book, alongside *chipukanana* (insect), *mhuka* (forest animal), *zvipfuyo* (livestock), and *vanhu* (people). Other principal actors will be similarly referred to in *chidzimbahwe* (the language of *vedzimbahwe*) because that is where I am standing philosophically as an analyst and storyteller.

The name *mhesvi* comes from a longer one, *mhesvamukono*, "the one that drives the *mukono* (bull) crazy." It is derived from two meanings. The first is from the ideophone (*nyaudzosingwi*) *pesva* (the plunging action of a sharp object like a needle or thorn into flesh), referring to *mhesvi*'s amazing ability to penetrate human and animal flesh with its probing mouth. The second is a reference to the effect of this piercing action, that of *kupesvedzera* or *kupesva* (inciting) the bitten animal or person. The person instinctively attempts to swat the fly. The cow's response inspired the adage *kwadzinorohwa matumbu ndiko kwadzinomhanyira* (where their stomachs are lashed, that is where they run to), referring to the tendency of *mombe* (cattle) to stray and sometimes even gallop ahead of the herd toward the tall sweet grasses that lash their stomachs. There, *mhesvi* waits; once bitten, *mhuka* succumbs to *chirwere che mhesvi* (sickness of *mhesvi*, which translates to *chirwere chemhesvi* or *n'gana*, the Zulu for that which makes cattle "useless" or "powerless") and eventually dies, but not before being incited to erratic, frenzied behavior.

To be clear, *mhesvi* cannot be confused with *uo* or *mbuwe* (also called *mbuwo*), which incoming *vachena* called the blind fly. Unlike *mbuwe*, *mhesvi* is hard to crush. If you swat it, it remains intact; you must squeeze hard to crush it. And it is stealthy; by the time you feel that itchy sensation on your skin, it has already finished feeding on your blood and is flying away. You can't catch it with your bare hands. To kill it, you must be steady when it lands, relax, and then patiently make a move. You must use a knife, called *mhesvi* (from the word *kupesva*, this time meaning sharp and thin); slowly move the blade toward the engorged fly, then slash its legs off.[1]

Two types of *mhesvi* have historically inhabited *dzimbahwe*. The first thrives in the forests (*matondo*). The second lives in savanna grassland. The third is generally found in vegetation along rivers or valleys. I found no distinction among *vedzimbahwe*, Hlengwe, Ndebele, Batswana, and other southern African groups concerning these three types; on the contrary, *vedzimbahwe* and Hlengwe saw one *chipukanana* that dwelled wherever it was conducive. These divisions are therefore analytical, as is the language I use to re-Africanize them.

Mhesvi is internationally known by the vexatious Latin surname that *varungu* gave it in the late-nineteenth century, *Glossina*, from the root *glōssa*, tongue. *Glōssana* (*G.*) thus means "of the tongue" and refers to the action of using the tongue to suck—in this specific case, the action of the bloodsucking fly. *Varungu* identified two principal groups of *mhesvi* in *dzimbahwe*: a savannah grassland variety they called *G. morsitans* and two riverine types, *G. pallidipes* and *G. palpalis*. *Morsitans* simply means "biting"; hence, *Glossina morsitans* is a biting *mhesvi*. *Pallidipes* is Latin for "pale-footed"—hence, a pale-footed *mhesvi*—while *palpalis* refers to having palps or feelers located near the mouth. Every time these terms are used, the reader must treat them as *vachena*'s categories, not words *vatema* used.

Otherwise, I will refer to the grassland *morsitans* by its *chidzimbahwe* derivative *rutondo*—hence, *mhesvirutondo*. *Rutondo* is deep *chidzimbahwe* for *dondo*—a tract of uncultivated and uninhabited land that is either a forest or a wide grassland. I will use the term *mhesvirupani* to reference a *mhesvi* that lives in the valley, inclusive of rivers—that is, *bani* (*rupani* in deep *chidzimbahwe*). *Mhesvirutondo* and *mhesvirupani* thus substitute for the ubiquitous *Glossina morsitans* (in the first instance) and *Glossina pallidipes* and *Glossina palpalis* (in the second), which, as chapter 2 shows, have served to erase the African footprints of such knowledge in *ruzivo rwevachena*. This is one instance in which, as *vanhu vatema*, we need not treat our languages simply as archives or idioms, the applications of which are limited to those things that our ancestors coined them to refer. Instead, we must invent new applications and forms that enable us to specify and clarify what we mean to say.

Mhesvi (shown in figure 0.1) looks innocuous, and it is quite similar in size to *nhunzi* (the housefly). However, that is as far as the similarities go. When it bites you, that is when you know that this fly, which lives in the bush, is not your run-of-the-mill housefly—*if* you ever find out what has bitten you. Suppose you do: By the time you feel the pain, the *chipukanana*

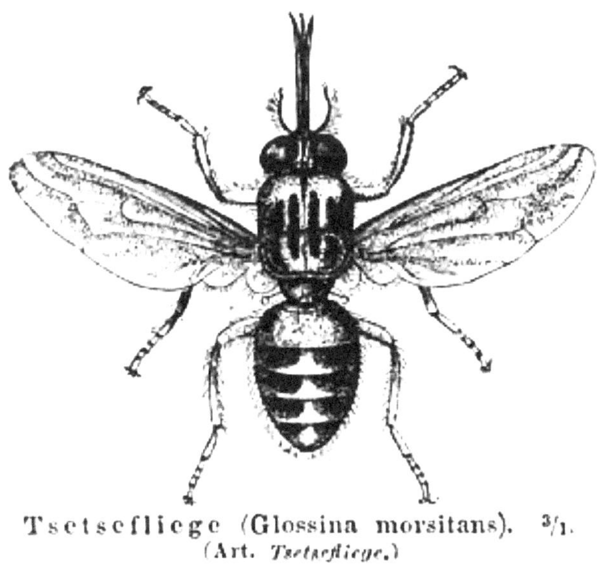

Figure 0.1
Mhesvi the tsetse fly.
Source: Wikicommons.

has finished eating its meal (drawing your blood) and is on its way. When you catch or kill it, look at it closely. It has an unusually long mouth, which explains why it bites so deeply and extracts blood. Like *mbuwe*, its mouth pierces through a shirt into the skin.

Let's assume you are very smart and alert and you have seen it alight on your skin or noticed it after it has already engorged: What's your next step? To swat it? That's another difference compared to *mbuwe*; *mhesvi* is as hard as a rock, and the force you apply to kill *mbuwe* would at most cause *mhesvi* a slight annoyance before it dusts itself off and flies away, as though mocking your stupidity.

Male or female, *mhesvi* is known to digest and assimilate ingested blood within two days. Most of it goes toward nutrition, with the surplus converted into fat that provides a reserve of energy in lean times. During feeding, *mhesvi* ingests the *hutachiwana* (trypanosomes), which attach themselves to the walls of the long mouth (with which it probes and sucks) and begin their developmental journey within *mhesvi*, ready for inoculation into its next bite victim.

Like several other *zvipukanana*, *mhesvi* has an internal protective lining in its middle stomach impenetrable to *hutachiwana*. Thus, it can carry infectious matter without being infected itself. As *mhesvi* pierces into flesh to draw blood, *hutachiwana* leave one carrier and enter others—from *mhesvi* into *nyati* (buffalo) into *vanhu* (people) into *chipfuyo* (domestic animal; plural *zvipfuyo* (Ford 1971, 77–78, 88).

The passenger most critical to this book is *hutachiwana*, the unseen thing that causes illness and/or death in people, animals, or plants. *Hutachiwana* literally means "the we have found it," in homage to the invisibility and elusiveness of the thing that cause death or illness. *Hutachiwana* is thus either contagious matter or poisonous material. On one hand, the name is ironic: a thing that can never be seen, never be found, difficult to confront and annihilate. On the other hand, the name is a declaration of the triumph of discovery: "*Tachiwana!*" (We have found it!). There is also a third dimension; after death, the immediate family of the deceased usually visits a healer to diagnose the cause, and will afterward remark: "*Tachiwana chauraya mufi*"(We have found what killed the deceased).

Two types of *hutachiwana* feature in this book. The first caused *gopé* or *hotsikotsi* (the disease of sleeping), which *vachena* called *sleeping sickness* or *human trypanosomiasis*. The second caused *hutachiwana hwen'gana* or *hutachiwana hwemhesvi* (also called *nagana*) in domestic animals; *vachena* called the disease *animal trypanosomiasis*, which translates to *chirwere chemhesvi* or *n'gana*, the Zulu for that which makes cattle "useless" or "powerless." Both types of *hutachiwana* would become to *vachena* the "trypanosome";

this book focuses exclusively on *hutachiwana hwen'gana*, shortened to just *hutachiwana* throughout, with *hutachiwana hwegopé* used when referring to sleeping sickness.

To put it another way, *hutachiwana* do not in themselves pose a threat to *zvipfuyo* or *vanhu* (people); they only do so when something transmits them. However, *hutachiwana* are engaged in microscopic journeys within *mhesvi* itself. In the most general terms, *hutachiwana* ingested in blood attach themselves to the walls of *mhesvi*'s long mouth and live there until they develop a tail, with which they swim into *mhesvi*'s throat, where they stay until ready to be inoculated into *mhuka* or *vanhu* during feeding. The younger the *mhesvi*, the more easily infected it is, especially when feeding on an infective host for its first meal (Ford 1971, 22–73; Bursell 1959). The site of that itchy *mhesvi* bite develops into a red spot, the periphery of which is pale. *Hutachiwana* stay here briefly before entering the bloodstream; the spot disappears gradually thereafter.

The first signs of infection when a person is bitten by *mhesvi* are the fever beginning, the heartbeat accelerating, and a temporary rash appearing. Next, the fever fluctuates, severe headaches begin, and glands begin swelling, becoming tender, smaller, and fibrous. Then, debility, insomnia, and edema of the eyelids take over, with frothing at the mouth, deep sleep by day, restlessness by night, and, rapidly thereafter, death (Ford 1971). *Mhuka* struck by *n'gana* first exhibit an intermittent fever, then anemia, edema, and lacrimation; the *mhuka*'s strained lymph nodes become enlarged, the cows abort pregnancies and struggle to become pregnant, and generally they lose appetite and rapidly lose weight, chronic diarrhea sets in, the nervous system collapses, and the *mhuka* dies.

By its movement or stillness, *mhuka* becomes (un)attractive and (in)visible as a host to *mhesvi*. Complacency or timidity under attack makes a perfect host—irritability (leg kicking, skin rippling, head shaking) a bad one.[2] *Nyati* the buffalo, with its love for chewing its cud under the cool breezy shade of a tree or in the shallow waters of the river, makes itself available to *mhesvi*. The little creature feeds as the *mhuka* feeds; "the slow, intermittent movements of grazing, browsing or foraging mammals allow large concentrations of 'following' flies to build up around the hosts" (Gatehouse 1972, 83).

Each *mhuka*'s behavior makes it tolerant, vulnerable, and even resistant to *mhesvi*. Regular hosts are larger browsers and grazers that permanently reside in one area all year; once they are eliminated or scattered, *mhesvi* cannot survive permanently on chance encounters with food hosts. Occasional hosts such as *mbeva* (mice), *majerenyenje* (anteaters), *tsoko* (monkeys), *makudo* (baboons; singular *gudo*), *mashuramurove* (storks), *magora* (vultures; singular *gora*), *nguruve dzemusango* (bushpigs), *nzou* (elephants), and *vanhu* are selective and migratory and therefore poor feeds. *Mhesvi* labors to get its meal from these "accessory food animals" in winter; in summer, vegetation overabundance disperses *mhuka* and encumbers *mhesvi*'s hunt for food (Trypanosomiasis Committee of Southern Rhodesia 1946, 13–14). *Mhesvi*'s blood diet and its feeding mobilities are ecologically contingent.

Stops on the *mhuka*'s pathways to/from *hufuro* (pastures, or *mafuro*) and waterholes become points at which *mhesvi* catch rides, feed, and become vehicles for *hutachiwana*. The beaten track through open ground renders *mhuka* easily visible to *mhesvi* taking cover from its own predators (*shiri* [birds]). When *mhuka* come to the pool to drink, the female *mhesvi* lands on them to feed, with the male flies in tow, evidently intent on mating. Only *mhuka* whose drink and bath times coincide with *mhesvi*'s active (daylight) hours become its food source (Ford 1971, 21–32).[3]

Mhesvi lands on the animal, is carried around as it feeds, fills its stomach with blood, then flies away to rest in the tree branches nearby. When hungry, it indiscriminately follows and probes any darkish moving object for blood—even car tires. The mobile nature of its food source (*mhuka*) means that *mhesvi* must travel with and on its meal, on terms and with an itinerary the animal dictates and by the diurnal rhythms of light, temperature, and humidity. *Mhesvi* is "quiescent at night," except warm, moonlit ones.[4] When *mhuka* stops at waterholes, rivers, or lakes, or returns to graze, browse, or chew its cud in the tree shade, the *chipukanana* pounces and feeds upon it. In general, *mhesvi* is more active later in the day, especially during or right after peak temperature and minimum humidity, when the light intensity is declining and the silhouette effect improving (Jack 1939; Pilson and Pilson 1967).[5]

As noted earlier, the male *mhesvi* usually comes for mating, not feeding. It is a master tactician, deploying itself strategically in the grass or woods in such a way that the flying females and their movements are silhouetted against the sky. As the female *mhesvi* descends on an animal and engorges itself with its long mouth, the male pounces on her. The male can be seen settling on a human or animal, taking off after females, and settling again. So long as the females are around, the males ignore sucking blood altogether. Once the female is fertilized, it reproduces without sex for the rest of its life (Swynnerton 1921a).[6]

Mhesvi is a *chipukanana* known for accompanying moving objects for significant distances. It feeds on the blood of *zvinosvosvoma* (reptiles), *shiri*, *mhuka*, *vanhu*, and *mhuka dzomurukova* (animals that live in water); it draws all its water from blood.[7] The *chipukanana* tends to catch a ride "on the backs of pedestrians and cyclists, and under the hoods, etc., of motor cars, though [it] commonly pursue[s] fast-moving objects for a considerable distance on the wing" (Jack 1930, 493). The faster the moving object or vehicle, the more *mhesvi* is attracted to it and the further it is carried.

Altitude and climate—specifically, the movement of air masses that determine whether it rains—also determine the presence of *mhesvi*. Relief rainfall and cold temperatures create vegetational conditions unsuitable for its survival. The *chipukanana* normally thrives under warm to hot conditions (Ford 1971, 461). The vegetation that gives *mhesvi* shelter and the valleys where it breeds depend on the low-pressure air mass that moves north, south, and back again in response to the seasonal change, drawing in southeasterly trade winds and bringing rain. *Vachena* called this the *intertropical convergence zone* (ITCZ; Ford 1971, 116).

Geological and climatic mobilities determine a third factor critical to the presence of *mhesvi*: vegetation. Rainfall distribution determines the distribution of plant life; plant life determines forest animal populations, some of them vehicles and food sources for *mhesvi*; and all *mhesvi* essentially are forest *zvipukanana*. They mainly inhabit the *mupani* and stream bank forests, all composed of close thicket. *Mhesvi*'s distribution in all four seasons varies according to leaf fall and releafing times; trees in good leaf offer cover from predators and the sun. The hotter the day, the closer *mhesvi* stays to the trees, only leaving the shade when temperatures decline (Jack 1971, 9).

Mhesvi in *Dzimbahwe*

Fewer *zvipukanana* terrorized the white man more than *mhesvi*. There are over twenty-three known species of this *chipukanana* throughout Africa. The United Nations' Food and Agriculture Organization (FAO) estimates the total value lost to infestation by *mhesvi* in Africa's so-called green deserts to be $4.5 billion per annum. *Gopė/hotsikotsi* infects an estimated fifty thousand to seventy thousand people each year. There is no vaccine, and the available drugs are toxic and ineffective (Enserink 2007, 311).

Two types of *mhesvi* terrorized Southern Rhodesia (now the Republic of Zimbabwe) throughout the duration of and even after British settler rule (1890–1980). The most widespread was *mhesvirutondo* (forest *mhesvi*), which *vachena* called *Glossina morsitans*, found in the country's Zambezi Valley and the southeastern borderlands of Zambia and Mozambique, respectively. The other type was *mhesvirupani* (valley *mhesvi*), which *vachena* divided into *Glossina pallidipes* and *Glossina brevipalpis*, found in the Sebungwe (to the northwest) and Chipinge (southeast) districts, respectively.[8] The historical limits of *mhesvi* are shown in figure 0.2.

For nearly two decades after the great *chirwere chemombe* (cattle plague) known in German as *rinderpest* (1896–1897), Southern Rhodesian authorities did not need to worry about *mhesvi*—and they didn't. *Vachena*'s ox wagons introduced a devastating *hutachiwana* against which *chipfuyo* (livestock) and *mhuka* had no natural immunity. They died en masse, denying *mhesvi* its most versatile means of transport and food source (Spinage 2003; cf. Crosby 1986). Only those *mhuka* in the remote borderlands along the Zambezi, Limpopo, and Savé River Valleys survived. In the vast acres of the now-*mhesvi*-free land, *vachena* established their *mapurazi* and mines (Mavhunga 2014).

As the herds of *mhuka* began recovering and returning to their old haunts, so did *mhesvi*. In 1909, two cases of *gopé* were confirmed in the

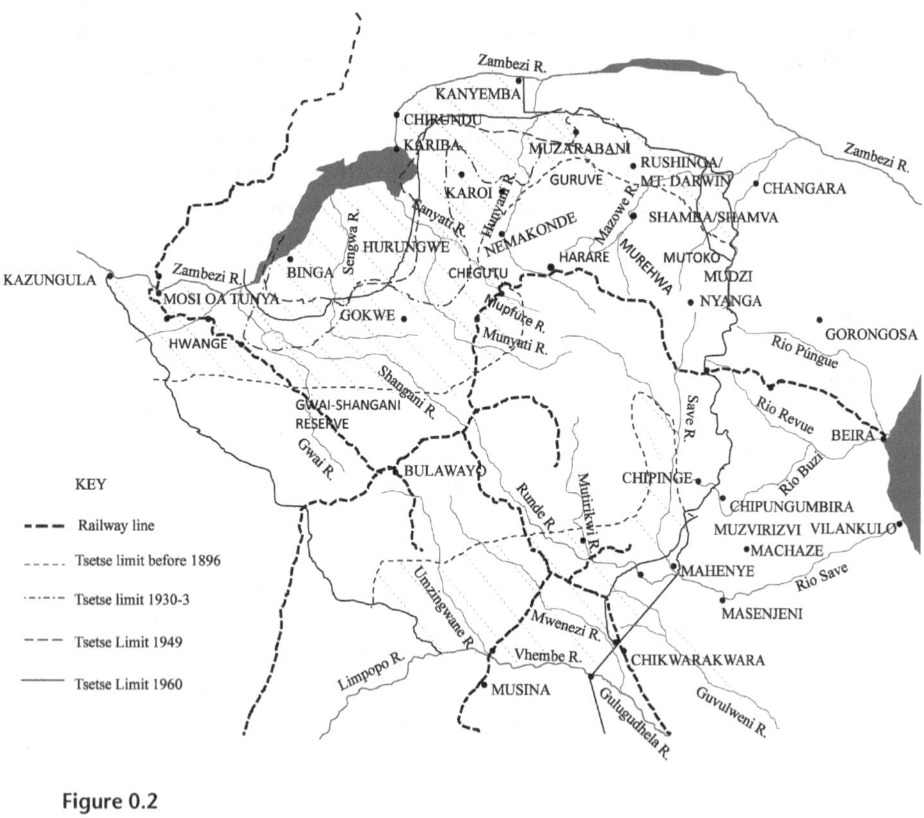

Figure 0.2

Known historical limits of *mhesvi* in what is now Zimbabwe.
Source: Mavhunga 2014.

district of Sebungwe (Fleming 1910, 1913; Stephens and Fantham 1910). The number of cases increased in the 1920s and 1930s (Blair 1939). In 1923, the chief entomologist warned that "the continued steady advance of this pest [was] ... creating a very grave situation."[9] By 1949, *mhesvi* had become "one of man's worst enemies over the greater part of Africa south of the Sahara."[10] Only with the introduction and intensification of spraying of *chepfu* (poisons) that *vachena* called *organochlorine pesticides* (OCPs) in the 1950s was the scourge finally brought under control.

During this period since 1900, extensive field research has been undertaken "to know how, when, why, where tsetse does what it does, and [apply] this *ruzivo* to practical control and combat" of *mhesvi* and the *hutachiwana* it spread (Phillips 1929, 438). The white men who studied insects in general called themselves *entomologists*; later, these *vachena* who focused their studies on *mhesvi* (*Glossina*) specified their job title as *glossinologist* (Jack 1930, 1933, 1934).[11]

The object of their experiments was to know in a thorough way the "life economy" of this *chipukanana*, the bite of which was deadly to *vanhu* and *zvipfuyo*. This entailed knowing not just *mhesvi* and its animal associates, but also the vegetation and physical environment within which it lived, refuged, bred, and hunted, thus providing a "scientific basis" for its control (Trypanosomiasis Committee of Southern Rhodesia 1946). The researchers talked about "a lifetime of affectionate study" of *mhesvi*—that is, "to live and breathe and think with it."[12]

Of course, these *vachena* were newcomers to *mhesvi*, and though they had political power, designed activities, and directed what, how, and where information was to be gathered, they could not personally *live with mhesvi* for sustained periods without falling victim to the bite of another *chipukanana*: *hutunga* (the one that gores), which *vachena* called *mosquito*. Only African men recruited as "flyboys" (fly catchers) could make the regular fly rounds or inspections of the targeted areas. They produced the information that the white entomologists compiled into "scientific reports," the object of which was to know the pestiferous mobilities of the insect.

Pestiferous Mobilities of *Mhesvi*

As a mobile *chipukanana*, *mhesvi* interacts with many different environments and actors that are subjects of multiple disciplines. First, it allows us to enter a conversation about *zvakatikomberedza* (surroundings) or *zvisikwa* (creations), not just referring to "nature" or "the environment" around us in *vachena*'s sense, but including social, economic, and political components as well (Ngulube 2017; Mutasa, Nyota, and Mapara 2008). *Dondo* or *sango* (forest) becomes the home of *mhesvi*, and *mhesvi* itself becomes a *nyongororo* (parasite) that pounces to feed on any *mhuka* that moves within *dondo*. A large body of scholarship has explored *sango* (and spaces within it) as a site of the hunt (Mavhunga 2014), resource conflict (Murombedzi 1994; Mutimukuru-Maravanyika 2010), warfare (Daneel 1995; Alexander, McGregor, and Ranger 2000; Le Billon 2001; Draulans and van Krunkelsven 2002), game reserves (Carruthers 1995), sacred space (Daneel 1970; Ranger 1999; Werbner 1989; Bourdillon 1978; Rennie 1978), and the source of veterinary disease that *vachena* solve (Brown and Gilfoyle 2010).

What passes as the history of science in Africa is still made up of social and political histories of disease and medicine. In most of these, *vatema* are either victims of *vachena*'s policies or those who must be saved from disease by Europe's medical advances. Although they illuminate the pathologicomedical aspects, these studies do not consider Africans as intellectual agents before or during *vachena*'s encounter with *mhesvi* (Lyons 1992; Hoppe 2003; Packard 2007; Herbert 1975). Scholars have also addressed comprehensively the disruptive effect of the movement of *vanhu* and *zvipfuyo*, but not mobility itself as a historical force producing such disruption (Giblin 1990; Lyons 1992).

This book addresses *sango* as a mobile *pabasa* or *nzvimboshandwa* (workspace) where what *vedzimbahwe* call *ruzivo rwemhesvi* (knowledge of tsetse) was produced and applied. *Mhesvi* was a resident of and a traveler within *sango*, and *sango* allows us to shift the conversation about *zvipukanana*, including *mhesvi* itself, from considering a mere epidemiological or environmental problem to the production of *ruzivo*. *Zvipukanana*'s presence and movement in and out of *sango* constitutes a serious risk to *vanhu* because of the *hutachiwana* it carries and transmits through its bite.

Mhesvi opens a space in which an African can say something about the increasing global circulation of biological material that presents serious health issues. The (likely) presence of deadly *zvipukanana* and *nyongororo* forces political, military, legal, scientific, and other bodies to come together and confront this issue (Mitchell 2002). Recent studies explore both the negative and positive aspects of deliberate usage of insects as weaponry. On one hand, security agencies worry about the potential of individuals or groups deliberately mobilizing *hutachiwana* for a harmful purpose (Ginsburg 2000). On the other, there is a positive deployment of *zvipukanana*, first sterilized and then released into *zvakatikomberedza* to exterminate their own kind or to act as early warning equipment when their kind invade (Beisel and Boëte 2013; Beisel, Kelly, and Tousignant 2013; Tousignant

2013). Still, a sense of foreboding exists: what if those insects (read "deadly biological weapons") get into the wrong hands?

The development of means to enhance the speed and distance of transportation and communication has increased capacities for the spread of *hutachiwana* such as SARS (severe acute respiratory syndrome) and electronic viruses such as various types of ransomware. Others have called such outbreaks "networked diseases" (Ryan and Glarum 2008; Wenger and Wollenmann 2007); I focus instead on the modes and work of mobilities—not just those that are human made, but especially mobilities concerned with natural transport, such as growing interest in transcontinental bird migrations and avian flu transmission has highlighted (Chen et al. 2005).

These mobilities have been understood as "mobile life" (Clark 2013) and "the biology of life on the move" (Dingle 1996) and their study as "multispecies ethnography" (Kirksey and Helmreich 2010). My investments in *mhesvi* lie elsewhere: the mobility of *vedzimbahwe*'s knowledge of the pestiferous movements of a peripatetic *chipukanana* that carries *hutachiwana* deadly to *vanhu* and *zvipfuyo*.

Pestiferous animals are not just *any* living beings; they are vermin beings (Mavhunga 2011). They can be "exterminate[d] ... with an easy conscience" (Mamdani 2001, 13): pests, weeds, terrorists—one living thing intruding into the space of others, becoming "matter out of place"—*dirt* (Douglas [1966] 2002, xvii, 2).

Dirt in *chidzimbahwe* is *tsvina*; the meaning is the same as in English. As a concept, dirt is the "bridge between our own contemporary culture and those other cultures where behaviour that blurs the great classifications of the universe is tabooed. We denounce it by calling it dirty and dangerous; they taboo it." The beholders invoke harm to whip the defiant into submission (Douglas [1966] 2002, xi). Because each kind has its own environment, those that don't fit in become "anomalous creatures" (xii–xvi). As Douglas says, "there is no such thing as dirt; no single item is dirty apart from a particular system of classification in which it does not fit." Dirt "exists in the eye of the beholder" (xvii, 2). "Dirt offends against order. Eliminating it is not a negative movement, but a positive effort to organize the environment" (2).

Dirt must be stopped, cleansed. *Nzira* (ways) and *maitiro* (means) must be found to clean the dirt. *Nzira nemaitiro* (ways and means) constitute *ruzivo* (knowledge) deployed to make (*kugadzira*) dirty spaces clean (*kuchenesa*). *Vanhu vatema* created numerous stratagems to manage and coexist with *mhesvi* that *vachena* later borrowed and deployed to control the insect. These included *moto* (fire), specifically late burning of forests; *manhuwe* (repellents); *miteyo* (traps); *kugarisika kwevanhu* (human settlement); *rusosa/ruzhowa* (fences); the use of *muchetura* (poisons) or *mishonga yezvipukanana* (pesticides); and so on. The advent of Rhodesian rule introduced a new dynamic to relations between *mhuka* and *vanhu*: from coexistence to exterminating the insect, along with any *mhuka* and *sango* associated with it.

Thingification

Writing *in Black Skin, White Masks* in 1952, Fanon compared the discrimination and persecution of Jewish and black people and their relegation to the role of pests in a most illuminating way. The Jew, he said, is suspect "because he wants to own the wealth or take over the positions of power. But the Negro is fixated at the genital. ... Two realms: the intellectual and the sexual. ... The Negro symbolizes the biological danger; the Jew, the intellectual danger. To suffer from a phobia of Negroes is to be afraid of the biological. For the Negro is only biological. The Negroes are animals. ... They go about naked" (Fanon [1952] 1967, 157–165).

When people are seen as vermin, the instruments designed for verminous animals also can be extended to them. The Nazi described genocide against the Jewish people as delousing, performing an act of hygiene, removing dirt (Raffles 2010, 141). When talking about *nyongororo* (parasites), the high-ranking Nazi Heinrich Himmler said in 1943: "They suck our blood (as do Jews). They carry disease (as do Jews). They enter our most intimate parts (as do Jews). They cause us harm without our knowing it (as do Jews). They signify filth (as do Jews). They are everywhere (as are Jews). They are disgusting. There is no reason they should live" (Raffles 2010, 144).

The Nazis called it *Rassenhygiene* (race hygiene) against *Judenfieber* (Jewish sickness), *nyongororo* to be eliminated in "delousing facilities" equipped for eliminating them not *like* but *as* actual lice and other bugs, with hydrogen cyanide in gas chambers (Raffles 2010, 155–157). The boundaries between human and animal collapse; the dehumanized is eliminated as a pest. It can be subjected to experiments, as in the Tuskegee syphilis case, in which infected blacks were deliberately not treated to study the effects and progression of the disease (Reverby 2009); and the Guatemala syphilis experiment of 1946–1948, in which doctors infected soldiers, prostitutes, prisoners, and mental patients with syphilis and other sexually transmitted diseases to examine the effects of specific drugs (Reverby 2011).

We often make the mistake of thinking in terms of comparison rather than connection; however, consider the case of locusts and insurgents as *terrorists*, acting not separately but together, creating overlapping terrors. Locusts that originate in places made inaccessible by war—in Mauritania, eastern Mali, northern Niger, northern Chad, Sudan, Somalia, Iraq, Afghanistan, and western Pakistan (Raffles 2010, 209–222).

The othered must be made to look threatening if violence against him/her is to be justified. The Guinea Bissau revolutionary Amílcar Cabral would sum up the black existence under what *vatema* called *hunhapwa* (slavery) at the hands of *vachena* thus: "To co-exist one must first of all exist" (Cabral 1974). Aime Césaire called this *thingification*: the transformation of the black person into a thing—in this case, a machine or "an instrument of production" (Césaire [1955] 2000, 42–43).

The fullest extent of thingification can be seen in the deployment of *vanhu vatema* as instruments to know and eliminate *mhesvi* (as flyboys, hunters, spraymen) and their exposure to *chepfu* in the name of antitsetse operations. No study to my knowledge has yet fully grappled with the harmful effects of such *chepfu* sprayed to eradicate either *mhesvi* or other pestiferous *zvipukanana*, like *hutunga*, *hwiza*, and *mhunduru* (swarming, green crop–destroying worms *vachena* called *armyworms*), as well as *zvimokoto* (swarming, small, grain-eating birds *vachena* called *quelea birds*).

Deintellectualization

Dehumanization is not merely about being made a tool or a pest; it is also a reduction to a nonthinking thing, or deintellectualization. The deliberations of scientific associations and scientists of the 1890s to the 1950s—all of them *vachena*—show *vatema* as objects of study, not agentive participants in the production of *ruzivo*. Anthropology became an instrument for revealing the structure of *vatema*'s societies and functions of custom so that the Rhodesian authorities and the church could better effect subservience. Because no university existed anywhere else in Southern Africa until the University College of Rhodesia and Nyasaland in 1957, South Africa becomes an important site for understanding the prejudiced intellectual milieu responsible for deformalizing and delegitimating knowledge produced by *vatema*.

The reduction of people into "species" in the Enlightenment tradition rests upon their gradation based on race, the white being deemed civilized and blacks savages. In the 1920s, Lévy-Bruhl's concept of the "primitive mentality" was particularly influential in the white anthropological community in South Africa. In his presidential address to the South African Association for the Advancement of Science (SAAAS) in 1926, J. D. Rheinallt-Jones cited Lévy-Bruhl's concept when referring to "the absence of the scientific point of view" among "the natives." He claimed that when faced with a situation, instead of following *vachena*'s course of "order and reason" and "calm and complete confidence in the immutability of natural laws," the primitive mind instead sought the guidance of "the mystic," "the occult," and "invisible power." Hence, Rheinallt-Jones said: "The intricate arrangement of a combination of methods appropriate to the end pursued, does not necessarily imply deliberate activity of the understanding, nor the possession of knowledge capable of being analysed, generalised, and adapted to unforeseen cases. It may be merely practical skill, formed and developed by use, and thus maintained—a skill comparable with that of a good billiard player who, without knowing anything either of geometry or mechanics, has acquired a ready and accurate intuition of the movement required in a given position, without needing to reflect upon his stroke" (Rheinallt-Jones 1926, 86). *Vatema*'s knowledge, according to this view, was merely "experiential," as opposed to formed from thought and abstraction.

Darwin's *On the Origin of Species* (1859) was also used to mark the African's place in the unilineal "development" of culture and thought, from simple to complex, "culminating in the full flower of Western European civilization" (Hoernlé 1933, 77). *Ruzivo rwevatema* thus became mere raw material for *vachena*, who produced *ruzivo chairwo* (true knowledge). At best, elements of *tsika dzevatema* (what *vachena* derogatorily called "native culture") mattered only as examples of early (primitive) stages of development toward *tsika dzevachena* (white civilization), which was pure (77). *Vatema* had modes of thought, yes, but ordered within—never independent of—"superior" culture (Driberg 1928).

The ability to express ideas in ways that only *vachena* adjudged as writing became the measure of a society's capacity to be scientific. In the 1920s, a distinction was arbitrarily drawn "between peoples whose culture includes records of their own past, and who ... are historically-minded and may have their own written history, and *recordless people* who are not historically-minded and whose history, if such it can be called, is little more than legendary tradition" (Hoernlé 1933, 79; my italics). "Proper" historians studied people with written history; ethnologists studied recordless people (79–80). The latter's task was to render "the natives" legible for *hurumende* (the government) to cook them (*kubika*) into governable subjects. "The natives" became samples, specimens, data, and, at best, informants, rather than intellectual agents in their own right.

Indeed, in his presidential address in 1920, the Swiss missionary and ethnologist, the Reverend Henri A. Junod, declared before the SAAAS that "for the present the white race has to rule and guide the black race" (Junod 1920, 76). No science could emerge from such a race; the reverend restated as a matter of fact what another reverend, Johannes Winter, had said in 1914: "Sentiment has no place in science. Science is cold and dry as the moon" (Winter 1914, 371). Junod was even blunter: "We Europeans of the twentieth century possess what I may call the *scientific spirit*, while Bantus are still plunged in the *magic conception of Nature*" (Junod 1920, 79). According to Junod, "the Bantu" did not ask who had made it the way it was; he just accepted it as it was. Only in times of illness or crisis did he seek to understand the sources of the malice, which "he at once believes ... are produced by spiritual agents like ... ancestor spirits [and] witchcraft," all cured by "magic" (79). To our pious reverend here, *vatema* equals magic; *vachena* equals knowledge.

The 1920s was also the era of eugenics in Southern Africa for this bigoted class of intellectuals. Harold B. Fantham, a prominent eugenicist professor of zoology at the University of the Witwatersrand (Wits), for example, was convinced that to maintain "racial fitness ... physically and mentally" in South Africa, the boundaries of intimacy between whites and blacks had to be policed ruthlessly (Fantham 1925, 1927). He considered *vanhu vatema* mentally unfit but physically fit and therefore good for menial labor, and whites mentally and physically fit and therefore intellectual, scientific, and civilized. The white gene pool had to be defended against contamination by the "feeble-minded," lest "a lower level of civilization" drag it down (Fantham 1925, 405).

Vatema were no more than zoological material, studied as primates—or something close thereto. Possibly the crassest racist writer of his generation was the Rhodes University College professor James Edwin Duerden, after whom the annual Duerden Lecture at the university is named. In a 1925 paper, Duerden made a call for "controlling the quality and quantity of humanity" using eugenics (Duerden 1925, 60). Referring to *vanhu vatema*, Duerden said: "Their hereditary factors appear to have reached their maximum response under their particular surroundings. ... They appear to have become stationary at a very low level of achievement regarded from a European standard" (67). "We are here to do our best for the Bantu, and to give him as many rights and privileges as he can wisely use," Duerden said, in a condescending, paternalistic tone, rejecting that racial prejudice had anything to do with his "scientific" observation. "We have given him peaceful settled conditions of life. ... Schools for elementary education and for industrial instruction, Schools of Agriculture, Training Colleges, and finally, a University College. He has everything which the European has."

One question now remained: "How will his hereditary factors respond to these changed conditions? Will he rise to the level of the white in mental, moral, and spiritual values?" (68–69).

"Bantu studies" (later sanitized as "African Studies") in Southern Africa was born as a eugenic project, to ensure that "the Bantu" or "the native" was "now for the first time being studied on the spot, not by theorists overseas," but by local, university-trained *vachena* in ways that addressed critical issues of "native behavior" that might affect their white lot. By 1925, Wits and University of Cape Town had Bantu Studies Departments, and Rhodes's was on its way. Duerden declared: "These should supply data to enable the white to understand the native in all his aspects, material and psychological" and aid him in "expressing his good will" to "the native" (Duerden 1925, 69).

The "high point" of *vachena*'s reduction of *vatema* into objects of study was the establishment in 1921 of *Bantu Studies, A Journal Devoted to the Scientific Study of Bantu, Hottentot & Bushman* (see figure 0.3). In 1942, when *Bantu Studies* became *African Studies*, only the linguistic hygiene improved; the subject and object of research remained the same (Biesheuvel 1952).

To repeat, *vatema* came to *vachena*'s "science" as objects of study, not producers of knowledge. Just as *vachena* had done using anthropology, they now extended what they called "*natural* sciences" to dehumanize and deintellectualize *vatema*. For a case in point, consider Lawrence Wells, Wits student and lecturer, and later chair of the University of Cape Town's Department of Anatomy, and then founding head of the University of the Western Cape's Anatomy Department. His master of science thesis was titled "The Foot of the South African Native," which he described as follows in a SAAAS presentation in 1929: "Undoubted atavism occurs in the following cases: the varieties of the flexor digitorum brevis in which a deep head is present, arising from the long flexor tendon (22 cases); the abductor hallucis slip of the extensor hallucis longus muscle (8 cases); the absence or extreme reduction of the peroneus tertius (10 cases); the insertion of the peroneus longus into the first metatarsal only (6 cases); the presence of vestiges of the peroneus digiti quinti and peroneus digiti quarti muscles (8 cases); and the disposition of the interossei about an axis passing through the third digit (2 cases)" (Wells 1929, 796; also Wells 1938, 1952).

Wells's contemporary, Simon Biesheuvel, summed up *vachena*'s perception and treatment of *vatema* as experimental objects a decade later: "African society ... provides a natural laboratory for students of the social

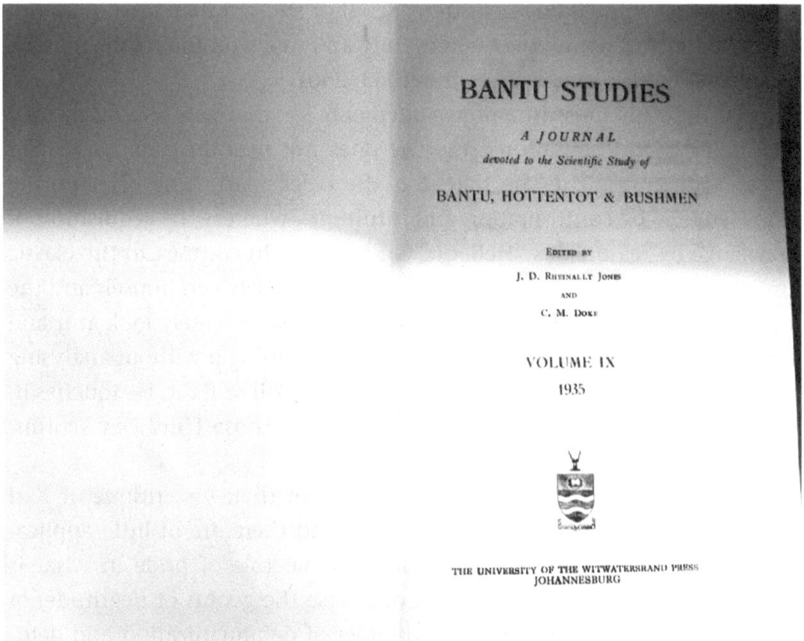

BANTU STUDIES

A JOURNAL

devoted to the Scientific Study of

BANTU, HOTTENTOT & BUSHMEN

EDITED BY

J. D. RHEINALLT JONES

AND

C. M. DOKE

VOLUME IX

1935

THE UNIVERSITY OF THE WITWATERSRAND PRESS
JOHANNESBURG

Figure 0.3
Bantu Studies, published as *African Studies* from 1943 on.
Source: Bantu Studies 9 (1935).

sciences" (Biesheuvel 1952, 45). Africa was "an imperial living laboratory," with white males from Europe front and center, and *vatema* as objects of study (Tilley 2011). *Vatema* could not be experimenters themselves.

Self-Reintellectualization

Strangely, even influential black intellectuals bought into this narrative of the unscientific African and found black pride elsewhere. Recall Rev. Johannes Winter's words from 1914: "Sentiment has no place in science. Science is cold and dry as the moon" (Winter 1914, 371). Leopold Senghor (intellectual and first president of Senegal) cast *vatema* in the same light—as full of the "warmth of human feeling—some joy and much pain" and "superior without invention and conquest"—while the whites were "hard and cold," with all their technology and science (Le Baron 1966, 268). His compatriot in negritude, David Diop, added that the black man "would teach rhythm to the world that has died of machines and cannons" (269).

In 1963 when enunciating negritude, Senghor stressed that "man does not live by millet and rice alone; he lives truly and solely on the myths that are his spiritual nourishment" (Senghor 1963, 250).

In perhaps his most infamous statement, Senghor said the black man is "a field of pure sensation"; that he does not measure or observe, but "lives" a situation. This, he said, was the black man's way of acquiring "knowledge"—by confrontation and intuition—whereas the acquisition of knowledge by reason was "Hellenic" and white: "In contrast to the classic European, the Negro-African does not draw a line between himself and the object, he does not hold it at a distance, nor does he merely look at it and analyse it. After holding it at a distance, after scanning it without analysing it, he takes it vibrant in his hands, careful not to kill or fix it. He touches it, feels it, smells it. The Negro-African is like one of those Third Day Worms, a pure field of sensations" (Senghor 1964, 72).

Senghor inspires me in every other way but that: his critique of Karl Marx as never having governed a country and therefore of little application to the African situation; the foundational role of pride in what is best about our African cultures; and of course the genius of negritude, of taking the black skin, the object and target of dehumanization and deintellectualization, as the ultimate symbol of pride, self-rehumanization, self-reintellectualization, and self-liberation. On the matter of science and technology, I declare my profuse disagreement. I am not alone. Ayi Kwei Armah certainly had Senghor in mind when lamenting *vachena*'s education system, which encourages us "to describe our society as thoughtless, rhythmic, playful, child-like, and irrational ... to make us incapable of reasoning" (Armah 2010).

Armah's ire was also directed at scholars like Abiola Irele, who declared in 1983 that the only future for Africa lay in following Western culture and civilization. "It is of no practical significance now to us," Irele said, "to be told that our forefathers constructed the Pyramids if today we can't build and maintain by ourselves the roads and bridges we require to facilitate communication between ourselves, if we still have to depend on the alien to provide us for necessities of modern civilization, and if we can't bring the required level of efficiency and imagination to the management of our environment" (Irele 1983, 3; echoing Towa 1971, 1979).

Armah unleashed his anger at such Irele-esque thinking:

> We have helped cover up the fact that the European claim of proprietary rights over rationality is ahistorical. ...Worse, there are African intellectuals [who] not only swallowed the hoax about 'Western rationality,' but actually took to arguing, straight-faced, that (a) since African culture was essentially irrational, and (b) since

life in the modern world, modernisation, depended on the mastery of rational routines, (c) therefore for Africa to move into the modern world, Africa would have to borrow rationality from the West—Europe and United States of America. It is a formula that consecrates the partnership of deceitful teaching and gullible apprenticeship. (Armah 2010)

Armah circles in on a theme he first developed in *The Beautyful Ones Are Not Yet Born* (1968): When we talk about liberation, "what, precisely, are we talking about?" He hears people celebrating Kwame Nkrumah, Kenneth Kaunda, Julius Nyerere, Patrice Lumumba, Modibo Keïta, and Sékou Touré as heroes who led *vanhu vatema* to freedom from foreign oppression—but he has one question: "If they freed us, why are we still talking about our liberation as if it still lay in the future?" (Armah 2010).

Ngugi wa Thiong'o answers that question rather well: "Unfortunately, the colonial phase of imperialism did produce an African elite with the mentality that was in harmony with the needs of the ruling classes of the imperialist countries. And often it was this African elite, nurtured in the womb of imperialism, with the cultural eyeglasses from Europe, that came to power or who held the reins of power during the neo-colonial phase of imperialism" (wa Thiong'o 1985, 18–19).

This book takes up Ngugi's challenge. *Self-reintellectualization* refers to *vanhu vatema* home and abroad researching, reinstalling, reasserting, and revaluing themselves not just as eaters of other people's knowledge, but producers of their own. In this way, conversations become possible with others who see from elsewhere, who should not be expected to see from our experiences and locations, whose perspectives we have no mastery over, nor they ours. That knowledge humility paves the way for a reconsideration of *vanhu vatema* as intellectual agents, engaged in their own journey of *ruzivo* and coming to the encounter with *vachena* not as blank slates, but as intellectual agents.

Mhesvi is one space from which I have chosen to embark on this journey, this reintellectualization of my own past, a declaration of independence, not just for myself, but every African who feels more comfortable thinking in their own language and keywords. I seek to reclaim my own humanity through an insect—seemingly innocuous, inconspicuous, grossly underestimated, and yet ... My point is a simple one: if a tiny insect can be that powerful, that unignorable, and that indefatigable, what more *vatema* whom *vachena* reduced to a thing just like the fly?

Zvipukanana (insects) have been reasonably well studied from other angles before. Some accounts meticulously detail the environmental management regimes that controlled pests with or without disturbing the natural environment, but then generalize *vachena*'s destruction of such indigenous practices when they could fruitfully engage with how such *ruzivo* became a critical ingredient (at the very least) and the foundation (as this book argues) of "*vachena*'s" pest control itself. Helge Kjekshus (1977) investigated indigenous modes of ecology control in his influential book *Ecology Control and Economic Development in East African History*. However, he may have moved too quickly to declare that the rule of *vachena* ended these indigenous forms of environmentally friendly management of *mhesvi*. At the level of practice, that is true in many cases; when the spaces to practice certain forms of *ruzivo* are removed, the knowledge dies. I did not see that in *Transient Workspaces*, where the hunt continued—retaining its essence, but elastic enough for inbound repertoires. I do not see that in this book, either, because *vachena* did not bring any readymade stratagems against *mhesvi*; they relied on *ruzivo rwevatema* both for survival and for the basic premises from which they began constructing "tsetse science."

The Mobile Workshop

How do we position *mhesvi* so that we can recover the trajectory of knowledge production concerning it? In other words, where can I start to see *mhesvi* from a *mudzimbahwe*'s point of view? How did people know *mhesvi*? What did they know about it? What happened to this *ruzivo*? Why is it important, at this moment, to know this information?

The Mobile Workshop is a serious consideration of the movement of ideas, of the things to be known, the people that seek to know and produce knowledge, and the materials with which they do so. In its basic form, *kufamba* (mobility) is to be in action or in motion, to act or to move (about). *Kufamba* and *vafambi* (travelers) refer to something more specific: the conveyance, transporting, re-moving, or re-placing of something from one thing or place to another, bringing into analytical relief the passenger carried, the vehicles, and the pathways, or pathlessness.

Positioned within mobility, *ruzivo* shows *that vachena*'s moment in power in Africa was never a monodirectional transfer of knowledge but a *kusangana kweruzivo* (knowledge encounter). *Kusangana* is the coming into contact of two things coming from different directions. Applied to *ruzivo, kusangana* captures two knowledge systems, the endogenous and the inbound, each engaged in its own itinerary, which is then fundamentally shaped, altered, diverted, or ended because of the encounter. In other words, the encounter provides a site where we can observe *kufamba kweruzivo* (the mobility of knowledge) between cultures. *Vachena*'s project in Africa—whether political, economic, or ecumenical—is often portrayed as a civilizing mission, as a massive tidal wave that swept in and flooded all indigenous *ruzivo*. The narrative of this historical moment as not as encounter (between peoples) but conquest (hence the privileging of *colonization* and *colonialism*)—wherein *kuziva kwevachena* (white knowing) replaced *kuziva kwevanhu* or *maziviro avatema* (how [black] people know) and black people under "colonial" rule simply followed—is a false narrative. Repositioned as *kusangana kweruzivo*, the moment of encounter represents instead a front-to-front engagement, albeit one that occurred within unequal relations of power.

This book is therefore a quest for an African spirit of *ruzivo* with *mhesvi* as a site of *kuziva* (knowing). It is a story of how a mobile *chipukanana* carrying deadly *hutachiwana* pulled all kinds of actors—physicians from Europe and North America, *vadzvanyiriri* and their government, and *vatema* and their *ruzivo*—to engage in the work of controlling and eradicating it, thereby producing a salubrious, *mhesvi*-free environment and *ruzivo*. This knowledge was a knowledge of *mhesvi*—specifically, the role of mobility in its pestiferous actions and, hence, *mhesvi* as a mobile workshop of pestilence.

Conventionally, a workshop is limited to *people*, more specifically to a group that meets to engage in intensive discussion and activity on a particular subject or project; or to a *room* or *building* in which people use specific skills to perform mechanical work, usually manufacturing; or to an unspecified *place* where the making or repairing of things happens. *Mobile workshop* usually refers to a portable or wheelable toolbox; a workshop on wheels, mounted on a truck; a meeting inside/on a moving automobile or bicycle; or a mobile clinic.

The element of *work* in *workshop* normally refers to people—hence, workers or laborers. This book extends the concept of work beyond people engaged in productive or operative activity, occupying a space of employment, or *people at work*—beyond work as exertion. It draws attention initially to the task itself or that which is subject to intellectual and physical exertion or labor—and, indeed, to work as a product of such exertion, labor, or activity. The word *works* also refers to working parts, usually an assemblage of machines (hence waterworks or steelworks), but in this book it is extended to organic body parts and systems.

Hence, in the first sense, the mobile workshop becomes an assembly of circulatory, digestive, endocrine, immune, lymphatic, muscular, nervous, reproductive, respiratory, skeletal, and urinary systems. What makes the body a workshop is not just the existence of interdependent systems but the fact of their constant mobility throughout life until the body is either comatose or clinically dead. The bodies in question are the *mhesvi*, *hutachiwana, vanhu, zvipfuyo,* and *mhuka,* as well as inanimate elements of the biotic and abiotic environments. These are *bodies at work.* An outbreak of *gopé/n'gana* is a sign of bodies working; work is also discovering what is going on and stamping out *hutachiwana, n'gana,* the *mhesvi* that transmit *hutachiwana,* and the infective *mhuka.*

In this second (operational) sense of work, a mobile workshop is an assembly and mobilization (i.e., making mobile) of men, skills, research, knowledge, tools, social and political relations, institutions, and faith to accomplish the objective of controlling the *mhesvi* and *hutachiwana,* which also determines where these elements are deployed. The *mhesvi* becomes a worksite, where men earn means of livelihood—hence the idea of work as employment (*basa*), a task or work to be done (*basa, mushando*), working or doing work (*kushanda* or *kuita basa*), and the workplace or being at work (*pabasa*).

The individual becomes part of a state machine to rid the countryside of a pestiferous *chipukanana*; the work of *kuziva*/knowing *mhesvi* (the fly rounds), starving it (by killing *mhuka* that it feeds on predominantly), and vegetation clearance (to deny it shelter) are just a few examples of mobility as work. *Working* (*kushanda*) denotes performing an activity, the moving of limbs or unfolding of thinking that results in knowledge. *To work* the gun, the sprayer, or the trap denotes the handling of the tools that operationalizes intent into actual execution. Results do not simply happen; they must be *worked for.* Sometimes, people do not just *do work*; they may have to *be worked*—that is, driven into *chibharo* (forced labor), as was often the case under *vachena.*

The *shop* in *workshop* conventionally stems from a store that sells things, a carpenter's shop, or any factory or workspace. Where *work* denotes the activity, *shop* denotes the place, the site of working. To shop is to go to this site looking for something, usually to buy goods stored or manufactured therein. In the past, shopping could only be performed by going or sending someone in person to the floor. That is no longer the case with the advent of online shopping platforms and door-to-door delivery services. The sense of *shop* examined in this book is that of a place where a specified type of work is performed. Workshop in this sense is not tethered to a fixed workplace, but is a transient *workspace*—a site of work produced through and because of movement.

The difference between *Transient Workspaces* and *The Mobile Workshop* is not only that the former focused on means and ways and the latter focuses on the mobility of knowledge (*kufamba kweruzivo*). More importantly, the one focused on the hunt as workspace, untethered to fixed physical space, whereas *The Mobile Workshop* takes the *mhesvi* itself as the primary one of several workshops that are mobile, the others being *hutachiwana, mhuka, zvipfuyo, vanhu,* and *zvakavakomberedza* (surroundings), with the term *vedzimbahwe* used to describe the environment. Operationally, that entails locating our analytical space in, on, and around the mobile *chipukanana* that bites infective *mhuka* and ingests, carries, and inoculates *hutachiwana* into *zvipfuyo* or *vanhu* as it bites them, leading to the outbreak of *n'gana* and *gopé*, respectively. *Vanhu*'s responses to *gopé/n'gana* cannot escape dealing with the transport and pestiferous work that an indefatigable, unignorable *chipukanana* performs through its mobilities. Once the mobile work *mhesvi* performs is understood, it becomes possible to see why the work of controlling *mhesvi* was organized and executed in a specific way and why so much focus was devoted to the *chipukanana*'s mobilities.

That story, told through and through as a story about stopping or managing the mobility of the *mhesvi* and its dangerous microscopic passenger, can now be told.

Outline of the Book Chapters

The Mobile Workshop is composed of fifteen chapters. The story begins in chapter 1, "How *Vanhu* Managed Tsetse," which shows that *vanhu vatema* understood mobility as the centerpiece of their interactions with the insect. It commences from a view of *vatema*'s management of *mhesvi* as a site of innovation, illustrating the centrality of mobility in interactions between *vanhu* and *zvipukanana*. The chapter strategically deploys the travel accounts of *vachena* writing in the nineteenth century about their encounters with people living with and despite *mhesvi*. The strategy herein is to read these travel accounts as acts of witnessing to, and confessions about, *ruzivo rwemhesvi* (knowledge of tsetse) among the people living in the lands between and along the Zambezi and Limpopo Rivers. The chapter will map *mhesvi*-infested areas and *mhesvi* management techniques—namely, forest clearance, selective culling of *mhuka*, strategic settlement of *vanhu*, use of repellents, movement by night, and inoculation. At the end of the day, African *mhesvi* management was about mobilities management.

Chapter 2, "Translation into Science and Policy," explores the processes of translation through which this *ruzivo rwevatema* (knowledge of black people) entered the pantheon of *ruzivo rwevarungu* (knowledge of *varungu* or *vachena*) and, later, state tsetse and trypanosomiasis control and research policy. This chapter first examines European travelogues, which show that such *ruzivo* and practices were the foundation of what became science and means and ways of tsetse control. It makes a more general statement: Certain *ruzivo rwevatema* and practices formed the foundations of what *vachena* then called *science*, even while dismissing *vatema* as only good at creating and peddling myths and legends. Empirically, the specific stratagems that *vachena* built on were controlled *moto* (fire), specifically, *kupisa sora* or burning grass, forest clearance, prophylactic settlement, erecting buffer zones, cleansing chambers, and tsetse gates. The white entomologists in charge of Southern Rhodesia's *mhesvi* and *n'gana* control effort say so themselves. The concept of *cleansing* is used in the *chidzimbahwe* sense of *kuchenura*, from the root word *chena* (clean, white), in contradistinction to *tsvina* (dirt) or *chakasviba* (dark).

The rationale for centralizing *vedzimbahwe* is not that they alone had *ruzivo rwemhesvi* or knew of it more than other *vanhu vatema*. It is rather a methodological choice for managing the archive; otherwise, the *ruzivo* and practices were not hermetically sealed in geopolitical or geolinguistic boundaries but widespread throughout the region. *Chidzimbahwe* serves as a linguistic archive and a thought space from which I am descended and to which I am heir, and therefore one among several optics from which this shared *ruzivo* can be explained. This book is not necessarily the history of *vedzimbahwe* as *vanhu*; it is an exploration of their *ruzivo rwemhesvi*.

Chapter 3, "Knowing a Fly," examines what one government official appropriately called "an intelligence system of tsetse"—a thoroughly intrusive infrastructure and procedure of knowing this *chipukanana* (principally its mobilities) in the most complete way possible. This anthropomorphic formula for intrusive knowing sought "to live and breathe and think with" *mhesvi*; to do so entailed "a lifetime of affectionate study."[13] This meant placing a peripatetic *chipukanana* under surveillance, to know how much time it spent in different parts of the habitat at different times of the year; how much time it spent feeding, sleeping, or simply in vigilant mode, waiting to pounce on anything that moved.[14] Maps—of where it slept, bred, roamed, ate; its boundaries; strong points and weak points—were essential to successful operations against it.

Chapter 4, "How to Trap a Fly," considers one of the stratagems developing out of an intelligence system of *mhesvi*: trapping systems. This approach was based on the underlying principle that the *chipukanana* had very small reproductive potential, that a slight reduction in the *chipukanana*'s reproductive rate or increase in its mortality rate was enough to control its entire population. To do so, *mhesvi* had to be attracted to artificial baits laced with killing or sterilizing agents. These "attractant studies" targeted *mhesvi*'s mobilities, sensory system, and feeding behavior; once attracted, the next step was to trap the flies and, once trapped, kill or sterilize them.

Chapter 5, "Attacking the Fly from Within: Parasitization and Sterilization," discusses the method of killing the *mhesvi* from within its body. The first section deals with research (from the 1920s to the 1970s) on *parasitization*—the destruction of the *mhesvi* through deliberately promoting the proliferation of *nyongororo* naturally found in its body. These *hutachiwana* either were naturally lethal to *mhesvi* or could be genetically engineered to be so. The second type of research focused on sterilizing the *chipukanana* through the capture and release of sterile males by means of chemical sterilants and gamma radiation.

Chapter 6, "Exposing the Fly to Its Enemies," considers two stratagems, both derived from past and prevailing practices of *vanhu vatema* of killing *mhesvi* and exposing it to its predators. One involved using *moto* (fire)—specifically, late-season burning—to achieve maximum destruction and expose to predation all *mhesvi* in their adult phase, their *zvikukwa* (the insect at its worm or pupa stage, what *vachena* called *puparia*; singular *chikukwa*), and their *zviguraura* (literally, "the one that has cut off its intestines," what *vachena* called *larva*; singular *chiguraura*). The second strategy was the mechanical clearance and chemical phytocides of the forest for the same purpose.

Chapter 7, "*Cordon Sanitaire*: Prophylactic Settlement," focuses on the use of fencing and forced resettlement of *vatema* as methods of "tsetse control." The argument is that *vatema* and their *zvipfuyo* were deployed as methods of pest control and to act as an outer ring of early warning systems to protect *vachena*'s cattle ranches (*mapurazi*, from the Portuguese word *prazos*). The chapter reflects on the meaning of a humanity (*hunhu*) experienced and lived under conditions of animalization, wherein *vatema* are dumped at the unhealthy margins, to live not just *like* but *with* other *mhuka* as *vachena* helped themselves to their healthy lands on the watershed.

Chapter 8, "Traffic Control: A Surveillance System for Unwanted Passengers," is concerned with the surveillance and cleansing infrastructure installed to stop "carried fly." On the surface, *traffic* might be interpreted as automobiles, bicycles, and foot movements—yet such movement is, at any other time, innocuous. What rendered it worth controlling was *mhesvi*, the real "traffic" that had to be controlled because it carried *hutachiwana*.

Chapter 9, "Starving the Fly," focuses on the deployment of huntsmen called *magocha* to eliminate *mhuka*, the primary bloodmeal source of *mhesvi*, and thus starve it. The chapter opens with an interview with one such *magocha*. These huntsmen were known in society as *magocha* (the ones who are always roasting meat) due to the massive amounts of meat at their disposal. Next, the chapter examines the relationship between *magocha* and the white tsetse field officers (TFOs) who supervised them and the work of indiscriminate and discriminate hunting. Because many technical aspects of the hunting itself have already been discussed elsewhere (Mavhunga 2014, 125–150), chapter 9 instead focuses on the perspectives of *vatema* covered only thinly or not covered in the earlier work.

Dirt is also a pollutant in the sense of chemical poisons, as chapters 10 to 13 show in their discussions of the deployment of organochlorine pesticides (OCPs) to destroy *mhesvi*. Chapter 10, "The Coming of the Organochlorine Pesticide," introduces the three OCPs that were most widely used in Rhodesia: DDT (dichlorodiphenyltrichloroethane), lindane (gamma-hexachlorocyclohexane), and dieldrin. Later, these were largely replaced with two other OCPs: thiodan (also called endosulfan) and deltamethrin. The chapter first explains what these chemicals were and how and for what they were originally designed in the United States and Europe, as well as the circumstances of their travel and deployment in *mhesvi*-occupied Africa, narrowing it down to Southern Rhodesia (later Zimbabwe).

Chapter 11, "Bombing Flies," explores the use of *ndege* (aircraft) to spray OCPs. The origins of the practice in KwaZulu, South Africa, are traced first. (*Ndege* is the *kiswahili* word for bird and thus is used throughout to refer alternately to both the flying animals and airplanes.) In the second section, the technical aspects of aerial spraying are examined as an example of the extension of methods designed in the United States for agricultural or military purposes to deal with *zvipukanana* and with conditions for which they were not originally designed. In the final sections—the bulk of the chapter—the deployment and performance of first fixed-wing *ndege* and then *zvikopokopo* (helicopters) are closely examined. *Vedzimbahwe* do not refer to a helicopter as *ndege* or *shiri*, but as *chikopokopo*—or, in deeper parlance, as *mukonikoni* (dragonfly) on account of its hovering and landing behavior.

The objective of chapter 12, "The Work of Ground Spraying: Incoming Machines in *Vatema*'s Hands," is self-evident. The chapter is organized into three sections. The first concerns the strategic deployment of inbound spraying equipment to perform or operationalize specific objectives and outcomes. The second section takes us inside the work of spraying, focusing on the meeting point between *mushonga wezvipukanana* (pesticide; hereafter just *mushonga*), *mushini* (spraying machine), and sprayman, known in the villages as *mafrayi* (fly man; singular *mufrayi*). The final section is a case study of a spraying campaign involving three neighboring countries: Southern

Rhodesia, Portuguese East Africa, and the Union of South Africa. The purpose of the campaign was to stop the advance of *mhesvi* from the Rio Savé region of Mozambique into the Savé-Runde area, potentially threatening northeastern South Africa.

The final two chapters deal with the fallout from chemicals deployed to deal with *n'gana* in cattle. Chapter 13, "DDT, Pollution, and *Gomarara*: A Muted Debate," begins to ask questions about the link between *vachena*'s OCP use and the high incidence of many types of cancer, a condition known in *chidzimbahwe* as *gomarara*. The word derives from *gomarara*, a plant that usually grows on other plants, deposited in the fecal droppings of birds roosting or stopping over. This *nyongororo* (parasite) takes over the tree, slowly replacing the branches and then the stem. Some *gomarara* kill the plant; others are just malignant. The condition of cancer is the same.

The rise in profile of *gomarara* in Zimbabwe raises two questions: Could the massive aerial and ground spraying of the Zimbabwean countryside be catching up with us? Is there a connection between *gomarara* and OCPs? The reader should note that this chapter is not a detailed treatment of the question; all it seeks is to bring together the global discussion of OCP carcinogenicity and the staggering statistics of *gomarara* in Zimbabwe, where the environmental pollution discussion is muted. First, a brief exploration of the state of *gomarara* in Zimbabwe is offered, drawing out the incidences of *gomarara* that are usually associated with OCPs. The second section reconstructs debates about OCPs as environmental pollutants, a discussion that I argue was muted at the height of the spraying campaigns of the 1950s and 1960s and is largely forgotten now. This is alarming given the banning of these chemicals globally, along with other synthetic products, such as lead-based paint and asbestos, that were once deemed very safe and now are known to be toxic. (In Africa, recalls are rare and class action lawsuits against toxic products virtually unheard of. Corporations get away with everything.) I then examine some of the investigations made into the environmental effects of OCPs elsewhere, marshalling that evidence to ask questions and to map and follow the itineraries of these pesticides in our bodies and those of *mhuka*.

As early as 1944, skeptics were already warning that these new chemical weapons against pests were "turning out to be double-edged weapons" that "may at the same time destroy both useful and harmful agricultural insects." Indeed, as Jane Stafford cautioned, "They may rid your dog of fleas, but insidiously ... damage his liver or paralyze him through nerve damage. They will rid your home of mosquitoes, flies and vermin, but the price may turn out to be high in human health and life" (Stafford 1944, 90).

Chapter 14, "Chemoprophylactics," addresses the use of trypanocidal drugs to cure or prevent *n'gana* in *mombe*. It first gives a historical overview of chemoprophylaxis in Southern Rhodesia, then turns to the problem of drug resistance and photosensitization, a clinical condition in which the skin's negative exposure and reaction to sunlight is heightened due to phototoxic drugs and chemicals. The chapter ends with some case studies of chemoprophylaxis operations in Southern Rhodesia, all showing how the early promise of chemoprophylaxis ended with unforeseen complications that poisoned instead of curing *zvipfuyo* of *n'gana*. The argument made is one about pollution of the most intimate kind: within the body, both of the animal and *hutachiwana* itself. The chapter shows a general pattern among all the drugs: They worked well initially before the animal either relapsed or exhibited signs of drug resistance, prompting the deployment of one drug to cure the effects of another.

Chapter 15, "Unleashed: *Mhesvi* in a Time of War," takes the discussion into the abandonment of "tsetse control operations" as the war of self-liberation intensified, into the fog of war in which the methods designed for *mhesvi* and other pests are extended to those *vatema* viewed as *varwi verusununguko* (freedom fighters) and those designated *magandanga* (terrorists). This does not mean all *vatema* and all *vachena* shared the same perspective or that all freedom fighters behaved consistently with that description but the majority did. This lumping together of "problem animals" and "problem people" into "vermin beings" justified the extension and slippage of instruments and methods from *zvipukanana* to the dehumanized *munhu*, whose elimination constituted a form of pest control.

Finally, in "Conclusion: *Vatema* as Intellectual Agents," I return to the question of *munhu mutema* as intellectual agent, reemphasizing that *mhesvi* is one venue in which *ruzivo* was applied, produced, and extended in different directions. What are the implications of knowing from the fly?

1 How *Vanhu* Managed Tsetse

Writing in 1932, the white South African antiquarian B. H. Dicke commented on how *mhesvi* had spectacularly saved people living north of the Limpopo River from certain conquest by the Boers, the Cape descendants of Dutch settlers. Locals warned two such Boers, Hendrik van Rensburg and Louis Trigardt, about the "dangerous" *mhesvi*-infested belt ahead. "That warning was not heeded," Dicke said:

> The tsetse fly … destroyed the trekkers' draught animals and forced them gradually to abandon all their wagons in the same manner as it caused the *voortrekkers* … to leave their wagons behind, one after another, on the road they were traveling to Delagoa Bay. Without wagons the van Rensburg group of *voortrekkers* was unable to avail itself of the usual Boer tactics of forming a *laager*, and … he was at the onset deprived of his chance of offering successful resistance by the tsetse-fly.
>
> In 1847 the Boers defeated [Mzilikazi] in Southern Rhodesia. In 1851 they defeated Sechele in Bechuanaland. In both instances the Boers took no advantage of the situation [because] the fly cut off the supply base, and would have isolated men who might have settled in the country of the defeated chiefs. … *It was the tsetse-fly that saved all those chiefs.*
>
> By the time the fly-infested areas contracted and passages opened through them, British influence had secured a footing in Bechuanaland and Southern Rhodesia. But, without the fly there would not have been an independent Lobengula [the Ndebele king] to grant concessions, there would not have been a Chartered Company, and South[ern] African history during the last forty-odd years would have run in entirely different channels. (Dicke 1932, 795–796; emphasis added)

Dicke's conclusion is that all people north of the Limpopo River owe *mhesvi* a huge debt of gratitude for saving them from certain conquest. This raises a question: How did *vanhu* (people) relate to *mhesvi* in general?

This chapter uses the accounts of *vachena* as eyewitness confessions to *vatema*'s understanding of *mhesvi*. This is a *vachena*-compiled archive, in which white travelers and settlers tell their own stories. I do not care much about everything else they say; I do take interest in their accounts as bearing witness and confessing to *vatema*'s *zvokwadi* (truths, facts; singular *chokwadi*) about *zvipukanana*, especially since these often were the only facts available to aid their survival in this tropical environment. What masquerades as a colonial library—an archive assembled by colonists and filtered through their biases and priorities (Mudimbe 1988)—turns out to be also *vachena*'s confessions of what *vatema* were doing. The "coloniality" of what Mudimbe called a "colonial library" turns out to be something of a ruse, even for such institutions as the church and later the "colonial state," forced by hardship to appropriate local idioms as opposed to handing out the fruits of so-called civilization to "the native."

It would have been easy for me simply to resort to ethnographic accounts and interviews of elderly people in their eighties and nineties (born in the 1930s–1940s) whose parents were already adults when *vachena* occupied *vatema*'s lands (in the 1890s). Their accounts have been used like this before: as "oral traditions" or sources of evidence that reveal *vatema*'s voices without filtering them through *vachena*'s subjectivities. Such a method would be perfect if I was looking for the "unfiltered African voice." I prefer instead to hear nineteenth-century *vachena* confessing on paper, in their own words, what they saw and experienced while traveling in the Southern African countryside. Indeed, the last half of the nineteenth century constitutes the critical moment of encounter between *vanhu vatema* and white hunters, missionaries, traders, and self-styled "explorers" that yielded the archive upon which this chapter is based. From the 1850s to *vachena*'s partition of *vatema*'s lands in the 1890s, these travelers published papers on *vatema*'s understandings of *mhesvi*, its movements, and its management in journals that professed to be scientific and professional.

Strategic Deployments

Where Dicke attributes to *mhesvi* the continued independence of people along and north and east of the Limpopo, I look for the *ruzivo* of the locals that enabled them to practice what I call *strategic deployment*. By this I mean the transformation of natural features such as steep mountainsides, swamps, and forests into pest-control infrastructure without even touching them—for example, people placing the *mhesvi* between themselves and their enemies. In that way, people turned *mhesvi* into a weapon against their aggressors (without touching it). Alternately, such *mhesvi*-infested areas were simply avoided.

The following are several examples. The first is that of "*bantu*"-speaking herders of *mombe* (cattle), whom archaeologists say had moved south through a strip of land between the Kgalagadi and upper Limpopo because this area was a *mhesvi*-free "corridor" in the period prior to 1 CE. They would have stayed north of the Limpopo, but they were pushed through the *mhesvi*-infested belt into the Transvaal by "bantu," migrants from the Cameroon-Nigeria borderlands who arrived in the region around 300 to 500 CE. *Mhesvi* ensured that no cattle-keepers could settle on the Limpopo Valley without losing their *zvipfuyo* (Dicke 1932, 793). We have no record—orally, ephemerally, or textually—of what these early peoples called this deadly *chipukanana*.

In a second example, the rulers of the sprawling Munhumutapa kingdom deliberately left an area near the Indian Ocean unsettled because of the presence of *mhesvi* and *hutunga*. Munhumutapa (the king) ruled over his *vedzimbahwe* subjects, among whom the *chipukanana* was (and is still) called *mhesvi*. The presence of *mhesvi* in the eastern territories of Munhumutapa comes from Portuguese documents. In 1569, two Portuguese travelers—the army captain Francisco Barreto and the Jesuit friar Father Francisco Monclaro—left for Munhumutapa at the head of a formidable expedition of five companies, each two hundred harquebusiers strong, backed by a corps of cavalry and cannon and twenty small ships. Trouble started when the mission reached Sena, 120 miles inland. The horses started dying—along with cattle and, soon, even the troopers. Monclaro concluded that the local Swahili guides had poisoned the grass and the waters and asked Barreto to see to their execution. Barreto was eager to oblige—but the deaths continued. Only fifty men survived. Barreto was not among them (Monclaro [1571] 1975). As it happens, the expedition at that time was trudging through thick *mhesvi* and *hutunga* country.

This instance of strategic deployment in the environment, where settlements were positioned in such a way that invaders or attackers encountered the deadly *zvipukanana* first before reaching the inhabitants, is not just a Southern African phenomenon. The historian and philosopher Lansiné Kaba chronicles a spectacular example from sixteenth-century Mali. It is 1591. The famed empire of Songhai is facing an unprecedented invasion from the forces of the Zargun Pasha of Morocco, who are armed to the teeth with the very latest guns. Songhai's beleaguered warriors are armed only with bows and arrows and spears. The forces of Songhai lure the Moroccan invaders into "an extremely unhealthy site on the Niger River infested with mosquitoes and tsetse flies." The war rages until 1595 as exhaustion, thirst, starvation, destitution, and sickness from *mhesvi* and *hutunga* destroy the horses and decimate the invaders. The Pasha requests six fresh army corps worth of reinforcements, "but they, too, fell victim to Songhai's attacks and tropical diseases" (Kaba 1981). Morocco wins but is mortally wounded.

Back in Southern Africa we find our fourth example: that of three northbound refugees fleeing from the Zulu king Shaka from the 1820s to 1840s, avoiding the thinly populated Limpopo Valley *mhesvi* belt (Dicke 1932, 794). One group—that of Zwangendaba—blundered into the *mhesvi*-infested Dande-Mutoko-Nyanga-Mudzi-Gorongosa-Tete area south of the Zimbabwe River and the equally heavily infested Luangwa and Shire Valleys north of it—with heavy consequences (Barnes 1951). The Kololo under Sebetwane suffered a similar fate before eventually arriving in the Linyati River-Mapfungautsi plateau area in 1840 and establishing their kingdom near the majestic *Mosi oa Tunya,* but with far fewer herd animals (Ford 1971, 335).

The third group under Mzilikazi was more knowledgeable about *mpukane* (*mhesvi*) presence and chose its routes more wisely. In the 1820s, this Ndebele group migrated north through a chasm between the windward slopes of the uKhahlamba Mountains in the west and the *mpukane*-infested Lebombo Mountains to the east to save their cattle herds from the deadly *mpukane*'s bite (Fuller 1923, 8). The further north they went, the less their knowledge applied. They eventually settled in the western part of *dzimbahwe* after realizing that the presence of *mpukane* was blocking their path north across the Zambezi, but only after severe losses of *inkomo* (or *izinkomo,* as cattle are called in *isindebele*). Mzilikazi had no choice but to settle around what later became *Bulawayo* (the place of the killed). West was the Kgalagadi desert, the northeastern and southeastern fringes bursting with hungry *mpukane* (Dicke 1932, 794). To the east stood the not inconsiderable power of *vedzimbahwe,* as well as the kingdom of Gaza, established by another of the Nguni kings, Manukusa Soshangane. It is from the Gaza that an account of the use of *cordons sanitaire* in the management of *inthesi,* which the locals called *ndedzi,* will later come.

The final example comes from *vedzimbahwe,* caught between the Gaza, Ndebele, and *mhesvi,* who chose to build their settlements on hills rather than retire to the safety of the *mhesvi*-infested lowlands, where they would live but lose their cattle. At least on the highlands, *vedzimbahwe* could spot their enemies from afar, use *moto* (fire) and *hutsi* (smoke) to alert (signal) others, and roll loose rocks on to the clambering attackers (figure 1.1). Altitude ensured that the damp, warmer, and pest-friendly valleys were avoided, except for short durations when *vedzimbahwe* came down in the dry winter to pasture their *zvipfuyo.*

Figure 1.1
Vedzimbabwe's nineteenth-century practice of building on hilltops to command the view of the country below and deploy light, smoke, and sound as sentinel systems against surprise attack.
Source: Zambezi Mission Record 1914.

Further south in the valleys of the Umfolozi and Hluhluwe Rivers, the Zulu successfully kept their cattle safe from *mpukane* through strategic cattle migration and grazing. In summer, they drove their herds into the highlands, returning them to the lowlands in the cold, dry winter. *Mpukane* coming out of *egobolondweni* (*zvikukwa* in *isizulu*; literally, "[that which lives] in the shell") under cold, wintry conditions did so in a labored way and found it difficult or impossible to transmit *hutachiwana*. Therefore, even if they emerged, *mpukane* were too weak to fly and bite and were virtually harmless. As Burtt (1946) and Fairbairn and Watson (1955) showed, the Zulu grazing management system was designed strategically to utilize lowland *idlelo* (pastures) in the season in which *mpukane* was free of *isihlungu* or *igciwane* (*hutachiwana* in *isizulu*), steering clear of them when *mpukane* was infected and deadly.

The Orma of Kenya used the same grazing system, evacuating cattle from the riverine grazing landscape in summer and returning in winter only when it was dry and tree foliage had dropped. Their grazing philosophy held that "when undergrowth and the canopy opened up the habitat of *mhesvi* would shrink and grazing become possible. In the areas such as the delta known to harbor mosquitoes and flies that bite cattle, the Orma burn the dry cattle dung to create smoke screen for repelling mosquitoes" (Oba 2009, 35). By contrast, *vachena*'s land tenure systems were fixed and left no flexibility to deploy flight as a way of avoiding *mhesvi*. This weakness of the extension of *vachena*'s enclosure system of individuated property rights is also reflected in the sleeping sickness disaster the Belgian authorities created in the Semliki Valley of Congo (van Hoof 1928).

Those *vanhu* and *zvipfuyo* inhabiting *mhesvi*- and *hutunga*-infested areas for sustained periods acquired physiological defenses or tolerance of *hutachiwana* (i.e., trypanotolerance), which they transmitted to their offspring. Elsewhere in Africa, *mombe* breeds like the Ndama and Muturu (Dwarf Shorthorn) of Nigeria and Ghana, and the Ngoni breeds near Lourenço Marques—both very short—were highly disease resistant owing to long exposure to tropical disease agents, including *mhesvi*, ticks, and varieties of flies (Chandler 1952, 1958). Nobbs (1927, 336) remarked that the *kasiri* of *vedzimbahwe* was, compared to *vachena*'s breeds, "more resistant to disease" and suffered "less than improved cattle from redwater and gallsickness, and epizootics appear[ed] sooner to lose their virulence." There is no reason to suggest that this variety of *mombe* had not been "salted" as well through centuries of the gene line's exposure to *mhesvi* bites.

Some people could be carriers of the sleeping sickness trypanosome without expressing any symptoms. This was so in the Zambezi Valley, where "healthy carriers" had existed in high proportion and visitors became infected in places where locals were perfectly healthy (Blair 1939). At different times between 1850 and 1890, Robert Moffat, David Livingstone, John Kirk, and Frederick Courteney Selous traveled along and crossed the Zambezi, and either they or their companions were struck with "fever." It is easy for us as readers to deduce that the fever was malaria; it is always easy to say *hutunga* caused the *nyong'o* (malaria) and killed people while *mhesvi* killed *mhuka*. Yet the fever occurred in what became the epicenters of *gopé* in the twentieth century. Assertions that *gopé* was a new disease in Southern Rhodesia (Fleming 1913) become suspect.

These few opening examples show that people across Africa did not always avoid *mhuka* or live separate from them, but instead learned to coexist with them. Suppression of thicket formation, shifting cultivation, grazing, and controlled late-season burning all were deployed to keep vegetation low and *mhesvi* at bay. To create a *mhesvi*-free zone, people simply hunted all *mhuka* in each area of land separating *sango* from *musha* (village). All *mombe* were banished to one side in summer when *mhesvi* was active, then brought back in winter when *mhesvi* was least active. In times of drought, these areas became *mafuro* (pastures) of last resort, avoided when seasons were normal.

Mhesvi Management Techniques: A Survey

Whereas *vachena*'s control of *mhesvi* was based on *exterminating* the *chipukanana*, that of *vatema* prior to the coming of *vachena* was based on *managing* it. *Vatema*'s approach was rooted within a deep spiritual ethic toward earth's endowments, not a lack of means of mass destruction (Mavhunga 2014, 23–40).

Edges were generally spaces of exile or refuge. Those defeated in war, running away from succession disputes, or dispossessed of their land were forced into flight. Unless they waged an assault to displace others or found refuge under powerful protectors, these fugitives were forced to *mhesvi-*, *hutunga-*, and drought-infested margins to eke out a challenging existence. The margins between powerful polities were not necessarily no-man's lands. The margin was also a space deliberately left unsettled as security against pestiferous intrusion, so that the deadly predators and *zvipukanana* would destroy or weaken the inhabitants of outlying areas before they reached the king or chief. With that in mind, the following are brief surveys of *ruzivo rwemhesvi* in different societies in the region.

Batswana

Vachena traveling in the upper Limpopo found Batswana using various methods of movement management or traffic control to mitigate the effects of *tsetse* on their herds. One was to travel by night while the *setshidinyana* (insect) was sleeping. In 1877, local Batswana guides told the Englishman Thomas Baines to "leave ... before dark so as to be able to get into a definite track, and yet not so soon as to rush into the fly until it has retired for the night" (Baines 1877, 65–66). He followed that advice and did not suffer molestation from the *setshidinyana*. *Basweu* (*vachena*) like Baines usually arrived in the Southern African winter (starting in May) and hunted until October, when the rains began (and fevers erupted), trees were leafing again (and spotting became difficult), and Batswana leaders enforced the closed season to allow *diphôlôgôlô* (*mhuka*) to breed and replenish. In the hot and humid summer months, *dikgomo* (cattle) grazed in *mafulô* (pastures) in which, barring known riverine fringes, *tsetse* was absent. The rules against taking *dikgomo* into the *diphôlôgôlô*-rich and *tsetse*-infested areas were strictly enforced in summer.

Then, in the dry winter months, as *mafulô* became depleted and crops in the uplands were harvested, *dikgomo* could be systematically driven through tsetse-infested areas to *mafulô* in *tsetse*-free fields. For this *dikgomo* movement, Batswana counted on the cold nights, while *tsetse*'s body and wings were numb with cold and the passage from *motse* (village) to the winter *dikgomo* posts was safe. The ground frost in late May through to early July made for perfect conditions to move *dikgomo*. Then, in September–October as the rains broke, it was a matter of waiting for the heaviest downpours or a cold day on which to drive the herds back from their winter grazing.

Batawana

What is now called the Kalahari is *vachena*'s corruption of an African name, based on a combination of ignorance and arrogance in pronouncing names *vatema* had given their neighbors, lands, rivers. Places. "These are the lands of Bakgalagadi," Batswana had told them. *Vachena* could only manage "Kalahari."[1]

On the northern fringes of the Kgalagadi lies Lake Ngami, an *endorheic lake* (a water body that does not flow into the sea) that formed a shallow belt of the Okavango Delta. Ngami fed off the seasonal waters of the Taughe River, tributary of the Okavango. It is one of the fragmented remnants of the ancient Lake Kgalagadi. When Ngami was full, the Taughe flowed east, and when the salt lakes were full, the Taughe flowed west. Being extremely shallow, the salt lakes evaporated quickly, exposing white salt, which *tlou* the elephant, *tau* the lion, *nare* the buffalo, and other *diphôlôgôlô* came to lick, bringing with them *tsetse* and attracting the hunter's interest.

This is the home of Batawana (the Tawana people), who are named after the cub (*tawana*, from *tau* [lion]). Locals—not just Batawana, but also Bakuba, Basarwa, Mbukushu, and Maxereku—strategically located their settlements in such a way that the unhealthy *tsetse*- and *monang* (*hutunga*, mosquito)-infested environment became a weapon against invaders; anyone invading their land from the east and south had to deal with these *setshidinyana* first (Holub 1880, 173; Livingston 1851, 23; Ashton 1937, 67).

The swampy conditions and fringe thickets made for good breeding and sheltering for both *tsetse* and *monang*. As a rule, the locals steered clear of such places filled with pestiferous flies to protect their own lives and those of their *diruiwa* (livestock), preferring instead to build on higher ground. They also seem to have developed natural resistance to the *setshidinyana*'s sickening bite, for they lived not far from the lake itself to have access to the fish; *diphôlôgôlô* that came to drink, swim, and cool off in the Ngami's shimmering waters; and farming in rich alluvium of the valleys. In addition to being accosted by *tsetse* and being a reservoir of *mogare* (pathogens), *nare* the buffalo was also a carrier of bovine pleuropneumonia and bovine tuberculosis. The coughing and pneumonia Livingstone recorded on the shores of Lake Ngami in 1850 suggest this.

Around Lake Ngami, we see the use of night movement as a way of moving *dikgomo* between home and *mafulô* through *tsetse*-infested country. Two sets of *kitsô* (knowledge, knowing) were essential for this movement. The first was an intimate understanding of *tsetse*-infested and *tsetse*-free areas critical to the siting of *motse*, *dikgomo* outposts, and *mafulô*. Batawana knew that *tsetse* lived in the bushes and reeds, rarely in open country; they were not found everywhere, but only in specific spots, seldom shifting habitat. It was thus possible to see *dikgomo* grazing on one side of a *tsetse*-free river or mountain range, even as *tsetse* swarmed on the other side. Second, Batawana had a thorough understanding of *tsetse* mobilities, which enabled them to time their own movement to occur when *tsetse* was immobile and avoid moving when *tsetse* was most mobile and virulent. One British traveler described how Batawana cheated *tsetse* by means of *traffic control*: "Should the natives, who are well acquainted with localities frequented by the fly, have occasion to change their cattle-posts, and are obliged to pass through tracts of country where it exists, they choose, I am told, a moonlit winter's night, as, during the hours of rest in the cold season, it does not bite" (Anderson 1856, 488–489). However, this was a misrepresentation of *kitsô*; dark nights, not moonlit ones, were effective.

In other words, Batawana had *kitsô* of the physiology of *tsetse* and could manipulate the effects of time and weather conditions on it, moving when those elements rendered the *setshidinyana* immobile. In reverse, they avoided movement of *dikgomo* when time and weather conditions favored *tsetse*. There is *kitsô* there, of strategic deployment—again, without touching the *setshidinyana*, its bloodmeal source, or its habitat—which turned the environment into a pest-management apparatus without touching or modifying it.

Ndebele

From the 1840s, when the Ndebele settled in the western half of *dzimbahwe*, they did not permit any settlement on the southern border of the Zambezi as a pest-control measure against *mpukane* (tsetse), *inyamazana* (forest animals), and *abantu* (people). They saw the river as "their natural frontier of defense against their enemies … [who] themselves do not settle in that country, in consequence of the bad fevers prevailing all along the riverbanks" (Pinto 1879, 486). The price the Ndebele paid for their initial ignorance of *mpukane*'s whereabouts is now immortalized in the Ndebele name for Nata River; the circumstance of its naming by the Ndebele was recounted vividly by Robert Moffat in 1854. To put this in context, the Ndebele monarch Mzilikazi ka Matshobane was traveling with a healer or *inyanga*, the king accompanying the priest, heading north to Inyati—toward the Zambezi and into *mpukane* country. Moffat write in his diary on September 6: "Moselekatse yesterday told the Mashona doctor that as we would soon pass all outposts, when sheep and goats would cease, he must return" (Wallis 1976a, 371).

He naïvely assumed that Mzilikazi was trying to avoid confrontation with the Kololo: far from it. As they reached the Nata River, the priest learned why:

> Having passed a village about two miles, we halted beside the river Nate (sic, Nata), or according to the Matabele, *Amatse-a-monyama* (black water, sic *Amatsheomnyama*). ... This is the river along which [Mzilikazi] and a large company descended when he was driven from the Bahurutse country by [Zulu king Dingane's] warriors. It was his determination at the time to have proceeded to the Zambezi, which he intended crossing and taking possession of a new country for himself beyond, but was arrested in his northward career by tsetse. ... He with his company and a great many cattle had no sooner entered the tsetse region when scores died. He instantly saw that advance without them and of course without food, would be impossible, when he commenced a retreat in a direct course to whereabouts he now lives. From their ignorance of the locality of the tsetse, it was some days before they got out from among them. The cattle died ... rapidly. (Wallis 1976a, 371)

After a decade in *dzimbahwe*, the Ndebele king had clearly learned the correlation between the presence and movement of big *inyamazana* and the drift of *mpukane* toward settlements. *Ndlovu* the elephant usually drifted during winter from the very dry Zambezi uplands to the better watered south. *Mpukane* followed. In his October 6, 1854, diary entry, Moffat notes how Mzilikazi addressed the problem: by instituting "game-laws, so that no one but his own people can hunt the elephant" (Wallis 1976a, 375). The Englishman Frederick Courteney Selous (whom *vedzimbahwe* called *Serowe*), hunting in Ndebele country in the 1870s during the reign of Mzilikazi's son and successor, Lobengula, bitterly lamented the king's refusal to allow hunting in mid-April. When finally allowed to hunt in June, he was restricted to an area west of the Gwai (Selous 1881, 55). Not without good reason: Hunting in the Zambezi Valley in 1872, *Serowe* had seen people struck by "fever and ague" and caught it himself. The attacks came only when he stopped and rested a few days (Selous 1893, 294).

Mzilikazi followed a policy called *ukulagisa,* under which those with many *izinkomo* may lend them out to relatives without. These kin keep and manage them in exchange for milking, manure, and draft power. Mzilikazi and later Lobengula had the royal herds driven to valleys along the big rivers for winter grazing. The herders set up temporary *izinkomo* posts, composed of stockades and temporary huts. With the onset of the summer rains, they rounded up their *izinkomo* and shepherded them toward healthier areas (Garlake 1978, 491–492). The *izinkomo* posts were usually located far from *ekhayeni* (villages), in *mpukane*-free areas. Parceling out the herds to kin or subjects located in different districts also insured against losses in the event of contagious diseases, while keeping the infected herds apart and away from *ekhayeni.*

Moreover, the bite of *mpukane* was never avoided completely; people took their *inja*/dogs into *mpukane*-infested areas to gain natural resistance through sustained exposure to bites. The Tonga preferred to mix and dry dead *mpukane* in herbs (leaves or bark), grind the mixture into a fine powder, and then administer it to *izinkomo, inja,* and *imbuzi* (goats) orally. *Izinkomo* suffered under *mpukane*'s bite—but not smaller stock. Says Moffat on July 27, 1854: "The scattered inhabitants have abundance of game and are able to keep flocks of sheep and goats, which do not suffer; and it is remarkable that this should be the case, for their hair or wool is thicker than that of other animals, [and] there are about them vulnerable parts which the tsetse can easily reach. Dogs immediately fall victims" (Wallis 1976a, 368; see also Livingstone [1857] 2001, 96).

Kololo and Barotse

In the 1840s, a new force arrived on the upper Zambezi: Bafokeng (Basotho) of chief Sebetwane, who founded the Kololo polity. The Kololo were forced immigrants who headed north after disturbances in the Eastern Cape associated with the rise of the Zulu kingdom under Shaka. They fought their way north through Hurutshe, Kgatla, Ngwaketse, and Batawana country, crossed the Zambezi, and deposed the Barotse dynasties in 1840. In the next decade, Sebetwane repulsed the Zulu and Mzilikazi's Ndebele; in fact, such stubborn resistance and the scourge of *zeze* (their name for *mhesvi*) along the Zambezi discouraged the Ndebele from further northern expansion or migration. In the 1860s, the kingdom Sebetwane had built fell victim to internecine wars, paving the way for Barotse—VaRozvi descendants from *dzimbahwe* under Riwanika (Lewanika)—to retake power. The history of the Ngami-Barotseland area is reasonably well documented (Flint 2003).

Vachena traveling in the Kololo's domains in the 1850s talk of a vibrant kingdom that Sebetwane held together despite *zeze* (Selous 1893, 302). David Livingstone and fellow Englishman William Cotton Oswell visited Sebetwane's kingdom in 1851 and found that *zeze* was not found everywhere, but only in particular spots (Oswell 1894). Downstream of the Katima Mulilo bend, the Zambezi passed through rocky terrain and formed a series of rapids and cataracts, the most ferocious being at Kalilabombwe, Nambwe, and Gonye. All through these rapids, the Zambezi's banks were clothed with riverine thickets full of *zeze*. Then, the *zeze*-infested belt ended abruptly as the high banks peeled away from the river toward the north-northwest and north-northeast, forming two parallel plateaus twenty to thirty miles apart. The intermediate space between these two ranges, one hundred miles long, was *zeze*-free and the core of Barotse land (Livingston 1854, 296).

The Kololo and Barotse believed, like their Batawana neighbors in the southwest, that *zeze* could indeed be cheated by moving at night. This was *ngamboto* (a matter of common knowledge) and not necessarily *butalifi* (special expertise)—especially for men, who from childhood were schooled in herding *likomu* (cattle; singular *komu*) and knew that their *likomu* would die if they entered certain areas during the day. Even the most *zeze*-infested places could be "crossed with safety by night if sufficiently narrow to allow of the cattle being driven before sunrise." In the 1850s to 1860s, *vazungu* (whites) tested this idea during their journeys and "found [it] also correct" (Kirk 1865, 154).

The locals also used *sinuka* (odorous grasses) to enable the safe movement of their *likomu* (a few of them) through *zeze* country in broad daylight. They noticed that *zeze* avoided *limbweletete* (human excrement), especially whenever a homestead had been built in a *zeze* habitat; the insects had withdrawn instead of swarming to the bushes where people relieved themselves or to *likomu* pens. The Kololo smeared each *komu*'s skin with mixtures of *mulaha* (liquid cow dung), *mabisi* (milk), and some medicines before setting off through *zeze*-infested belts. They also burned the roots of a certain shrub underneath the bellies of standing cows, the smoke whiffing and wafting into the skins to act as *sutelezi* (repellent; Kirk 1865, 154; Livingstone [1857] 2001, 82–83).

The Kololo also used other forms of *sutelezi*. For instance, *vazungu* found the locals passing *likomu* through smoke "made from sun-dried cattle dung [*lisu* in *silozi*]. Fires burnt in the cattle kraals during the nights generated a very strong smell from which the cattle received a certain degree of protection" (Kjekshus [1977] 1996, 54). Among the local Kololo of the upper

Zambezi, in 1857 Livingstone observed: "The well-known disgust which the tsetse shows to animal excreta, as exhibited when a village is placed in its habitat, has been observed and turned to account by some of the [witch] doctors. They mix droppings of animals, human milk, and some medicines together, and smear the animals that are about to pass through a tsetse district; but this, though it proves a preventive at the time, is not permanent" (Livingstone [1857] 2001, 96). Of course it was never meant to be!

Ten years later, another *muchena*, Benjamin Bradshaw, remarked that *zeze* did not stay long in a camp after *mulilo* (fire) was lit, even though they would be everywhere beyond it. When "much troubled" by them in camp, the local guide told him: "Make a fire and they will go away." About which, Bradshaw admitted, "I found the experiment to succeed" (Bradshaw 1877–1878, 52). The reference to an "experiment" was a way of claiming scientific credentials among his Royal Geographical Society audience of naturalists. The real scientist was the African practitioner!

Korekore

Further down the Zambezi, in the area between the Batoka and Kariva gorges and their frontage from Lupane through Gokwe to Hurungwe and Guruve, the use of *manhuwe* (repellents) was quite pronounced among the Korekore people. They used *zumbani, sumba* (both indigenous mints), *mungezi*, and *mubhubhunu* as *manhuwe* against *mhesvi* and *hutunga* (Kazembe and Nkomo 2010; Gudhlanga and Makaudze 2012, 75–76). These were rubbed on the skin to repel the *chipukanana* as it landed, applied as internal medicine against *gopé* (sleeping sickness) and *nyong'o* (malaria), and used as a fumigant. After internal treatment, hot charcoals were placed in a small potsherd, over which the dry roots (or powders thereof) were placed. The smoldering potsherd was then put underneath the standing cow for it to be soaked in *manhuwe* and provide an anti-*mhesvi* deterrent to further bites.

The procedure was continued over several weeks, for as long as the symptoms of poisoning remained, and terminated once they were gone. Of course, not all bitten *mombe* would live, but the ingested medicine was deemed to build immunity in the event of future *mhesvi* attacks. The *manhuwe* was intended to see the herds safely through a *mhesvi*-infested belt (C. Livingstone 1861–1862, 34). The roots and leaves of *muvengahonye* (literally, "the maggot hater") were crushed and smeared around a kill site when skinning a carcass so as to repel flies (especially *mhesvi*) that secrete maggots into meat.[2]

Tokaleya

We have already seen how settlement choice enabled people to manage *mhesvi* by avoiding it. A classic example is found in the area surrounding *Mosi oa Tunya* (The Smoke that Thunders, which *vachena* called Victoria Falls), home to two interrelated people: the Toka or Batoka, and the Leya or Baleya, who speak *chitonga*. It is after the former that the Batoka Gorge downstream of the falls is named. The people living near the gorge settled there because of its elevated lands, a fine healthy climate, and a vantage point from which to defend themselves against both pestiferous *vanhu* and *zvipukanana*. Here, fever (*n'gana* and *nyong'o*) was unknown, large herds of *mombe* "furnished an abundance of milk," and "the rich soil largely repaid the labour of the husbandsman" (C. Livingstone 1861–1862, 32).

Living in an area surrounded by *mhesvi* to the southwest, south, and east, Batoka (and their neighbors the Nanzva, under Chief Hwange, VaRozvi emigrants from the south) developed a number of stratagems to deal with the pest. These communities, living between the Gwai and Deka rivers, had through their close settlements and cattle grazing maintained a *mhesvi*-free space. The Ndebele raids pushed them across the Zambezi, leaving the country uninhabited: "Overrun with game, [the area had] become one of the great strongholds of tsetse, extending from the Zambesi river for at least 60 or 70 miles in a southerly direction" (Bradshaw 1877–1878, 52).

As Bradshaw observed, one remedy for *mhesvi*-struck cattle among the locals was a mixture of herbs and dead *mhesvi*, which they used as an inoculant. The first ingredient was the dried root of an unidentified plant. The second was composed of a dozen or so *mhesvi*, which also were dried. Some of the dried roots were then mixed with the *mhesvi*, the mixture being ground together into a fine powder. The mouth of the cow was forced open so that the mixture could be administered internally. Inoculation extended to other *zvipfuyo*. Dogs and *mbudzi* also gained natural resistance to *mhesvi* bites through living through and being exposed to continued attacks. Either they were taken into *mhesvi* country and exposed, or *mhesvi* were caught, brought home, and fed to *zvipfuyo*. Batoka caught *mhesvi* by "slipping the blade of a knife or edge of an assegai on their legs and then, turning it carefully over, crush[ing] them beneath it" (Bradshaw 1877–1878, 52). Any stock brought in suddenly from outside of *mhesvi*-infested areas usually died within days. Subsequent generations acquired immunity from their parents, even if a few died after birth (Selous 1881, 131). It is possible that Batoka, like other peoples, also extended these traditions of inoculation to deal with the smallpox outbreak, as reported in the Gaza and other cases (e.g., Junod 1918; Gelfand 1964; Herbert 1975; Apffel-Marglin 1990).

Guruve, Dande, and Beyond

Inoculation was a widespread practice in the eastern and southeastern areas of Dande, Mutoko, and Gorongosa, all areas inhabited by *vedzimbahwe*. In the Limpopo Valley, Mauch observed the practice of inoculation as the only remedy against *mhesvi* in 1868:

> Only one remedy appears to help which rests on the "homeopathical" principle. The actual fly, taken internally, renders the stings innocuous, as I experienced with a dog that, after such a treatment, I took along with me to the lowest part of the Zambesi and which I sent back with my companions in a completely healthy condition. In 1868, when I had an ox, a female donkey and a dog with me and experimented with dissolved ammonia, the ox and the dog perished, while the female donkey, to which I had not applied the mixture of this salt, [survived]. (Bernhard 1971, 233)

As this example shows, these local stratagems inspired traveling *vachena* to also embark on their own experiments—with mixed results.

In the northeastern parts of *dzimbahwe*, the *mhesvi*-infested belt stretched from Guruve to Centenary, Muzarabani, Rushinga, southeast through Mazowe, Mutoko, Mudzi, Nyanga, and east through the Gorongosa mountains, all the way to Mutarara. Here as elsewhere, *mhesvi* was not just a target of medication but an ingredient in the medicine against *gopé* and *n'gana*. In northern Nyanga, it is not men but women who are mentioned in Montagu Kerr's account: drying quantities of *mhesvi*, pulverizing them with bark of a certain unnamed root, then mixing the contents in water, which they then fed their *makwayi* (sheep), *mbudzi* (goats), and *imbwa* (dogs) through the mouth. *Mombe* did not survive here even with this medicine (Kerr 1886a, 74; Kerr 1886b, 33). Edward Maund's narrative is not specific about which *vedzimbahwe* or what specific area, but in 1891 he observed that "the Mashonas dry and pound the fly, and give it to their dogs, a fly a day, as a safeguard against the effect of tsetse" (Maund 1890, 653; 1891, 12).

Gaza

Charles Swynnerton was twenty-three when he arrived at Gungunyana Farm, abutting Chirinda Forest in Chipinge, to begin his new job as a farm manager in 1900. (For map orientation in this entire section, see figure 1.2). For the next two decades, he undertook studies of the local *svifufunhunhu* (insects) and *sviharhi* (animals), tapping into the *vutivi* (knowledge)

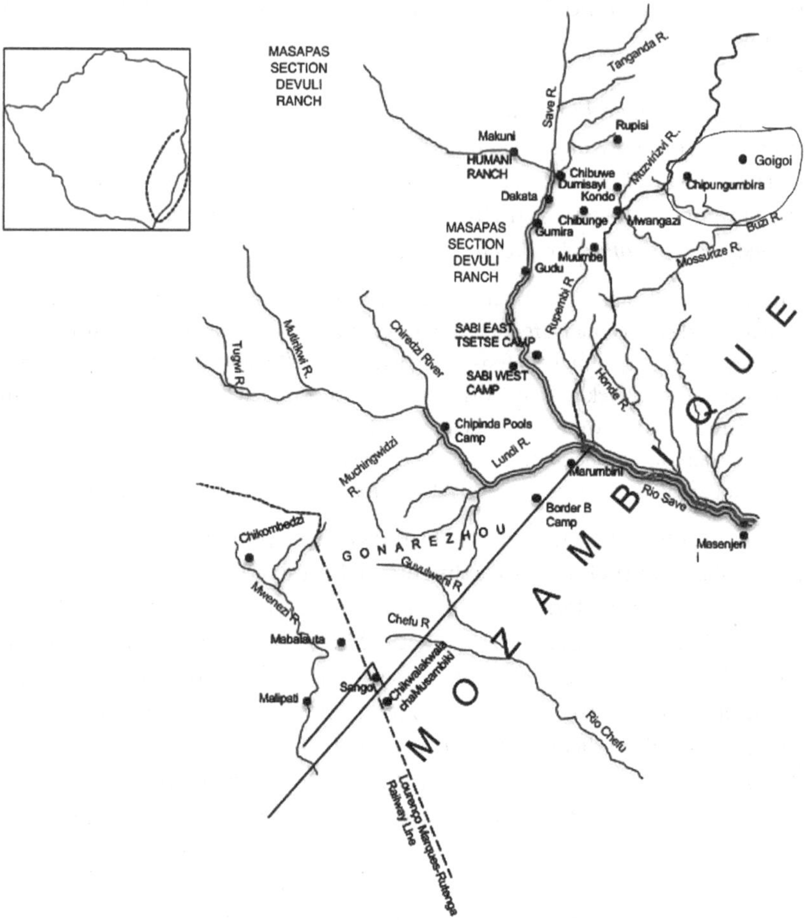

Figure 1.2
The southeastern *ndedzi* belt, showing the general area east of the Muzvirizvi and, later, the southeastern *ndedzi* area of the Savé-Runde, Guvulweni, and Rio Savé.
Source: Author.

of local Tshangana, Ndau, and Tsonga people (Marshall 1938).[3] He was told how Gaza king Mzila had set about combining game elimination, forest clearance, and prophylactic settlement to deal with *ndedzi* (*mhesvi*) and was able to keep *tihomu* (cattle) within a fly belt. This is Swynnerton's account:

From Gandwa, Umzila sent an order to "*sondela enkosini*" (draw near to the King). Thereupon an immense compulsory movement of the population took place. The

> country to the east of the Sitatonga Hills, particularly in and south of Gunye's, was at that time more fully populated than that to their west, but almost the whole of this population was deported, territorial chiefs and all, to the lower parts of the tract between the Sitatongas and the present British border, to Spungabera [*sic*; Chipungumbira] and Gwenzi's country, and, to the Umswirizwi [*sic*; Muzvirizvi] Valley, to Sinjumbo's Hills and Chimbeya's, and even eventually to the Sabi [*sic*; Savé] east of this area. (Swynnerton 1921a, 315)

According to Swynnerton, *sondela enkosini* was not meant to address only the problem of *ndedzi* and would have happened regardless of its presence. What is important about it is not its purpose or intentionality, he says, but the effects on *ndedzi*, according to surviving subjects of the Gaza king. Mzila maintained three large *tihomu* herds at Umpombo's, at Dongonda, and west of the Sitatongas. There is no doubt that the *ndedzi* problem—in addition to security needed due to surprise attacks from his brother and challenger to the throne Mawewe's still-fighting troops—was a primary concern.

To be clear, forest clearance was not the only strategy the Gaza were using. Like other neighboring peoples, they also fed dead *ndedzi* to *timbyana* (dogs) and *timbuti* (goats). Nor was such a practice limited to *ndedzi* or *sviharhi*, as we saw earlier in other communities. However, the logic of clearing a portion of forest was to draw a clear boundary and buffer zone between these *tihomu* and *sviharhi*, a project that marauding *nghala*/lions tested time and again when attacking the settlements. Water was also a problem, as was the feud between Chiefs Makuyana and Gogoi, subordinates to the Gaza. Regardless, Mzila heavily settled the buffer zone and suppressed any recovery or intrusions inside it through regular hunts (Swynnerton 1921a, 335).

Bush clearing, whether on its own or followed by settlement, deprived *ndedzi* of its habitat and cover from predators. Tree felling itself was a process of applying technology, with its many tools—not least *xihloka* (axe), composed of a blade fixed to a wooden handle. *Svihloka* (plural of *xihloka*) were themselves products of two industrial processes: First, that of turning earth into metal, and metal into blade, depended on skilled, well-reasoned manipulations and syntheses of air, ore, and fire; there was nothing accidental there at all. The *démó* (blade) was made from iron mined in the mountains and underground, smelted in the *svindlu svasvifuri* (forge), and finished with the *svigalanyundzu* (hammer) and *ndzilo* (fire) of *svifuri* (smiths). The wooden handle was the handiwork of *muvatli* (carpenter or sculptor; anyone except small children and women could be one); it required thorough *vutivi* of specific trees with hard wood resistant to cracking and boring *svifufunhunhu*.

Felling was not a haphazard process. If intending to destroy the tree for good, it was cut in such a way that the stump was left flat-topped. If intending the tree to regenerate shoots, the stump was cut at a thatch-roof elevation (or the white man's 45° angle). The bush generally regenerated very slowly over many years, while clearance followed by settlement gave permanent protection from reinvasion, if the area was large enough.[4] The effect on wooding was profound: Bush vanished, leaving bare country, save for the concentrated *tiko* (village) and gardens. The measure was confined to the lowveld; toward the hilly areas of the Pwizizi and Mtshanedzi rivers, the bush was not completely cleared, but was "surrounded on three sides by a broad cleared cordon, and on the other, backed by the highlands" (Swynnerton 1921a, 333). Beyond the Mtshanedzi and Budzi (Busi) neighborhoods, the buffer zone Mzila created "remained completely uninhabited and uncleared." The Zinyumbo and Gogoi areas were completely cleared right up to Mwangazi, Gwenzi's country, the Muzvirizvi River Valley, and portions of the Savé, which were populated with gardens (333).

The effect on *sviharhi*, carriers of *ndedzi*, was thorough:

> Large mammals became very scarce—not merely big game, but pigs and baboons. ... Drives with nets were organized across the entire country, and game, pigs and baboons were thus killed wholesale. If a herd of buffalos was reported subsequently anywhere west of the Sitatongas, it was at once hunted; if pigs appeared in a garden, they were at once tracked down to their retreat ... and killed. Except on its fringes the "Oblong"' ... was a great uninhabited game reserve. The game in it was thinned, it is true, and was kept well driven within its borders, but there still remained enough to attract ... hunting parties. In the heavily settled areas a few bushbucks, duikers and pigs were still to be found throughout the period. (Swynnerton 1921a, 333)

It was a very thorough clearance. The overall effect was such that *ndedzi* and bigger *sviharhi* could not cross where they were not wanted.

Tihomu (cattle) were absolutely banned from the cleared areas; any required for ceremonial slaughter were brought in from safe areas as needed. Countless attempts to introduce them in Gunye's country and south of the Budzi from the Mwangazi eastward resulted in losses, the numbers lost being replenished from the safe areas and through raiding *vedzimbahwe*, who called the Gaza *madzviti* (pillagers). *Tihomu* thrived in Zinyumbo's hills, on the Mwangazi, along the Muzvirizvi and toward Chipungumbira. In the Gogoi-Makuyana stretch, people kept *tihomu* "right under the Sitatongas both at and opposite the Rupisi and from the great bend of the Mtshanedzi to its source, also in the hills behind the cleared guard-area between the Mtshanedzi and Pwizizi and up to and beyond [what later became] the ... British border" (334; also Dube 2009).

On the Savé, elders who had participated in *sondela enkosini* told Swynnerton:

> This had previously constituted a separate fly-belt, which was eventually almost completely wiped out by native cultivation. The rinderpest ... may have given the *coup de grace* to the surviving remnant or two ... but at any rate cattle were already being placed and kept successfully all over the old fly area in the seven years between Gungunyana's [*sic*; Ngungunyana's] departure ... and the advent of the rinderpest. ... When the country was closely settled, cattle were kept successfully in places where they had always died before; and when the settlement was well established, they succeeded where in its earlier days they failed, though fluctuations still took place with successive shifting of the population. It is true that herds actually abutting on the fly still suffered small and occasional losses, as they are doing today to a greater extent on the present fly boundary. (Swynnerton 1921a, 334)

With the withdrawal of Ngungunyana to Bileni in 1889, *nhoveni* (forest) returned, and so did *ndedzi* (Selous 1893, 304; Millais 1895, 142–143). This story has echoes in Kjekshus ([1977] 1996). The difference is that the physical infrastructure died because of *valungu*'s (whites') regimes of ecological management, but the ideas that undergirded it did not. What happened to them?

2 Translation into Science and Policy

By settling on the central watershed, first the Ndebele (1838–1840) and then *vachena* (1890–1893) built their settlements based on *ruzivo* (knowledge) and millennia of strategic deployment in and modification of the environment by local *vedzimbahwe*. I argue that it is impossible and ahistorical to consider the project of creating a Rhodesia absent the *ruzivo* upon which its establishment was founded. The same is true for the Ndebele before *vachena*. They had displaced VaRozvi. It is also true of all incoming *vedzimbahwe*, who had scattered and/or subjugated Basarwa (the San). That is not to say these newcomers added nothing to environmental management knowledge, but to caution against a tendency to blow the transformative power of incoming things out of proportion. Instead, encounters between incoming and local *vanhu* are repositioned as *kusangana kweruzivo* (knowledge encounters).

It was not by accident that the *dzimbahwe* plateau became white man's country, while *vatema* were squeezed out to crowded reserves and the *hutunga*- and *mhesvi*-infested lowlands that *vedzimbahwe* had strategically deployed themselves against and away from on the watershed (Ford 1971). Reserves like Gwai and Shangani in Ndebele country, Hurungwe in the north, and Matibi II in the southeast were drought-prone, *mhesvi*-prone, and infested with *mhuka*. People felt that they were being treated as *mhuka*, evacuated to live with other *mhuka*: *nzou* the elephant, *shumba* the lion, *gudo* the baboon, and other creatures of the forest. Forced resettlement in such areas inevitably set up confrontation between *hurumende yevadzvanyiriri* and those calling themselves "African nationalists" (who popularized the labels of *vanhu*/*vatema* as "Africans") from the 1940s onward (McGregor and Ranger 2000). The point is that the originators of the *ruzivo* that made the plateau into livable space were being thrown to the undesirable margins.

Methodologically, *vapambevhu*'s theft of land made livable by *vanhu* serves as a call to carefully explore the encounter between *muchena* and *mutema* as one about *ruzivo*. The better starting point is not the moment of partition but the encounter between *vachena* traveling in the late nineteenth century and publishing, for example, in the *Journal of the Royal Geographical Society* (principally), and local custodians and practitioners of this *ruzivo*. Why did these *vapambevhu* (abductors of land) defer to *ruzivo rwevatema* (black people's knowledge)? Why did they write about it so candidly?

It's not that *vaive vanhu vakanaka chaizvo* (they were wonderful, objective human beings). The pain, loss, and sorrow that *mhesvi* inflicted upon these writers forced them to write about their experiences. With no experience of *mhesvi* overseas, these itinerant *vachena* had no choice but to defer to local inhabitants, who told them about and taught them their own means and ways of dealing with the insect. The itinerants then described these encounters in their memoirs and submissions to peer-reviewed journals, for which travel and experience were critical credentials of natural history.

Vachena's mobilities were disruptive by their very nature: Hunting with guns, wounding *mhuka* that then fled across the country, was disruptive. Even more disruptive was geographic exploration, which always entailed moving across—against—long-respected boundaries between *misha*, *sango*, and *hufuro*.

Equally, the local environment and its constituent elements were also facilitative of *vachena*'s mobility. Hospitality and hostility, availability or paucity of water, and the presence or absence of threats to good health were just three of many critical determinants of the success or failure of a journey into the Southern African interior. As carriers of a deadly *hutachiwana* (pathogen), *mhesvi* and *hutunga* often coexisted in the same places through which itinerant *vachena* passed.

As in the last, this chapter methodologically continues to read traveling *vachena*'s accounts in search of confessions—this time, first about the pain *mhesvi* exacts on them as they explore (according to them), hunt, trade, and preach, and then about the stratagems they learned from local people and wrote about in their journals and memoirs.

Translation: How Local Knowledge Moved into Western Science

In the nineteenth century, the *vachena* traveling in the Southern African hinterland to collect, write about, and paint *zvipukanana* were zoologists, specialized entomologists, and part-timers. They were not journeying in *terra incognita* or places without political jurisdiction. They might have been on their own some of the time, but they were never alone; all were under the jurisdiction of one form of authority or another, political as well as spiritual (Mavhunga 2014). Fewer white travelers understood protocol more than Emil Holub, whose account illustrates the role of etiquette as a lubricant to mobility.

Writing in 1881 at the end of nearly a decade of staying and traveling in the regions between the Vaal and the Zambezi, the Bohemian implored the aspiring white explorer to make acquaintances among "the natives," entering their *misha* and seeking permission from their kings to pass through. The visitor had to purchase food from the local inhabitants and hire them as "servants." It was not scientific enough to merely "list names of tribes and countries" and describe their "most interesting customs" and just breeze through the countryside, spend a night or two, and then vamoose. The traveler, Holub said, could only understand "the natives" if he stayed months, even years, among them, learning their language, practicing their customs, observing how they related to each other and to other "tribes" and to the white men. Therefore, Holub elected to go into places that were "not yet in any way civilized," having also lived among "tribes living among the white men," so that he could "notice the difference between those who enjoy the benefits of civilisation and those who do not" (Holub 1881, 3).

Hence the ease with which itinerant *vachena* accessed *ruzivo* and practices about *mhuka, mhesvi,* and their actions. In this chapter, you will see traveling *vachena* appropriating this *ruzivo*—sometimes acknowledging its sources, usually representing it as commonsense, and sometimes considering it as a "myth" they now put under "experiment." In this way, these travelers installed themselves firmly as the producers of true *ruzivo,* derived from "scientific" method, whereas "the native" was content with "myths" and "legends."

Some traveling *vachena* acknowledged their local sources, whereas others did not, obviously to claim the credit of "discovery" and exaggerate the pain they had endured to "explore" and how their ingenuity and improvisation had saved them from certain death. Experience was a very good teacher, if only complementary to knowledge already shared with them by local inhabitants or gained as a direct consequence of heeding or disregarding "native advice" and entering into certain encounters they could have avoided.

To restate the question, then: Why did traveling *vachena* appropriate and write about *ruzivo rwevatema* (knowledge of black people) about *mhesvi?* The answer is very simple: They had no means of controlling *mhesvi* and there was nothing to work with except locally produced ideas and practices. Otherwise, they risked losing all their oxen or horses and having to walk and carry their loads themselves—or spend heavily on local porters.

William Cornwallis Harris was one of the first Englishmen to suffer the pain of *mhesvi* enough to write about it. In 1836, after the *chipukanana* had terrorized him greatly on the upper Limpopo, he was trying to describe it to his audience of *vachena*. He called this strange *chipukanana* "a large species of gad-fly, nearly the size of a honey-bee." Some early travelers called this "horrible" *chipukanana* a "warble fly" (Clark 1857), others a "poison-fly" (Chapman 1868, 174).

It was Batswana that revealed the name *tsetse* to the Englishman Roualeyn Gordon-Cumming as he hunted in the upper Limpopo environs between 1843 and 1849. The distinction between *ntsi* (the ordinary house-fly; plural *lintsi*) and *tsetse* (*mhesvi*) was very clear, because the latter's bite spelled doom for a Motswana's *dikgomo* wealth. Gordon-Cumming (1850) did not take long to bear witness to this: "When under the [Modimolle] mountains I met with the famous fly called 'Tsetse,' and the next day (17th August, 1846) one of my stud died of *tsetse*. He had been bitten under the mountain range lying to the south of this fountain" (227). The Englishman observed the head and body of the stricken animal to swell up "in a most distressing manner before he died. His eyes were so swollen that he could not see, and in darkness he neighed for his comrades who stood feeding beside him" (227). Meanwhile, two other *vachena*, Cotton Oswell and Major Frank Vardon, were hunting in the Marico-Limpopo area. The latter apprehended one *mhesvi* that he took on his person to England (Oswell 1894, 113). Traveling through the Letaba-Lepalale stretch, crossing the Limpopo, and out north toward the watershed, the German geologist Karl Mauch decried the presence of "the Tsetse (*Glossina*), this pest and scourge for anyone who does not travel on foot." Its bite on his right wrist had caused an hour-long inflammation, but the infliction on the oxen was fatal (Bernhard 1971, 99).

Traveling *vachena* usually "escaped almost unscathed," but the horses and *mombe* did not. They encountered *mhesvi* in the bush, "or among the reeds," ready at a stir's notice to pounce on the wayfarer, "but not rarely in the open country." They noticed while walking, through being bitten, that the *chipukanana* was confined to particular spots; *vanhu* that had lived long locally knew the *mhesvi* not to shift its haunts (Anderson 1856, 488). The local chiefs and commoners told *vafambi vachena* (white travelers) things about *mhesvi*, guided, carried loads, and tracked for them.

After all, the local men these travelers employed as guides were the mobile sentinels who "reconnoitered in front, so as to announce the appearance of the poisonous insect at once" (Mohr 1876, 290). Countless times Moffat and Livingstone (figure 2.1) had been led by Basarwa guides through *mhesvi* country (Wallis 1976 vol. 2, 163). The men displayed encyclopedic *ruzivo* on what dangers inhabited which stretch of country and ways to avoid them. When the wayfarers outspanned, they always posted one of these local men "to watch every insect that approached them" (Baines 1877, 63). These men were usually moving well forward and, upon sighting *mhesvi*, tracked back to warn the main party of *vachena* further behind. Local knowledge defined the itinerary (Mohr 1876, 290).

The *mhesvi* was also an annoyance with its "incessant persecutions," as this account from Thomas Baines, a veritable artist, shows: "At the moment, perhaps, when one requires the utmost steadiness and delicacy of hand, a dozen of these little pests take advantage of his stillness, and simultaneously

Figure 2.1
David Livingstone (left) and Robert Moffat (right).
Source: National Archives of Zimbabwe.

plunge their preparatory lancets into the neck, wrists, and the tenderest parts of the body; one or more cunning fellows actually selecting the places where the lines of fortune radiate or cross, with a skill in palmistry that would do honour to an experienced gipsy" (Baines 1864, 511).

Anyone who has seen *mhesvi* will know that this *chipukanana* is a persistent, if uninvited, companion of anything that moves. "They accompany us on the march, often buzzing round our heads like a swarm of bees," noted David and Charles Livingstone (1865, 205) as they proceeded up the valley leading to the Mburuma pass along the Zambezi. John Kirk's account of an encounter with *mhesvi* near the Zambezi-Kafue confluence is particularly hilarious: "While walking along the river-bank in search of game, under flat-topped acacias, I heard a buzzing sound, and saw a cloud of bees, I ran off, while they followed. On looking back I found it was only 'Tsetse'; so, arming myself with a leafy branch, I kept them off and continued my journey; they accompanied me for some distance however. I have never again seen them congregate in this manner" (Kirk 1865, 155).

Mhesvi was a "cunning" companion with "intention" to feed. Hence, "when intending to bite," the flies landed "so gently that their presence [was] not perceived till they thrust in their lance-like proboscis [long mouth]." The pain was "acute" but brief, giving way to "the disagreeable itch of the mosquito's bite" (Livingstone and Livingstone 1865, 207; Chapman 1868, 175–177). In 1857, Moffat remarked on the state of Mzilikazi's bare-skinned messengers to David Livingstone, "after passing through the tsetse and arriving with skins stung till they are rough as a file" (Wallis 1976b, 109). (For a portrait of Mzilikazi, see figure 2.2a.)

What types of locally generated knowledge of *mhesvi* did these white writers translate into written text for their white audiences? The answer: any information and stratagems that people shared with them verbally and in practice. Take their use of night movement to cheat *mhesvi* as an opening example of *vachena* using *ruzivo rwevatema*. Writing on the move in the Sesheke area of the upper Zambezi, David Livingstone ([1857] 2001) tells of traversing the "20 miles infested by the tsetse during the night ... so pitchy dark [that] we could only see by the frequent gleams of lightning" (353). Later, as he trudged through the upper Limpopo *mhesvi* belt, Baines (1877) was grateful that "the night was dark and cloudy, preventing any observation for latitude, but affording us additional security against the insect pest" (66). Advising his all-*vachena* public about the road from Pretoria to Delagoa Bay via the Lebombo Mountains, Baines (1877) warned of a forty-mile strip of unhealthy country that needed to be crossed "as rapidly as possible, and at night or during a cold day, when the fly is dormant" (108–109). The distance from the Lebombo mountains to the port of Delagoa Bay was thirty to forty miles, low country with a "not undeserved reputation for unhealthiness," a significant portion of it *mhesvi*-infested but "sufficiently narrow to be passed through in one night" (108–109). Here we see how travelers manipulated temperature and the darkness of the night to "cheat" *mhesvi* as *vatema* had educated them to do.

These wayfarers took this *ruzivo* to Europe, where, as geographical explorers, they presented their papers before scientific associations such as the Royal Geographical Society and the Linnaean Society. They published in the *Journal of the Royal Geographical Society*, *The Field*, *Country Life*, *Fortnightly Review*, and other forums in which explorers and other empire publics "congregated." They also deposited trophies they brought from the colonies into museums of natural history, live *mhuka* in zoos, and plant varieties in botanical gardens (Austen 1908). The most influential south to north transfer of *ruzivo rwemhesvi* and its forest animal hosts occurred during the past five hundred years, peaking in the last two hundred.

Take, for example, John Kirk, who in his address to the Royal Geographical Society in 1865 "forgot" to acknowledge who had educated him about *ndedzi* when talking about his movement by night to cheat *mhesvi* in the Zambezi Valley. Clearly, Kirk is building on this local practice as he tries to "scientifically" explain night movement as a strategy against *mhesvi* bite: "In the morning while the dew hangs on the grass, and before the heat of the rising sun has warmed the air, the 'Tsetse' is dull and sluggish, resting on the underside of some leaf or blade of grass; when forced to take wing they may then be easily caught. ... By night I have never been bitten by 'Tsetse,' nor do they fly about after sunset" (Kirk 1865, 150).

In his diary entry of October 23, 1871, Baines too does not acknowledge the source of his wisdom, but takes two precautions against *mhesvi* that we earlier discussed people using. He says he "sent word to have the oxen kept away till after dark"; then, with his local staff, he "set fire to the grass and to heaps of rubbish to drive away the Tsetse, a few of which we saw." Curiously, Baines acknowledges the Boers for cutting through the Limpopo *mhesvi*-infested belt from the Transvaal into Ndebele territory. But who taught the Boer farmers this knowledge of *mhesvi*? *Vanhu vatema*, as shown in the previous chapter. Boer farmers like Theunis de Klerk told Baines that they "knew where to ride their horses with safety between the patches of fly; they also have safe or inoculated oxen and even ride their horses in; they will not tell their medicine, but charge an ox for making a horse safe" (Baines 1877, 61, 68). Baines found that *mhesvi*, "though occupying large tracts of country, does not completely overspread it, but leaves parts which are known to various hunters, and which serve as channels by which a course may be steered with some chance of escape from the deadly insect." Of course, as Baines discovered, the ability to steer through these patches depended on whether one was able "to obtain a skillful pilot" (80). A farmer named Andreas Duvenage told Baines of "a safe passage through the fly, between Blauwberg and Zoutpansberg." Devenaar lived eighteen miles north of Marabastad and was reputed to have the best-known road through the *mhesvi*-infested belt. He crossed the Limpopo at Commando Drift, west of Musina, "meeting only one patch of fly, which he rides through in the night" (81, 84).

In 1881, a young *muchena* named Humphrey gave an account of *mhesvi* in the same area in which Baines had found it, and he described the distribution of the *chipukanana* and ways of cheating it through tactful mobilities: "Leaving the last halting-place free from tsetse in the evening, they travelled all night to avoid the insect, and before morning reached a narrow strip of country free from fly but without water, though there was grass for the oxen. The next night a shorter march brought them to the river in time for the oxen to drink and return back to the spot free from fly before daylight" (cited by Frere 1881, 15). It bears repeating that, as I showed in the previous chapter, it was the standard practice of *vatema* to travel by night to cheat *mhesvi* all across east, central, and southern Africa.

By 1888, the scientific position in Britain was based on *ruzivo rwevatema*:

> At present no cure is known for the bite, nor does inoculation seem to afford any protection. The fly is said to avoid animal excreta, and in some parts a paste composed of milk and manure is smeared on cattle which are about to pass through the 'fly-belts.' This affords a certain amount of protection. Lion fat is used in the same way, and is said to be efficacious. The fly is found as a rule in the neighbourhood of water, and its habitat is usually sharply defined. Often it occurs on one side of a stream, but not on the other. The limits of the 'fly-belts' are well known to the natives, and travelers can ensure comparative safety to their cattle by passing through these districts after sundown. (Encyclopaedia Britannica 1888, s.v. "Tsetse-fly [*Glossina morsitans*]")

This was the *Encyclopedia Britannica* entry on the tsetse that year—copied directly out of *ruzivo rwevatema* and noting the surest way to avoid trouble from *mhesvi*.

The nineteenth-century evidence indicates that people in practically all *mhesvi*-infested areas believed that the *chipukanana* was always present wherever there was *nyati* the buffalo—and big *mhuka* generally. These people had long lived near and hunted within *nyati*-inhabited country—some for centuries, the indigenous San for millennia. The belief among the San (the original inhabitants of Southern Africa) was that *mhesvi* fell pregnant and gave birth to whitish worms. Said Bradshaw (1877–1878): "The Bushmen have told me that the fly breeds in the buffalo droppings, and it seems as if there was some truth in it, because where the buffalos have been driven away in certain tracts, the fly has almost disappeared" (52). The Kololo believed that *mhesvi* laid eggs, reddish in color, on *mopane* tree leaves, on twigs scattered in the ground, and in the cow dung of *nyati* the buffalo.

Muchena proved *vatema*'s facts about the *mhesvi-mhuka* association through the barrel of his gun. By the 1850s, *nyati* the buffalo, *nzou* the elephant, and *ndunguza* the antelope had vanished, along with *mhesvi*, from areas along the Southern African coastline (Livingstone [1857] 2001, 82–83). In the 1870s, *Serowe* and another Englishman and William Finaughty observed this correlation between the extinction of *nyati* and the disappearance of *mhesvi* on the Linyanti (Chobe) River and Shashe-Limpopo confluence, respectively (Selous 1881, 190–203; Selous 1893, 294, 298; Finaughty 1916, 175–176). Other travelers noticed the same thing across the once *mhesvi*-infested stretch between the Lebombo Mountains and Lourenço Marques (Maputo) (Frere 1881, 19; Swynnerton 1921a, 335–336).

There were many figures engaged in this movement of *ruzivo* from Africa into Europe and North America through peer-reviewed journal publication, but the central one in the context of *mhesvi* was without doubt *Serowe* (Selous). Arriving in the Ndebele kingdom in 1871 aged nineteen years, the Englishman would cut his teeth as a hunter under local mentors who tracked for, guided, and showed him how to kill *ndlovu* the elephant and other big *inyamazana* that made him famous (Selous 1881, 51). Twenty years later, he would betray the trust of these people who had been good to him when he enlisted as the chief scout of the Pioneer Column. This was the occupying force that British capitalist and Cape politician Cecil John Rhodes organized to occupy land that his British South Africa Company (BSAC) had fraudulently acquired from Mzilikazi's son and successor, Lobengula (figure 2.2b), in 1889 under the Rudd Concession. After annexing *vedzimbahwe*'s lands to the east, in 1893 the BSAC invaded the Ndebele kingdom. The territory became to *vachena* the "colony" of Southern Rhodesia in honor of Rhodes.

Ten years after the annexation of the Ndebele kingdom, Ernest Edward Austen's *A Monograph of the Tsetse Flies* was published. In it are found

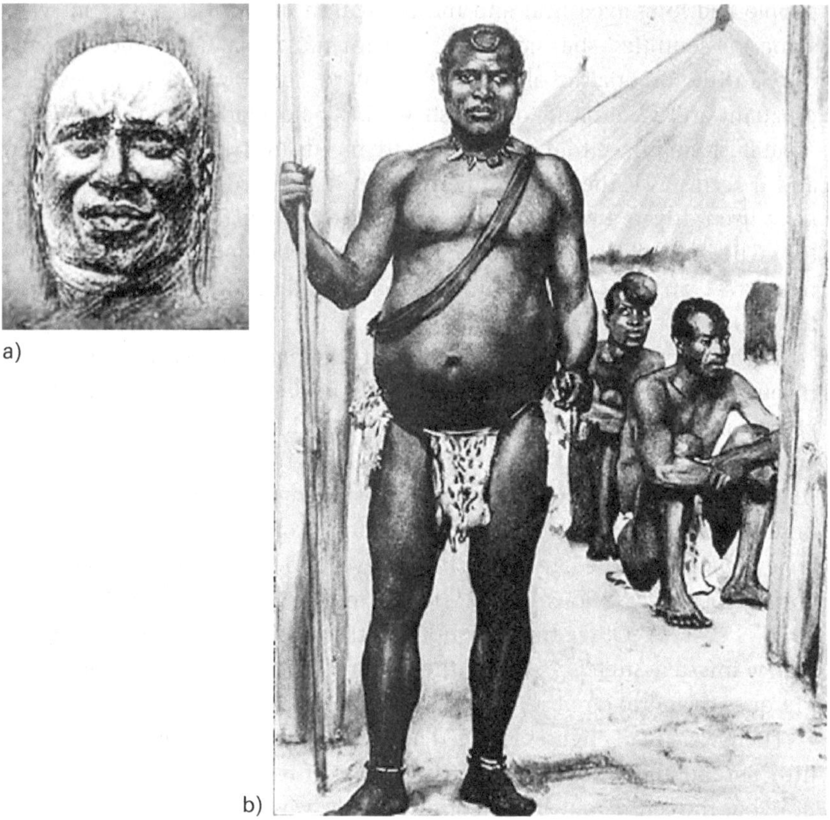

a)

b)

Figure 2.2a, b
Mzilikazi (left) and Lobengula (right).
Source: National Archives of Zimbabwe.

excerpts of *ruzivo rwevatema* about *mhesvi* and nineteenth-century white travelers' affirmations and applications of the same, complete with experiments on the move as the travelers encamped, walked, hunted, and decamped. Major Austen, a bacteriologist specializing in blood-sucking flies at the British Museum of Natural History in London, initially endorsed *Serowe*'s ideas in his 1903 intervention concerning the *mhesvi-nyati* connections, but the discord started not long afterward. That exchange (summarized here) is significant because it marked a shift in the basis of *hunyanzvi* (expertise) on tsetse derived from *experience* (*Serowe* the "big game hunter," imbiber of "native testimony," witness to its veracity through experience) to *experiment* (Austen, lab scientist, trained bacteriologist, using experiment to prove fact or falsehood). Austen was suggesting a paradigm shift: that neither *nyati* the buffalo specifically nor *mhuka* in general was necessary for the survival of *mhesvi*; that instead of taking the *mhuka-mhesvi* association for granted, attention must be focused on the latter's relationship to the environment. Knowing through experiment should now become the only permissible route to facts; *kuziva* (knowing) *mhesvi* through the experience of losing oxen to it no longer counted (Mavhunga 2007).

Austen's call for a "scientific" approach to *mhesvi* and *Serowe*'s insistence on the role of big *mhuka* and *nyati* inaugurated serious research interest in at least four elements of mobilities—namely, of *mhesvi* itself; of *mhuka*, whose blood constituted its food; of *hutachiwana hwen'gana* contained in the blood; and of *zvipfuyo* and *vanhu* it bit, infected, and rode on. The goal of this research was to design ways to detect, catch, suppress, or eliminate *mhesvi*,[1] to target both the surroundings and *n'gana*-spreading mobilities.

Swynnerton (see figure 2.3), whose writings about *sondela enkosini* we discussed in chapter 1, was born December 3, 1877, in Lowestoft, England. At an early age, he spent time in India with his army chaplain father before returning to attend school at Lancing College, Sussex, where his natural history interest blossomed. In 1897, aged twenty years, he was admitted into Oxford University but elected instead to migrate to Natal, South Africa. There, he met a naturalist named Guy Marshall, who encouraged him to accompany him to settle in Southern Rhodesia, where Marshall had been already well established in Salisbury since 1893. Swynnerton briefly worked in a store that Marshall partially owned, but the job was very boring. The young Englishman removed to the Melsetter (Chipinge) district, where he found work as a farm manager. In those days, a person just needed to be a white male and prepared to withstand living in remote areas to get such a job. In 1900, the Englishman moved to Gungunyana Farm, abutting Chirinda Forest, again as a farm manager.

Gungunyana is a corruption of *Ngungunyana*, the son and successor of Mzila and the last Gaza king, who in 1889 migrated to Bilene to avoid Anglo-Portuguese encirclement. Swynnerton arrived just a decade after this emigration. In 1902, his boss and friend Marshall bought the farm. Swynnerton would spend the next two decades at Gungunyana Farm. During that time, he undertook detailed studies of the local ecology, tapping into the *vutivi* (knowledge in *xitsonga*) of local people—that is, what remained of the epicenter of the Gaza kingdom. In 1919, aged forty-two years, he relocated to Tanganyika to take up a new position as a game warden. In

Figure 2.3
Charles Francis Massy Swynnerton.
Source: http://rayhewlett.org.

1928, he was appointed the director of Tanganyika's Department of Tsetse Research at Shinyanga. Swynnerton died in a plane crash in 1938.

Whereas Marshall thrived as *xitivi* (an expert) on weevils and rose to become the director of the Commonwealth Institute of Entomology in London, Swynnerton carved out his early natural history career on Gungunyana Farm collecting plants in 1903 and with some intensity in 1905 and 1906. The large numbers of herbarium specimens he sent to the British Museum in London drew praises as demonstrating "a precision in localisation and notes on economic uses which made this collection a model one" (Marshall 1938, 39). Swynnerton made an immense contribution to the botany of Southern Rhodesia, to the extent that British Museum botanists even named "new" species after him (Rendle et al. 1911). In 1906–1908, he wrote an amazing description of plants in the Chimanimani mountains (Goodier and Phipps 1961). As a farm manager at Gungunyana, he spent considerable time researching the plant life of Chirinda Forest (Mullin 1994). Swynnerton was not only interested in vegetation; he published on *svinyenyana* (birds) and *sviharhi* (wild animals) as well (Swynnerton 1908a, 1908b).

His major contribution was without doubt on *svifufunhunhu* (insects). Many of his extensive specimens found their way to the British Museum; in 1907, he was elected into the Linnaean Society. Health complications forced him to give up farming in 1918, and he now ventured into investigations on the habits and distribution of *ndedzi* (*mhesvi*) in northern Muzvirizvi (Mossurize) District. This area is located on the borderlands surrounding the Sitatonga Hills inside Mozambique. After taking up the wardenship in Tanganyika in 1919, he explored *sviharhi-ndedzi* relations even further, taking the ideas gathered at Gungunyana Farm and Muzvirizvi a step further into his experiments as director at Shinyanga, where he eventually published *The Tsetse Flies of East Africa* (Swynnerton 1936). When Swynnerton died in 1938, Marshall wrote a moving obituary in the US science journal *Nature*, omitting from his tributes the people who had taught the deceased man so much about the plants, *svinyengana*, and *svifufunhunhu*—and not least *ndedzi*.

Most of the elderly people who had taken part in *sondela enkosini* were still alive. They told Swynnerton exactly what had happened during and since Mzila's initiative. The local people who provided valuable *vutivi* upon which he based his "scientific" claims sometimes appear merely as "my native informants" in his text. Their *vutivi* to him also seems at times mere "information" and "native testimonies." By contrast, all *vachena* who barely contributed a thing to collecting the *vutivi* are recognized by their full names and their statements noted as concrete scientific positions. Regardless, Swynnerton's reason for turning to *vutivi* was that "the study of any successful campaign, by whomsoever carried out, is bound to add usefully to our general knowledge on the subject of controlling tsetse." Coming as a "botanist and ecologist rather than the unaided entomologist," he sought to understand from local *svitivi* (experts) the types of woodlands in which they had seen specific types of *ndedzi* (Swynnerton 1936, 317).

The Sitatonga study best demonstrates the huge debt Swynnerton owed his local "informants," as he called them. To his credit, he acknowledged it. His journeys with them—often led by them—into the local lives of *ndedzi* began in June 1918. For three months, he conducted a "preliminary investigation" to determine *ndedzi* habits and distribution. He had intended to focus on the Gogoi area, but it yielded little, so he relocated to the country east of the Sitatonga mountains, where he had caught considerable numbers of flies in 1900. He says the project was conducted "with the help of my own farm natives only" and local people through the offices of a local

Portuguese official named Senhor Lanne. The area west of the Sitatonga Hills had been the "*scene of a particularly fine experiment* in the banishment of tsetse" six decades earlier—by Mzila, the Gaza king (Swynnerton 1921a, 316; my emphasis). Indeed, it was a vast open laboratory.

Hlengwe locals were unanimous about the connections between big forest animals and *ndedzi* presence. *Mbavala* (bushbuck), *honci-nhova* (warthog), *khumba* (bushpigs), and *nhungu* (kudu) did not wander far and were "the fly's most reliable food-supply in its permanent haunts" (Swynnerton 1921a, 342). By contrast, *mhopfu* the eland, *ndlopfu* the elephant, and *nyarhi* the buffalo were wanderers, the latter two moving in great herds, covering extensive grazing grounds, and moving back and forth between them, splitting into smaller groups or ostracized into bachelor herds or lone bulls (342). *Nyarhi* the buffalo combined both the capacity to be a reliable feeder of *ndedzi* and an excellent organic vehicle for it as it moved through the countryside. Hence, Swynnerton found the problem in Muzvirizvi to be "not so much that [*nyarhi*] feeds the fly (which would be fed and contaminated in any case by the pigs) but that it carries tsetse far and wide in the rainy season and so brings it into contact with the cattle" (Swynnerton 1921a, 316).

Migration and vehicular role aside, the local mobilities and habits of *sviharhi* were themselves critical to explaining which kinds *ndedzi* best preferred for blood. *Ndlopfu* used the same path to the waterhole and were predictable to *ndedzi*. *Mvubu* (hippopotami), by contrast, spent the whole afternoon basking in the sun poolside; *ndedzi* had something of a field day with them. The "natives of the *morsitans* area," Swynnerton remarked, were "unanimous" in stating that *ndedzi* fed on *mfenhe* the baboon and that "wherever you find baboons you will also find fly" (Swynnerton 1921a, 336). They gave him countless examples "in which baboons driven from their gardens had left numerous replete flies behind and others in which flies were attracted in numbers to baboons that were killed" (336). *Nkawu* the monkey also raided their gardens, but was not as readily found with flies as *mfenhe*, which were slower, bigger, and more visible to *ndedzi*. Occasionally, they also found *ndedzi* on cane rats, diurnal rodents of which there was a surfeit. They also "universally incriminated" *khumba*. Swynnerton concluded that "any attempt ... to destroy the fly by starving it in its permanent haunts is doomed to failure if the bushpigs, and perhaps the baboons also, are not destroyed; and the destruction of the pigs in this type of country is not easy" (337). The same *xiharhi* (save *mfenhe*) would be targeted under Rhodesia's selective or discriminative game elimination from 1956 onward.

Ndedzi also drew on numerous *svinyenyana, svikokovi* (reptiles), and *dewulana* (bats) for blood. Slow-moving ground feeders included *mhangele* (guinea fowl), *nghwari* (crested francolin), *gumba* (white stork), and *ntsutsu* (egret). *Ngwenya* (crocodiles), *nyoka* (snakes), and *ngwahle* (iguana) were the most common reptile food hosts (Swynnerton 1921a, 328–329). Add to that *mhunti*, a little duiker that usually rests under logs and between tree buttresses, favorite resting places for *ndedzi*.

Locals disclosed to Swynnerton that smaller *sviharhi* also supported *ndedzi* populations in lieu of the more docile, slow-moving *nyarhi*. One of his guides, "a very observant native, was particularly convinced of it." When faced with a choice between *nyarhi* and other *sviharhi* that *ndedzi* followed, the guide said: "The buck are much more restless under its attentions than the buffalo, the hartebeests especially keeping up their dance when tsetse are about them; so that the fly can feed more easily on the buffalo" (Swynnerton 1921a, 339). Swynnerton subsequently concludes: "It follows *also from my observations* on this expedition that the old idea that tsetses possess a preference for *nyarhi* may be perfectly correct, though it will show itself strongly only where, and while, buffalos are so abundant as to make the fly comparatively independent of less favoured food" (340; my emphasis).

How then can we explain *ndedzi*'s preference for *nyarhi* the buffalo? Was it because the bovine was black? Yes, according to a local old-timer named Mabuzana, who lived close to the Mtshanedzi River south of Gogoi. Mabuzana told Swynnerton that *ndedzi* was especially attracted by a black coat, and an interesting conversation ensued. "How do you know that?" Swynnerton asked. "Because I have one!" Mabuzana replied (Swynnerton 1921a, 339). Four decades later, the Branch of Tsetse and Trypanosomiasis Control (BTTC) would embark on what it called "attractant studies" that found *ndedzi* to be attracted to dark colors.

The whole idea of "carried fly" was also something commonsensical and experiential. Locals had reported an increase in flies in their *tiko* (village) that were being brought from other areas on the bodies of *vanhu* and *sviharhi*. These people lived along the rivers where water remained as uplands dried up, so *sviharhi* would be coming to the river to drink. *Ndedzi* gathered at the waterholes in ambush. *Sviharhi* or people going to the river or passing through deposited flies picked in the uplands that were drying up, increasing as the dry Southern Africa winter intensified (Swynnerton 1921a, 370).

Yet it was not always the case that *sviharhi* brought *ndedzi*; in fact, local *vanhu* in the Mpapa area told Swynnerton that the *ndedzi* presence predated the arrival and concentration of *nyarhi* herds. At Masando, *vanhu* said the herds were too small to qualify as "concentrations." At Kanyezi, the large collection of *mahlomalavisi* (pupa) showed that *ndedzi* concentration already existed prior to the temporary presence of *nyarhi*. As far as *vanhu* were concerned, it was common *vutivi* that *ndedzi* concentration was "permanent though variable in numbers, that it had been there before the buffalos came (as it had also survived their departure), and that the animals had stayed a very few days only" (Swynnerton 1921a, 371). Swynnerton's guide Kanyezi lived close to the valley where *nyarhi* had concentrated. He had lived with the flies and seen the *nyarhi* herds when they arrived and believed unequivocally that the concentrations preceded this bovine movement (371).

Another local named Gundoda told of a concentration that took place in the *tiko* during the rains in areas where *ndedzi* was abundant. He was basing his *vutivi* on what he had experienced in his own *tiko*, which not only had good shade but was located on the edges of a well-watered valley—a rather "tempting dismounting-place for fly." He described to Swynnerton that in spring and at the onset of the rains "every newcomer or passer-by would bring with him an accession of flies till their numbers became unbearable" (Swynnerton 1921a, 372). Some of the flies would follow these travelers, but many more stayed, accumulating with each passerby. To ward off these *svifufunhunhu*'s attentions, the itinerant carried a leafy switch, thrashing any that approached or bit into him.

The Language of Translation

Vanhu vantima (black people), *mirhi* (trees), *sviharhi* (wild animals), and *svifufunhunhu* (insects) did not enter *vutivi bya valungu* (the knowledge of whites) with their names but with those that *valungu* arbitrarily gave them. Even rivers lost the names that *vantima* had given them as a prerequisite for entering *vutivi bya valungu*, as did *vantima*'s names as a condition for getting birth certificates or baptism. In the process, such resources and the *vutivi* that *vantima* had volunteered to these writers was hidden in strange Latin or botanic names, *vachena*'s linguistic translations, and the written text. *Vutivi* was no longer recognizable to the very same people who had pointed it out to these strangers.

The movement of *vutivi* from *vantima* to *valungu* happened through not just any translation but what became known as *fanakaló* or *fanagaló*—a hybrid language composed of some English, Nyanja, and *chidzimbahwe*. Also known derogatorily as *kitchen-kafir, mine-kafir, pidgin bantu, isikula* ("coolie" or Indian), *chirooroo, fanikaroo*, and *chiraparapa* (*silapalapa*), *fanakaló* emerged as an expedient language of communication in the Eastern Cape and Natal between mostly English-speaking *vachena* and local Zulu and Xhosa inhabitants in the early 1800s; it thrived in the diamond mines of Kimberley and the gold mines of Johannesburg later (Cole 1953; Hopkin-Jenkins 1948, vii; Bold 1952, 6; Lloyd 1950, 3). As workers returned home and as whites settled in Southern Rhodesia, *fanakaló* increasingly became the halfway language "spoken wherever black meets white from the Cape Peninsula to the Great Lakes of Africa" (Bold 1949, 77). Initially, entomologists such as Swynnerton spoke no local languages; they supervised and relied on workers with no English-language skills.

Swynnerton also began the process of mystifying and alienating the knowledge of *ndedzi* generated by local people through renaming. The mystification began with the new names for *mhesvi* itself. What locals knew as *ndedzi* (*xitsonga*) and *inthesi* (*xitshangana*) Swynnerton now called "tse-tse," which, as noted earlier, is Setswana. Even *tsetse* was not considered scientific enough, thus leading to *Glossina* (shortened to *G.*), a Latin word for bloodsucking flies (which now became *diptera*). *Glossina* was further subdivided into three "subspecies": two gigantic *svindedzi sva nkova* (*mhesvi-rupani* in *xitsonga*) or river-loving types *valungu* now called *G. pallidipes* and *G. palpalis*, and the small *svindedzi sva nhoveni* (*mhesvirutondo* or savannah-loving in *xitsonga*), which *valungu* now called *G. morsitans*.

In the *morsitans* category were *Glossina morsitans Westwood*, named after John Obadiah Westwood (in 1850); *G. austeni* and *G. pallidipes*, "discovered" by Ernest Edward Austen (1903); and *G. swynnertoni*, by Charles Swynnerton (1923). The twelve *G. fusca* "species" (because *vachena* said now they were; *vatema* had their own categories) were not found in *dzimbahwe* and have no bearing on this book. Of the riverine "species," two *G. palpalis*, "discovered" by Robineau-Desvoidy in 1830, and *Glossina palpalis gambiensis* were present, especially in the southeastern areas. There was also what *vachena* now called *G. brevipalpis* and *G. longipenis*.

Which *munhu mutema* would now recognize, let alone pronounce, these names? At least they could still pronounce *tsetse*, a Setswana name now universalized across all of Africa and even *ruzivo rwevachena*. The language of translation erased the tracks of the *vutivi* and hid it from its originators.

3 Knowing a Fly

A *chipukanana* (insect) that threatened Rhodesia's foremost economic sector—agriculture—had to be studied and taken seriously. In 1961, the entomologist Edward Bursell called it *"an intelligence system of tsetse,"* an infrastructure and system of knowledge production so thorough as to know *mhesvi* in the most intimate way. The entomologist was talking about an anthropomorphic kind of *kuziva* (knowing), getting into a *mhesvi*'s intimate life, "to live and breathe and think with it." The goal was no longer simply studying *mhesvi*, but to engage in "a lifetime of affectionate study."[1]

Such intimate knowing required placing *mhesvi* under surveillance, to know how much time it spent in different parts of the habitat at different times of the year, how much time it spent feeding, sleeping, or simply waiting to pounce on anything that moved.[2] It also required mapping *mhesvi* (which *vachena* now called *Glossina*), establishing its boundaries, and knowing where to mount defenses and offenses against it.

That way of *kuziva mhesvi* had begun less surely and developed in a meandering way. Its journey is the subject of this chapter, starting with the institutional structure for *kuziva mhesvi* and then moving to knowing how to find it, what it eats, and what to do with such *ruzivo* (knowledge).

To know *mhesvi* in order to control it, *hurumende* established what V. Y. Mudimbe (1988) calls the "colonizing structure," albeit with important caveats. Mudimbe meant by this concept "the procedures of acquiring, distributing, and exploiting lands in colonies; the policies of domesticating natives; and the manner of managing ancient organizations and implementing new modes of production" (3). That definition fits the establishment of an institution upgraded from a division to a branch to a department and downgraded as needs must. I replace "domesticating natives" with "destroying *mhesvi*," and "new modes of production" with "the production of knowledge" while cautioning against Mudimbe's readiness to see such production as anything "new." The reason is simple: Mudimbe bought too readily into and overestimated the overwhelming powers of *hurumende* to define what became *ruzivo rwevatema* (the knowledge of black people) while grossly underestimating the role of *vatema* in that which he assumes to be *ruzivo rwevachena* (white science).

Mudimbe says that the colonizing structure effects "the domination of physical space, the reformation of "native" minds, and the integration of local economic histories into the Western perspective" (Mudimbe 1988, 3). The domination of physical space remains the core of the book's focus, but all three elements of Mudimbe's formulation are top-down readings of white rule and open to further explorations. The "reformation of natives' minds" masked a vast appropriation—theft—of *ruzivo rwevatema* that profoundly defined the *vachena*'s (never quite successful) efforts to dominate space. As a case in point, the work of Mhoze Chikowero (2015, 19–79) illustrates how white missionaries with their very soporific acapella music drew little interest from would-be congregants, their fortunes changing only when they allowed *vatema* to bring their own musical inventions into the church. *Vachena* spent the first sixty to seventy years battling to destroy a deadly *chipukanana* that *vatema* had coexisted with for millennia; just as victory was on the horizon, *vanhu vatema* rose in rebellion to demand their independence back. The "integration of local economies into the Western perspective" is only half the story and depends on one's analytical location: It is true when a person is *looking from vachena*'s perspective. However, seen from the rural village and *vatema*'s everyday lives, a person also sees *vatema*'s integration of Western economies into the local perspective, thus completely appending Mudimbe. Hence, his is only one part of the story; in fact, such a top-down approach masks a subtler reality: that of the colonizing structure being founded on *ruzivo rwevatema* and their means and ways of doing things.

Finally, Mudimbe's colonizing structure "completely embraces the physical, human, and spiritual aspects of the colonizing experience" (Mudimbe 1988, 2). I add a fourth element that science, technology, and society (STS) can bring richly into Africa studies: attention to inanimate and animate things as actors. In this case, the focus is on the thing to be known and what it does to deserve being known, as well as the things deployed in order to know it. The thing to be known is also a thing that is known by its deadly mobilities. It becomes the venue, the meeting point, of those seeking to know, their knowledge traditions, and the means (tools) of *kuziva* (knowing), tools that are not only human fabricated, but also dehumanized people—*vatema*.

Why did *hurumende* become so interested in *kuziva mhesvi*? What infrastructures were needed to enable *hurumende* to know *mhesvi*, and how do we account for their evolution? What was to be known about *mhesvi*? How did *vachena* go about the process of *kuziva*/knowing it? And, what do we do with and learn from that way of doing things that made *hudzvanyiriri* (oppression) strong, that made *mulumbeti* (the devil in *xitsonga*) strong?

Why Did *Hurumende* Become Interested in Knowing *Mhesvi*?

The rationale for the authorities' interest in *mhesvi* lay squarely in the *chipukanana*'s mobilities and transmission of *hutachiwana* (pathogens), the local varieties of which were deadly to *mombe* (cattle). The beef industry itself is an important example of how Southern Rhodesia's economy was built upon *ruzivo rwevatema*—in this case, *ruzivo rwemombe* (cattle knowledge).

The roaring entrepreneurial successes white people from all over the world experienced in Gauteng (the Rand) during the gold rush of the 1880s convinced them that a second "rand" lay north of the Limpopo. When governments could not sponsor the Rhodesian occupation, individuals formed venture capital companies to do so themselves on the strength of mineral concessions fraudulently obtained from local leaders.

Cecil John Rhodes was one of those men. In 1890, his British South Africa Company (BSAC) sent a Pioneer Column to occupy the land east of the Ndebele kingdom, which now became, to the British Government, the settler "colony" of Mashonaland. The idea of the "northern goldfields" as a land with footpaths paved in gold turned out to be a huge disappointment. Pioneer farm rights, quoted at £100 each when the Pioneer Column was disbanded, were selling for £55 to £60 each by April 1893, while just three hundred *vapambevhu* were occupying farms.

The "failed settlers" of Mashonaland now sought reprieve in the West. Three years after the occupation of Mashonaland, after mounting tensions, the BSAC invaded and subdued the Ndebele and renamed their lands Matabeleland. Not surprisingly, the lure of a share of Lobengula's *izinkomo* and fertile land in Matabeleland attracted 922 *vachena* as volunteers; the war ended relatively quickly. Mashonaland and Matabeleland, together constituting all the land between the Zambezi (north), Limpopo (south), Kgalagadi (west), and Nyangani Mountains (east), became Southern Rhodesia.

Here too is a clear illustration of *vachena* building upon local resources—albeit ones the Ndebele had mostly pillaged from their neighbors in raids into *dzimbahwe*, Tswana, and Tawana territory. Following the war, some 948 land rights were issued. Those who had taken part were given first preference to any claim until April 26, 1894, by which time many were struggling to sell and get out. Like Mashonaland, Matabeleland had no pavements of gold. The quest to appease the pioneers with Lobengula's *izinkomo* also stemmed from knowledge of a well-developed trade in *izinkomo* that had existed with the Damara, Namaqua, and Ovambo since the early 1870s, fed in part by the wagon traffic from Cape Town and Kimberley. A man named Balane had made several two-year expeditions from 1869 to 1886 to the territories, buying eight hundred to one thousand beasts for £1 each on his first journey, which he sold for £7 to £8 each in Kimberley. Another man, Ericksen, at one point bought twelve hundred oxen from the Damaraland, which he pushed south to Kimberley via Lake Ngami (Stigger 1971, 16).

The Ndebele, meanwhile, sold their *izinkomo* (cattle), *imvu* (sheep), and *imbuzi* (goats) to local Jesuit Missionaries for handkerchiefs, blankets, and cotton cloth, even though prices fell as missionaries gained local *ruzivo*. *Serowe* also bought *izinkomo* from the Ndebele and delivered them south in December 1882, while in late 1889, a man named Dreyer took delivery of 750 Ndebele *izinkomo*, having supplied part of the Rudd Concession rifles promised. As Stigger shows, by July 1893, *izinkomo* trading was said to be "the most important business engaged in by traders among the Ndebele, pieces of calico forming the principal item demanded in exchange, although guns, powder in flasks, beads, blankets, coats, waistcoats and trousers were welcomed" (Stigger 1971, 16). In that year, *izinkomo* trade between Matabeleland and *vapambevhu* of Mashonaland was quite pronounced. For instance, in March, one H. J. Hill and two *vachena* returned to Salisbury with three hundred *izinkomo* and four hundred *imvu* and *imbuzi*. Hill was one of several Mashonaland settlers—including notably C. M. Acutt, Colenbrander, Dawson, and "Matabele" Wilson—who were also involved in this trade (16).

Vanhu vatema in Mashonaland were also prepared to trade *izinkomo*, with those going to look for work in town or going to boarding school often passing through mission stations to sell goods and get calico or gold (see figure 3.1). They usually rode on *mikono* (bulls), which upon arrival or in transit they sold to *vachena*. For instance, in 1891, others in Manyika (now Manicaland) and the Zezuru around Salisbury (Harare) freely traded *mombe* for salt, beads, brass wire, calico, and old uniforms "to itinerant traders

Figure 3.1
Men and boys riding on *mikono*/bulls. In addition to being wealth and being important ingredients in ancestral spiritual sacrifices, *mombe* were deployed as transport to work in the cities and deployed to mission schools, sold or slaughtered once there for upkeep and dried meat. This picture illustrates the multifaceted roles of *mombe* in *dzimbahwe* society.
Source: Zambezi Mission Record.

visiting their kraals." They also traded at the auction in Harare, some for around £6 per ox. Together, Shona and Ndebele cattle satisfied local demand: "When taxation was first introduced into Mashonaland in 1893, any tendency for *vanhu vatema* to part with stock is said to have been discouraged because of the trouble involved in selling their beasts" (Stigger 1971, 18). *Vachena*'s purchase of cattle was largely consumer-driven.

From 1894 on, *vachena* became more arbitrary and aggressive than before. After the defeat of the Ndebele, a Loot Committee was established to plunder Ndebele wealth. Initially, it set out to impound thirty thousand *izinkomo* for distribution to *vapambepfumi*, but by December 1895 it had seized sixty-five thousand. These are only the reported seizures; tens of thousands more were seized but not accounted for. A settler named John Meikle in particular built his Meikles Stores and Meikles Hotel empire out of looted *izinkomo* and thousands of hectares he held for speculative purposes (Stigger 1971, 18).

Still, even with so much pillage, Southern Rhodesia was anything but booming by 1896. The entire settler community was in a state of total farming stagnation (Stigger 1971). That year, a devastating cattle plague (rinderpest) that had started in British Somaliland in 1888 and followed the wagon trails of British East Africa (Kenya and Uganda) and British Central Africa (Malawi and Zambia) arrived. In a typical case of "ecological imperialism" (Crosby 1986), *vachena*'s ox wagons had introduced a devastating *nyongororo* against which *vatema*'s *zvipfuyo* and *mhuka* had no natural immunity. They died en masse (Spinage 2003). The extermination of *mhuka* denied *mhesvi* its most versatile means of transport and food source. Only those *mhuka* in the remote borderlands along the Zambezi, Limpopo, and Savé River Valleys survived. In the vast acres of the now tsetse-free land, *vachena* established their *mapurazi* (cattle ranches, farms) and mines (Mavhunga 2014).

The rinderpest (and *vatema*'s risings from 1896 to 1897) capped a depressed atmosphere in Mashonaland that had begun with the failure to find gold; most *vapambepfumi* had left for Bulawayo after 1893. When the risings broke out, they finished much earlier in Matabeleland and dragged on in Mashonaland to the end of 1897. Then, a commercial depression and unemployment set in among *vachena* now reduced to paupers and miners taking refuge in town. Meanwhile, the social depression caused by sending women and children to safety out of Mashonaland, combined with locust invasions and the drought, worsened. From October to November 1897 alone, there were eight suicides among *vapambevhu* (Kosmin 1971). While *vachena* were deep in depression, *vatema* dominated crop and *mombe*/cattle production in Southern Rhodesia (Machingaidze 1980, 282).

When *vachena* turned to ranching, they could only (initially) build upon *mombe* they had forcibly taken from *vatema* between 1893 and 1898 as spoils of war (Samasuwo 2003, 489; Phimister 1978). These were draught-oxen and disease-resistant cattle varieties, like those that *vachena* called the "Nkone," "Mangwato," "Matabele," "Mashona" (also called the "Hard Mashona"), and "Tuli" breeds (Government of Southern Rhodesia 1924, 30). The settling *vachena* were cash-strapped and could not import any breeds from Europe (Machingaidze 1980, 285).

The foundations of settler society were not only built with means and ways (technologies) of *vanhu vatema*; in the early period, the settling *vachena* also played second fiddle to *vatema* in terms of productivity on the land. Before 1923, "beef production in the colony was severely hampered by the lack of capital on the part of most *vapambevhu*, crude ranching techniques, rampant cattle diseases, lack of transport facilities and … the lack of remunerative markets" (Samasuwo 2003, 490). World War I offered temporary respite, but by 1921, the prices of beef had slumped again; the local market was too small. The veterinarians had no choice but to slaughter *mombe*, keep carcasses in cold storage, and can beef.[3]

The process that would lead to a complete shift in fortunes started in 1923, when a commission of inquiry recommended the establishment of an abattoir to pave the way for producing frozen and chilled beef for overseas export. Negotiations began with the South Africa–based Imperial Cold Storage and Supply Company Ltd to set up abattoirs in Rhodesia.[4] The Rhodesian Export and Cold Storage Company Ltd (RECSCO) was formed in 1927 as a subsidiary of the South Africa–based parent company.[5]

It was not a rosy beginning. RECSCO's first ten years in business posted staggering losses, beef prices remained depressed, and the subsidy the Rhodesian government was paying only fed the parent company's bottom line, instead of benefitting Rhodesian farmers (Mlambo 1996, 57). Faced with a choice to turn RECSCO into a parastatal or become completely private, the government chose the former.[6] Arguments were made that "once a concern is under state control, efficiency goes by the board" and that the entire situation would "end in disaster."[7] They were waved away.

Out of RECSCO was born the Cold Storage Commission (CSC; also known as COSCO) in 1938. The parastatal was established as a guaranteed market for white ranchers and to rescue them from the grinding effects of the Great Depression through the forced acquisition of *vatema*'s *mombe* on the cheap.[8] In 1941, CSC turned its fortunes around so much so that it upgraded its Bulawayo abattoir; built three more in Salisbury in 1943, Umtali (now Mutare) in 1946, and Fort Victoria (Masvingo) in 1951; and installed cold storage facilities in Que Que (Kwekwe) in 1946 and Gwelo (Gweru) in 1947.[9] In addition to totally dominating the domestic market, CSC also expanded its exports into the United Kingdom, Northern Rhodesia, South Africa, and the Congo.[10]

Rhodesians (as *vapambevhu* came to define themselves) knew how to look out for each other—particularly in the 1930s, when white and black cattle owners had lost 200,000 and 250,000 head of *mombe*, respectively, due to Great Depression–induced hardship. CSC's "guaranteed prices and markets" policy for whites could not have come at a better time in 1938. The prices were good. The floor price in 1950 was 70 shillings per one hundred pounds, cold-dressed weight, increasing to 97 shillings in 1955 and 113 shillings four years later. The national herd also increased in the postwar period, from below 3.6 million head in 1948 to 4.2 million in 1955 and 4.75 million twenty years later.[11] In 1960, CSC extended its operations to Northern Rhodesia and Nyasaland.[12]

The CSC was created to guarantee white farmers' success. It was a racially discriminatory affirmative action infrastructure, a structure to create an industry to buy *mombe* from whites and guarantee them good prices, a market, and a sustainable settler economy. It was a government mechanism to deliberately give *vachena* an unfair competitive advantage and force *vatema* to sell their cattle. Before World War II, *hurumende* did not bother to organize the marketing system for *vatema*; rather, individual white traders, white farmers, speculators, and agents of CSC and Liebig's Ranch (property of the beef giant, Liebig's Extract of Meat Company) went around buying *mombe* in villages.[13] No weights were used; prices were negotiated on the spot (Mlambo 1996, 64).

The CSC not only became an apparatus for prospering *vachena*; it also took its place at the center of disenfranchising *vanhu vatema*—not through exclusion from its market, but by forcibly seizing their *mombe* and giving them to white ranchers. One thing had become very clear by 1941: Despite their "excessive herds," *vatema* were not selling (enough), and Rhodesia was failing to supply fighting troops at the front with tinned beef. The view of the Native Department was that *vatema* should be forced to sell their *mombe* and that their herds should be capped and the rest sold or forfeited to *hurumende*.[14] Under the Natural Resources Act No. 9 of 1941, Section 36, the government decreed that every household herd be limited to "prevent overstocking." The Native Department was charged with organizing *mombe* sales at designated markets. The 1942–1951 period was the most intensive one for forced destocking, perhaps equivalent only to the looting of Ndebele *izinkomo* after 1893.[15] To ensure that *vatema* had no alternative market, all potential buyers were supposed to show a Cattle Sales Permit Order (Government Notice No. 603), issued for two purposes only: buying for slaughter or farming purposes. The noose would further tighten in 1947 with the Native Cattle Marketing Act (Act No. 23/1947), under which the Minister of Native Affairs was granted power to prescribe methods of sale, venues, and who could or could not buy. Only the CSC, Liebig's, and butchers the CSC did not supply could buy. No such restrictions applied to white *mombe* owners.

Evidence showing the existence of this discrimination was suppressed. According to Mlambo (1996, 67), "Africans in their areas were being forced to sell cattle against their own wishes, ... the prices paid for the cattle were very low, ... cattle belonging to Africans who were absent at the time of sale were sold without their knowledge or consent and ... those Africans who refused to sell their cattle were punished or were threatened with punishment." A subsequent commission of inquiry was a sham; quite contrary to its own evidence, including Native Commissioners assaulting *vatema* who refused to sell, the inquiry dismissed these charges as lies and disclosed just how happy *vatema* were to destock. In Gutu District, for example, *vatema* told the Native Commissioner they would rather much slaughter their beasts than sell them—whereupon the official ordered the brushes of the cattle's tails cut as a sign they had been designated for slaughter and would be confiscated if seen again, incurring a huge fine for the owners.[16] Having stolen *vatema*'s land in the 1890s and almost two hundred thousand of their cattle from 1893 to 1896, *vapambepfumi* continued their looting spree into the 1950s. The CSC bought the cattle at arbitrary, giveaway prices, then passed them on to white farmers to fatten and sell for slaughter to its abattoirs—at monstrous profit.[17]

The cattle plundered from *vatema* built CSC—and *mapurazi* more generally. The looters normally targeted drought time, when *mombe* were just *matore* (walking carcasses), their prices rock bottom, and bought them for less than a song. They struck when the market was oversupplied and prices depressed and selected *mombe* still not yet fully grown, the prices of which were a fraction of adult values. *Munhu mutema* would never get a better price than *muchena*; there was a price for *vachena* (8 shillings 4 pence per 100 lb. weight) and another for *vatema* (4 shillings 7 pence). In 1951, *nhimura* (the Native Land Husbandry Act) introduced even more destocking measures. Free competition was only introduced in 1956, by which time *vachena* had made their money.[18]

What Kind of Infrastructure Did Knowing *Mhesvi* Entail?

It was *mhesvi*'s threat to the emerging cattle industry that forced *hurumende* to build a branch dedicated to swatting this fly. The shape, character, and composition of this branch can only be understood through the pestiferous mobilities of the problem insect. The constant changes—from a branch, to a department, to a branch, and in personnel and job descriptions—are consistent with the status of the war against the indefatigable *mhesvi* and its unending ebbs and flows.

In 1909, as the insect (which *vachena* called "tsetse fly") threatened to get out of control in the Chegutu-Kadoma districts, the government of Southern Rhodesia set up a Division of Entomology within the Department of Agriculture. There was no operational—let alone research—infrastructure when Rupert Jack (see figure 3.2a) was appointed government entomologist; in fact, only with the arrival of James Keswall Chorley (see figure 3.2b) to take the post of assistant entomologist was Jack's title elevated to chief

a) b)

Figure 3.2a, b
Rupert Wellstood Jack (left) and James Keswall Chorley (right).
Source: S2A3 Bibliographical Database of Southern African Science (left image); *Proceedings and Transactions of the Rhodesia Scientific Association* 1967 (right image).

entomologist. By 1921, these men had performed so little work that the division's annual report for that year was only half a page long. Throughout the 1920s, experimental work was conducted by the two men and a few farmers, and even that research focused only on trying to understand *mhesvi* breeding and the mechanical transmission of both flies and *hutachiwana* (which *vachena* now called "trypanosome") through on-foot traversal. So thin-staffed was the division that when both Jack and Chorley went out for fieldwork, the headquarters virtually ground to a halt.[19]

To share and better coordinate the burden of controlling *mhesvi* and give policy advice to government, the Southern Rhodesia Trypanosomiasis Committee (SRTC) was established in 1928.[20] Its chair was Llewelyn Bevan, under whose stewardship SRTC worked tirelessly to pull together different government departments and white farmers' associations to offer unified resistance to *mhesvi*.[21] In the 1930s, Jack and Chorley worked with SRTC to build up an effective research and operational unit.[22] As a starting point, a departmental committee of inquiry was established in 1931 to audit the state of the Division of Entomology and its tsetse operations.[23] The recommendation to place operations under the Department of Agriculture's administrative officer while restricting the entomologists to research *mhesvi* was a disaster in practice. By 1932, the entomological officer assumed the direct control of tsetse operations under the authority of the chief entomologist.[24] The key lesson learned was that operations and research were inextricably bound together and should never be separated.

With Bevan's death in 1938, the SRTC fell apart. Jack had also left by 1942, leaving Chorley—now chief entomologist—to reconstruct the committee and equip it to coordinate the many departments involved in the control of *mhesvi* and *n'gana* (Cockbill 1968).[25] Throughout Chorley's tenure, the work of combating *mhesvi*—let alone knowing it well enough to offer the right kind of response—suffered from serious logistical limitations. As he remarked in his 1951 annual report, "The only vehicles which are really suitable for the work in most of the areas are those with four-wheel drive. Of these the branch now has five, of which one is very old and constantly gives trouble. Apart from these the branch now has four 3-tonners and 12 smaller vehicles."[26] The division was, in summary, still a rickety, ragtag affair.

Two important reorganizations took place in 1951 that fundamentally redefined the control of knowledge production focusing on *mhesvi* and *n'gana*. The first was the reorganization of the SRTC into a smaller unit, composed of "only those officers directly concerned with tsetse fly and its effects."[27] Second, the Division of Entomology was transformed into a full-fledged Branch of Tsetse and Trypanosomiasis Control (BTTC) within the Ministry of Agriculture. It now had its own director answerable to the minister of agriculture, a senior entomologist, three entomologists, seventeen tsetse fly officers (TFOs) stationed in operational areas, and two clerical staff. The department hired Dr. Gerald Cockbill, a trained entomologist, zoologist, and botanist born in Cardiff, as a senior entomologist under Chorley, who now became the director.[28] The TFO's job was to supervise all aspects of field operations—initially game elimination, forest clearance, and fence construction, and later the spraying of OCPs. The entomologists were responsible for all research and were stationed at the central laboratory in Causeway, in Salisbury (now Harare), or at several field research stations—principally Rekomichi, Sengwa, and Lusulu, all located within the Zambezi's *mhesvi*-infested belt.

In 1953, another commission of inquiry was conducted to review tsetse operations, this time zeroing in on the long-running program of indiscriminate game elimination. The commission's report was submitted to the minister in 1954, recommending, among other things, the elevation of the BTTC into a full-fledged Department of Tsetse Fly Control and Reclamation (DTTRC).[29] The new department was established on July 1, 1956. It was, at best, an enlarged BTTC, with a director, a senior entomologist, seven entomologists, five senior field officers, twenty-five TFOs, clerical staff, African field assistants, and twelve hundred to fifteen hundred *mago-cha* and general workers.[30]

Chorley had already retired by March, and Gerald Cockbill took over as acting director of DTTRC.[31] He held the position until John Ford, previously director of the East African Trypanosomiasis Research Organization, took over as substantive director on September 27, 1957, with Cockbill as his deputy—designated senior entomologist initially, then chief entomologist.[32]

The idea of knowing *mhesvi* thoroughly as a prerequisite for killing it reached its climax during Ford's tenure. To achieve this, he threw everything into improving the research capacity of the DTTCR, aggressively recruiting qualified personnel capable of initiating and sustaining research and operations even when resignations, retirements, and other loses of staff occurred.[33] At the height of his powers as director in 1957, Ford's complement of research and operational staff is shown in table 3.1 (my own tabulation).[34]

Two key appointments reflect Ford's bias toward academically qualified laboratory personnel with hands-on, in-house training in field skills such as marksmanship, on one hand, and administrators with an understanding of entomological and operational work, on the other. He insisted, for example, that his deputy director must be a scientist with experience in running

Table 3.1
The Department of Tsetse and Trypanosomiasis Control's staff establishment in 1957

	Present establishment	Proposed establishment
HQ	Director Director	
	Senior entomologist (senior personnel officer [SPO I])	Deputy director
	Administrative and executive officer	Senior entomologist (SPO I)
		Technical assistant
		Administrative and executive officer
Field	1 entomologist (SPO II)	1 entomologist (SPO II)
	6 entomologists (PO)	5 entomologists (PO)
	4 senior field officers	1 survey ecologist (SPO II)
	26 field officers	5 senior field officers
	25 Africans, branch IV, grades I and III	18 field officers
		50 Africans, branch IV, grades I and III

tsetse operations. To pay for the post, he would cut two entomologist positions.[35] His senior entomologist would be a full-fledged field officer rather than simply an "Administrative officer sitting at Salisbury headquarters—which had led to failure to obtain data on operational progress, costing the Department money."[36] Finally, Ford also pushed for and got (in 1959) the appointment of a survey ecologist, which he viewed as a prerequisite to enabling the department to produce its own original maps of tsetse, game, and vegetation distribution, thus promoting greater accuracy and saving field officers time they would otherwise spend producing their own maps. The following year, the cartographic section was expanded with additional staff and provisions for their accommodation.[37]

The days of DTTRC as a department were numbered, however. On April 1, 1961, it was incorporated as a branch of the Department of Veterinary Services, with the assistant director of Veterinary Services (Tsetse and Trypanosomiasis Control) now the most senior official of the branch.[38] Two years later, the Federation of Rhodesia and Nyasaland came to an end, and in 1964 the position of director was formally abolished. Ford chose to retire instead of staying on to work for Rhodesia (formerly Southern Rhodesia), because its white minority defied Britain and declared unilateral independence rather than grant independence to the black majority.[39]

Gerald Cockbill assumed the reins from Ford at a time of escalating *mhesvi* menace. He consulted with senior staff and concluded there was a need for a second review of the *mhesvi* and *n'gana* position, which the government accepted in late 1964. It involved the reintroduction of game elimination, "albeit in a much-modified form, now known as selective game elimination, supported by selective application of residual OCPs to dry season resting and refuge sites of *mhesvi*."[40] Cockbill continued the research momentum gained during Ford's tenure. Even more emphasis was now placed on in-service staff training.[41] Hence, it was "felt strongly that any Senior TFO, Entomologist, or Senior Animal Health Inspector being transferred to a district where trypanosomiasis is a problem—lacking specialist knowledge—should attend a special course at the Central Laboratory in Salisbury."[42] In the past, efficient control measures had suffered from the lack of professional staff, especially entomologists.[43]

Chief Veterinary Officer (Trypanosomiasis) William Boyt lauded the incorporation of BTTC into the Department of Veterinary Services as "an unqualified success." Liaisons between the animal health inspectors and TFOs on the ground and among entomological staff had shown "a steady improvement." District veterinary officers and entomologists now understood each other's roles and opinions, thus eliminating past "danger of friction and delay." The cooperation with other departments and ministries was still very poor, however: "Not all Provincial Administrators or officials of the Southern Rhodesia Department of Agriculture and the Lands Department appreciate the necessity for the restrictions and facilities requested."[44]

Among the most important appointment was that of immunologist, based at the Salisbury lab, a Food and Agricultural Organization (FAO) position that was always occupied by a woman; when she was not there, the experiments most critical to knowing the trypanosome passenger of *mhesvi*—and, indeed, its bloodmeal—ground to a halt.[45] This Immunology Research Section of BTTC continually experienced high staff turnover. First, M. A. Bolton terminated her appointment as Immunologist in June 1965 on getting married, having laid "a sound foundation for further immunological studies."[46] She had also developed techniques and procedures to uncover changes that occurred in the blood of *mhuka* in response to changes in trypanosome presence and trypanicidal drug treatment. Although the senior veterinary officer had acquired "some of the more specialized techniques," the laboratory technician Bolton had trained had departed for 4.5 months of military training.[47] The branch's luck seemed to be turning in 1970, but the new immunologist, E. M. Steinberg, also quit her job to get married. That vacancy was filled by another woman, V. W. Emslie, "who within a few weeks successfully took over the intricate routine." By the end of September, she had "almost eliminated the backlog of samples awaiting identification."[48] Then she too left.

On April 17, 1972, Dr. Gerald F. Cockbill retired as assistant director of Veterinary Services (Tsetse and Trypanosomiasis Control), and Desmond F. Lovemore took over. Other significant appointments that year included R. D. Pilson as chief glossinologist, Glyn Vale as regional glossinologist (research stations and field investigations), and A. Marks, replacing T. J. Casewell as a regional glossinologist.[49]

Cockbill seems to have left a vibrant department still capable of attracting qualified staff despite the raging liberation war. Branch staff were particularly active in publishing, with Boyt, Davison, Hursey, Lovemore, Pilson, Robertson, and Vale contributing papers to local science journals such as the *Rhodesia Science News* and international ones such as the *Bulletin of Entomological Research*.[50] Meanwhile, in 1971, Dr. R. J. Phelps, an experienced University of Rhodesia glossinologist who was on leave, agreed to assist the branch in research. Two glossinologists and twenty-one TFOs were also hired, and one TFO was promoted to a vacant senior field officer position. Eleven new posts were also authorized, effective July 1, 1973: two

TFOs, eight learner/tsetse field assistants, and a clerical assistant. A twelfth position, senior TFO, was approved for an August 1, 1973, hire.[51] Another high point was Regional Glossinologist (Research Stations and Field Investigations) Vale's completion of his doctoral thesis, entitled "The Responses of Tsetse Flies to Their Host Animals" in 1973.[52]

Then the war escalated, control became increasingly difficult, and an infrastructure that had begun to take shape started unraveling. When we rejoin the archival record again, it is 1982, two years after independence. In the 1981–1982 operational year, the BTTC experienced further losses of experienced personnel who were either retiring or resigning, not least of which was Assistant Director Peter Napier Bax, who quit rather suddenly in April 1982. Gerald Davison took over as assistant director, with two deputies—Vale as chief glossinologist (research) and Hursey as chief glossinologist (operations).[53] More white officers quit in the 1983-1984 operational year, citing "the more attractive salaries and conditions offered by the private and commercial sectors."[54] It was also common in this period for whites to leave because of the uncertainty of transition to, and being ruled by, a black government. Many whites were struck by a fear of the unknown. They had never been in this position before.

Putting a Finger on the Pulse of the Fly

How do you gather data about a highly mobile *chipukanana*? And, assuming you successfully reduce *mhesvi*'s mobilities to "glossinological data," how do you make it actionable intelligence of *mhesvi*? This section discusses two aspects of the production of knowledge about *mhesvi*, focusing first on the flyround and then on experiments at three field stations.

The Flyround

Earlier, we discussed an old-timer named Mabuzana, south of Gogoi, who told Swynnerton in 1918 that *ndedzi* was attracted to him because of the black coat he was wearing. By 1928, a method of inspecting the presence or absence of tsetse in an area had emerged, in large measure thanks to Swynnerton's publication of the 1918 research from Muzvirizvi and his subsequent experiments at Shinyanga, Tanzania. This method involved walking a black "bait ox" along the footpaths and *tiko* (villages) located near traffic routes. Out of these beginnings, a systematic method of detecting *ndedzi* presence was born.

The *flyround* was a path created through the bush that a party traversed periodically, catching any *ndedzi* they encountered and recording data about them. The word *round* referred to the early practice of visiting selected areas to make "timed catches" of tsetse at "stations" (markings) along the path. The round was divided into sectors fifty yards long or corresponding with vegetation communities to be traversed (EATTRRO 1953, 30; Potts 1930).

In the rural areas of Chibwedziva (Savé-Rundé area) and Nembudziya (Gokwe), the men who traversed these flyrounds catching *ndedzi* were known in the *misha/tiko* (village) as *mafrayi* (the fly people). "*Kwaiva navanhu vaifamba nemombe* (There were people who moved about with black oxen)," recalls Willias Chabata, who was born in 1936. "They carried *zvihuka zviya izvi* (these nets). They hooked the *mhesvi* trying to bite the *jon'osi* (ox) with the net. They used to walk here, following behind the ox."[55] Said one man: "They would set off from here—one person with a net plus his companion moving ahead, holding the rope tied around the black oxen's horns. Right now [around 2 pm] they would still not yet have returned. They would walk and walk, then stop, for a while, because the *mhesvi* would know that there is food there, I better head that way if I'm to survive. And surely you'd see it land on the ox. Then the *mufrayi* (singular of *mafrayi*) following behind would swing his net and catch it."[56]

Dense thicket meant a path had to be cut through or hoed, but the normal practice was to follow existing paths wherever possible. The path was marked using a wheel, a chain and post, or trees marked at designated intervals (ten to fifteen yards) or "stations," the "sectors" in between clearly numbered to act as guides. The start of a round was marked 0, and subsequent sectors were marked with ascending numbers (EATTRRO 1953, 30; Ford et al. 1959).

The catching team was composed of two catchers—one recording, the other leading the ox (EATTRRO 1953, 9, 32). On approaching the start point, the catching party stopped about one hundred yards before the first station; every member had to ensure that no *mhesvi* was on him. On reaching the path, the *mufrayi* in charge wrote at the top of his notebook page the flyround name, date, and other details, such as colors for marking *mhesvi* for release, time of day, and so forth. Normally, the sectors were fifty to one hundred yards long; *mafrayi* walked at normal pace to the first station, stopped, checked for following *mhesvi*, caught any in pursuit or engorged on them or the *jon'osi*, and recorded them vis-à-vis the station post (32). Once details for each *mhesvi* were recorded, it was either killed and thrown away or marked and released. These catches were entered according to the number of the station; a weather note for that sector (J for *Jua* [sunny]; M for *Mawingo* [dull]; V for *Mvua* [raining]); the number assigned to *mhesvi* during marking if previously marked; male or female; whether teneral (soft, immature, and pale) or mature; the hunger stage if a nonteneral male; the location caught; flies *in copula* (entangled, mating); *mhuka* and fresh spoor seen; and other changes (grass burning, flooding, damage by *nzou*, etc.).

The coding system was based on *dzimbahwe* modes of meteorology and time, the actual words *Zuva* and *Mvura* being corrupted into *fanakalo*—that is, *jua* and *mvua*. *Mafrayi* also searched for *zvikukwa*, putting them into boxes and recording the number collected per searcher per standard working day and sorting them into whole or empty shells. Finally, flyrounds occurred at the same time every day, starting and ending at the time "when the length of a man's shadow equals his height" in the morning and in the afternoon (EATTRRO 1953, 33–34). This is yet another example of the appropriation of *dzimbahwe* ways of calculating time, as opposed to using clocks, which *vatema* did not have.

The role of black juveniles in the collection of *zvikukwa* cannot be understated. "A team of African juveniles varying from eleven to fifteen in number and working under an African field assistant," disclosed Desmond Lovemore, entomologist for Sebungwe District, "have been making regular collections of *G. morsitans* puparia since May in the Gwababa-Mbelele river area." By September 30, 1958, they had collected 14,786 "apparently live puparia," the majority going to the Salisbury laboratory "with a view to building up a laboratory population," with a small proportion handed over to Dr. Robert Barrass of the University College of Rhodesia for "work on parasitism."[57] This was no isolated incident; another team of five juveniles under the supervision of a black survey assistant had been collecting *zvikukwa* and shells in the same area since early August. They collected 947 "apparently live" *zvikukwa* and 4,397 shells.[58]

It is not an accident that juveniles were sent to collect *zvikukwa*. In communities in which fishing exists, the adults always dispatch the boys in advance to the alluvial riverbanks to dig up or scrape the ground for *nyongorosi* (earthworms) or *zvikukwa* to use as hook bait. I performed this chore as a boy; I sent my nephews as an adult. One acquires a certain *hunyanzvi* and exact *ruzivo* of where to look, what signs to look for.

Locals in Nembudziya still have memories of *hufrayi* (the work of fly-catching). Reuben Mavenge still remembers *mafrayi* using "*chimumbure* (net), the same one they used at the gate."[59] Another local resident still remembers *mafrayi* at the business center, "going out to look for *mhesvi*" early in the morning, returning at dusk. He used to hear his own father looking forward to catching more *mhesvi* so that his month's wage would increase. Many *mafrayi* "did sensible things with the *mhesvi* money"—or

Figure 3.3a, b
The flyround (left); searching for *zvikukwa* (right).
Source: http://www.fao.org/docrep/010/ah809E06.htm (left image); National Archives of Zimbabwe (right image).

simply spent it on beer and prostitutes at Gokwe, arriving home empty-handed. After pay, they would first go to Gokwe Center to surrender equipment, then go home. These *mafrayi* were not just anonymous; they were physical and social people with names—like Fema Ngonda, African assistant, flycatcher, and *mugocha*.[60] He is no longer the faceless, nameless, and invisible "fly-boy" of the entomologist's or TFO's field report. He and other *mafrayi* like him were makers of knowledge of *mhesvi* that *vachena* called "glossinology" (*glossina* knowledge), and they created this knowledge in a physical and intellectual way.

Sengwa Field Research Station

One of the recommendations coming out of the 1954 review was for a need to redouble efforts to understand *mhesvi* and *n'gana*. To do this entailed establishing more research facilities to enable the collection of more reliable data. Yet it was only after Unilateral Declaration of Independence (UDI) from Britain, with Southern Rhodesia becoming Rhodesia and international sanctions kicking in, that the establishment of these stations and localization of laboratory testing really took off.

Established in 1965, the Sengwa Field Research Station focused principally on the *mhesvi-mhuka* relationship, seeking to understand the habits, distribution, and dynamics of *njiri* the warthog, *dzoma* the bushbuck, *nhoro* the kudu (see figure 3.4a, b, c), and *nguruve yemusango* the

a) b) c)

Figure 3.4a, b, c

Njiri the warthog (left), *dzoma* the bushbuck (center), and *nhoro* the kudu (right), northern plains of Gorongosa National Park, Mozambique, in October 2011.
Source: Author.

bushpig—the favored hosts of the *mhesvi* that *vachena* called *G. morsitans*. The investigation—a joint undertaking between research officers from Lusulu and the Department of National Parks and Wild Life Management (DNPWLM) area in Gokwe—was to be conducted in the Sengwa area.[61] It complemented a similar investigation in East Africa on two *mhesvirupani* that *vachena* now called *G. swynnertoni* (after Swynnerton) and *G. pallidipes* (pale-footed *mhesvi*) as distinct from *mhesvirutondo* (now *G. morsitans*; Isherwood et al. 1961).

At its inception, Sengwa was given a mandate to address four broad points. First, it needed to determine the host preference of *mhesvi*, through the collection of blood meals from pairs of plots within each of five distinct vegetation types numbered from R1 to R10. Monthly samples of male *mhesvirutondo* were taken from each plot using flyround parties to determine seasonal variations in the fat content, size, and age of *mhesvi* populations. The goal was to "provide a basis for the interpretation of data on the game and *mhesvi* populations and composition of tsetse blood-meals of each plot."[62] The research complemented that in East and West Africa that dissected both *mhesvi* and their *zvikukwa* for clues (Bursell 1959; Potts 1933; Jackson 1949).

The second point was to determine how changes in the population of *mhuka* may influence that of *mhesvi*. Sampling in the two blocks, four square miles each, showed that numbers, distribution, nutrition level, and diet of *mhesvi* were similar in both. In August 1969, the DNPWLM research unit conducted a trial within Lutope to determine the feasibility of removing *njiri* the warthog by capturing them as they came out of their burrows. The *mhesvi* were marked with distinctive colors and released "so that a rough estimate of the effectiveness of the capture technique could be obtained from the relative numbers of marked and unmarked warthogs seen subsequently. ... The capture technique would be a feasible means of removing warthogs provided that two catching teams were used, and that the four-square mile area was surrounded by a pig-mesh fence to prevent re-invasion."[63] The routine sampling in Lutope and Sengwa intensified in October as entomologists sought to establish any changes in tsetse population associated with capturing *njiri*. Preliminary results indicated that the disturbance had no obvious influence on tsetse quantities apprehended on flyrounds and coming to bait oxen.

The third point, to determine the means by which host preferences operate, entailed two types of work: (1) distribution and behavior studies to detect any special relationships between distribution and behavior of *mhesvi* and its preferred host and (2) studies to determine characteristics that rendered preferred hosts attractive to *mhesvi*. Samples of *mhesvi* from flyrounds and of engorged *mhesvi* from stationary bait oxen were used in experiments focused on developing alternative sampling.

The fourth point involved the development of devices to track *mhuka* and thus map their movement patterns. Two self-supporting antenna masts—thirty-five feet and seventy feet, respectively—were installed to act as base stations for the radio tracking of *njiri* the warthog. Priority was placed on tracking radio-tagged *mhuka* and detailed observation of the daily activities of tame, free-ranging *njiri*, especially in the Sengwa Gorge camp area. The successful development of radio-tracking techniques in 1968 marked a major breakthrough in the study of *njiri* dispersion and behavior. Dr. A. E. Rogers was the point man in this experiment, and with his team of black field assistants he deployed equipment with "ranges of four to five miles when using fixed receiving stations, two to three miles with a portable Yagi antenna, and about three-quarters of a mile with a hand-held loop-antenna." These devices made it possible to "home-in rapidly on tagged animals in order to observe them, and also to plot their movements from a distance by taking successive bearings on them from two or more points."[64]

Five *njiri* were tagged with transmitters, and thirty-eight were tagged with plastic collars. The plastic collar was made from a PVC tennis court marker, on which symbols were painted with PVC paint. They served as valuable temporary tags for immature *mhuka*, "lasting until the animal overgrows them, when they drop off." In all, 163 *njiri* were tagged by the end of 1968, with recaptures showing that ear tags easily shed off and resulted in "fairly high" loss of information. Ear-notching was suggested instead.[65]

Encouraged, the research station extended this approach to *dzoma* the bushbuck at the end of 1968. In 1969, however, the radio collar experiment encountered a setback due to the short transmitter battery life, which was limited to about six weeks. Only toward the end of the year did the project resume, following the availability of a more suitable, longer-lasting battery.[66] The field collars attached to *njiri* continued to transmit beyond six months; the laboratory ones continued transmitting after fourteen months.[67] Battery life was critical, because these devices would be deployed in remote areas.

Finally, it is worth pointing out that up until the declaration of UDI in 1965, all blood meal lab tests were performed at Lister Institute in London, through the good office of Dr. Bernard Weitz. Sera were usually frozen at and transported in frozen state from Rekomichi, Lusulu, Sengwa, and other field stations to the airport, then flown overseas for testing. The difficulty of this process spurred Rhodesia to build a blood meal identification laboratory at Salisbury Experimental Station. By 1969, the lab had experimented with three kinds of hosts: *vanhu*, *zvipfuyo*, and *mhuka*.[68]

Conclusion

The chapter has shown that economics drove *hurumende yevadzvanyiriri*'s need to know *mhesvi*—an acknowledgment of the danger this *hutachiwana*-carrying *chipukanana*'s mobilities posed to the cattle (specifically, beef) industry. The industry itself is another example of *ruzivo rwevatema* as the foundation of the *hutongi hwavachena* (white rule); it must not be seen in isolation, but within a larger context of *pfungwa dzavatema* (idioms of black people) that *vachena* appropriated to survive and thrive as settlers. As the white occupation began, white writers submerged these appropriations under the propaganda that any credible knowledge they found *vatema* possessing and practicing (especially of stone architecture, metallurgy, and agriculture) had been brought in or built by an intellectually superior, more civilized people (Bent 1992; Selous 1893).

Some of these myths have been decisively debunked elsewhere on the African continent. For example, between 1500 and 1850, blacksmiths' strong, high-carbon iron outcompeted *vachena*'s inferior product, which had high sulfur content and was brittle (Thornton 1990–1991; Goucher

1981). The Portuguese (1470s–1600s) intended to bring their own engineers to revolutionize how metallurgy was performed and set about establishing a foundry in the part of the Kongo kingdom that is now northern Angola—but it was not to be. So inferior and uncompetitive was their steel that they ended up employing Kongolese smiths in their foundry in Angola and buying locally made iron for resale along the western coastline. Black merchants and middlemen also extended their indigenous entrepreneurial traditions to exert control over trade between the white merchants and the hinterland (Diouf 2000; Dumett 1983, 663; Henige 1977). Similar examples include the decisive role of locally domesticated crops (Brooks 1975; Bowman 1987; Clarence-Smith 1994), mining and metallurgy (Garlake 1978; Phimister 1978), and medicinal plants (Osseo-Asare 2014) in the development of the United States' and Europe's industry and science from the sixteenth century on.

We can now return to Abiola Irele's controversial statement discussed in the introduction, in which he once said that the only future for Africa lies in turning and following Western culture and civilization. "It is of no practical significance now to us," Irele said, "to be told that our forefathers constructed the Pyramids if today we can't build and maintain by ourselves the roads and bridges we require to facilitate communication between ourselves, if we still have to depend on the alien to provide us for necessities of modern civilization, and if we can't bring the required level of efficiency and imagination to the management of our environment" (Irele 1983, 3; echoing Towa 1971, 1979).

This view has its critics, who reject a total sublimation of *tsika dzevatema* or *chivanhu* (black culture) into an imitation of *tsika dzevachena* (white culture), because nobody can predict the identity or desirability of the outcome. They instead urge *vatema* to take all the positives they can get from outside, while maximizing the strengths of their own innovations (e.g., Gyekye 1997; Falola 2008). The example of *mombe* has shown how *vachena*, by force, expropriated one product of *ruzivo rwevatema*—*mombe*—to anchor their enterprise. The answer to Irele also lies in a tour of *vachena*'s early iron, gold, and copper mines throughout southern Africa, if not the entire continent, which were established precisely where *vatema* used to mine, smelt, and process such ores.

The rationale for the *mombe* discussion was to put into relief why *vachena* had to study *mhesvi* so thoroughly and intrusively. The rest of the chapter explored how it did this, taking the reader first through the establishment of a "colonizing structure" for the control and eradication of *mhesvi* and *gopé/n'gana*. The critique of Mudimbe—and by extension Africa scholarship—lies in the absence of a serious consideration of things (other than people) as actors in need of space in the historical narrative. When Mudimbe talks about the physical, people, and the spiritual elements, he is serving notice for the impossibility of merely talking of "the thing in itself," be it means and ways of doing things or *zvinhu* (things; singular *chinhu*) outside the political economy of contextual meaning.

Increasingly since the 1980s, some Western scholars of science and technology have insisted on treating humans and nonhuman things as actors or agents, what one such scholar calls a "parliament of things" (Latour 1993). This utopian vision takes for granted the status of being *munhu* (human), not considering the possibility that one minute, someone is *munhu*, and the next that person is designated *chinhu* and subject to treatment as an instrument for eliminating other *zvinhu*. Read from *dzimbahwe*, Harry Collins and Steven Yearley's (1992) dismissal of Callon and Latour's formulation of human/nonhuman agency as a game of epistemological chicken that purports to respect the latter's agency even as such *nonhumanness* is assigned from "a human-centered universe" (311) is both reasonable and problematic—reasonable because Latour/Callon's descriptions of such agency were far over the top, absent a thorough engagement with how those designated *zvinhu* speak or not; problematic because there are less controversial, more empirically grounded, context-specific ways of accounting for the agency of things designated *zvinhu*.

Mhesvi is one of those less controversial instantiations of insect agency. It posed a real danger through its pestiferous actions and caused the establishment of an entire structure and set of procedures dedicated to putting a finger on its pulse, with the ultimate goal of destroying it. This is how its location became a site of work, or a laboratory. The flyround shows *mhesvi*'s location as a site of mobile work. Where *mhuka* that *mhesvi* most fed on roamed, field stations (static infrastructures) were established. The major work, however, was at transient sites of work and involved transient work, with researchers tracking *mhesvi* ever alert and lurching onto anything that moved for food, while *hutachiwana* moved with and inside *mhuka* like *njiri*, *nguruve dzemusango*, and *mhesvi* itself.

I started from *chidzimbahwe* with culturally grounded understandings of *zvipukanana* and *mhesvi*. I have entered *vachena*'s language and ways of producing knowledge, in which *mhesvi* became *Glossina*. I started from traditions that deployed organic, environmentally friendly means and ways of managing and coexisting with *mhesvi* and *hutunga*; I am now taking the reader into an environmentally destructive, pollution-intensive moment in *mhesvi control*.

4 How to Trap a Fly

At the end of the pan-African Agricultural and Veterinary Conference in Pretoria in August 1929, Chief Entomologist Rupert Jack visited the KwaZulu area affected by *mpukane*. He spent ten days with R. H. Harris, the man in charge of tsetse operations there, learning about the Harris flytrap then in trials. In November of the next year, Jack again returned to KwaZulu to witness its demonstration. As he put it, the traps "function[ed] remarkably" well and the demonstration was "decidedly impressive."[1]

The Harris flytrap experiments began in the Nyambani valley of Kadoma in 1930, with various modifications alongside other models to determine what was best for Southern Rhodesia. Each was painted in a different color to determine which one yielded the most *mhesvi* catches. Dark blue and black turned out to be the most attractive, while results on other colors proved "disappointing in many respects." (Potts 1930; Nash 2007; Lloyd 1935; Swynnerton 1933; Jack 1939; Barrass 1960). Effectiveness varied considerably in the twenty-two traps according to position, amount of movement within the vicinity, seasonal and daily weather conditions, and *mhesvi* concentration. Most of the flies were caught from August to September, with the numbers decreasing in October and becoming "almost negligible" as the rains started. By contrast, a single *mufrayi* caught twice as many flies with a net in one hour as the best trap caught in a month.[2]

In 1932, Jack emphasized that any significant catch only took place during a small portion of the year "and even then has been too small to afford promise of practical benefit from the use of the traps." As he put it, "The reason for the failure of these fly traps in Southern Rhodesia lies deeper than in details of design—it lies, in fact, with the principle itself."[3] Chorley explained in 1937: "Any trap depending on shade as an attractant must give poor results with *G. morsitans* which is an open savannah loving fly, not so dependent on shade as *G. palpalis* or *G. pallidipes*."[4] Precisely the difference with the KwaZulu experiment! The trap could only be effective with a thorough knowledge of the bionomics of *mhesvi*. Deployed against *mhesvirupani* in KwaZulu, the Harris trap was very effective, but it was generally ineffective against *mhesvirutondo* elsewhere in Africa.[5]

K. R. S. and M. G. Morris's investigations in the Gold Coast (Ghana) in 1949 yielded a trap that would become significant in Southern Rhodesia. Fewer examples illustrate the enmeshment of *vachena*'s means and ways of controlling *mhesvi*, on one hand, and *vatema*'s ways of life and spiritualities on the other. In 1949, the two researchers sought "a means of controlling *G. palpalis* and *G. tachinoides* in sacred groves," pernicious foci for the concentration and dissemination of *mhesvi* and *n'gana* and frequently visited by local *vanhu*. The investigators defined the problem thus: "Rigorous taboo opposes the cutting of any vegetation within them, and one or two early attempts to introduce clearing methods resulted in the loss of confidence and cooperation of the people" (Morris and Morris 1949, 494; Morris 1950). The methods—the means and ways—had to be tailored around the cultural dynamics. "Thus," Morris and Morris (1949) concluded, "J. K. Chorley's Crinoline trap might be mistaken by a tsetse for the body of a man; small forms of Harris or Swynnerton's Screen traps [developed in Tanganyika] might represent the local dwarf breeds of cattle, but both were too big to resemble what are by far the most numerous of the food hosts along the banks of rivers: sheep, goats and, in the more thinly populated parts, duikers" (494).

Morris and Morris therefore designed a new "animal trap" (see figure 4.1a, b) resembling a *mbudzi* (goat-)-sized animal, which they trialed alongside Chorley's eighteen-inch diameter, twenty-four-inch long Crinoline traps made out of khaki and black material; Harris traps two-thirds standard size; and Swynnerton's awning screen box (ASB) traps with screens 4 ft. wide, 2 ft. long, and 2 ft. deep (Morris and Morris 1949, 494). The attractant was "a cylindrical body, 2 feet long by 1 foot in diameter, standing with the top or shoulder 2 feet from the ground, and covered with ordinary light brown hessian or burlap" (Morris 1961, 905).

By 1960, many traps had been designed (Swynnerton 1933, 1936; Chorley 1933; Lewillon 1945; Morris and Morris 1949; Morris 1960). All showed that a trap that worked well against river-hugging *mhesvirupani* (*vachena*'s *G. palpalis* and *G. pallidipes*) might perform badly against savannah-loving *mhesvirutondo* (what they called *G. morsitans*). Both Swynnerton's ASB trap and Langridge's box screen trap rated higher than the Morris trap (Moloo 1973, 231)—quite the opposite of *mhesvirutondo*-dominated Rhodesia!

The interest in traps seemed to have died down in the immediate postwar period and the 1950s. However, as game elimination and concerns

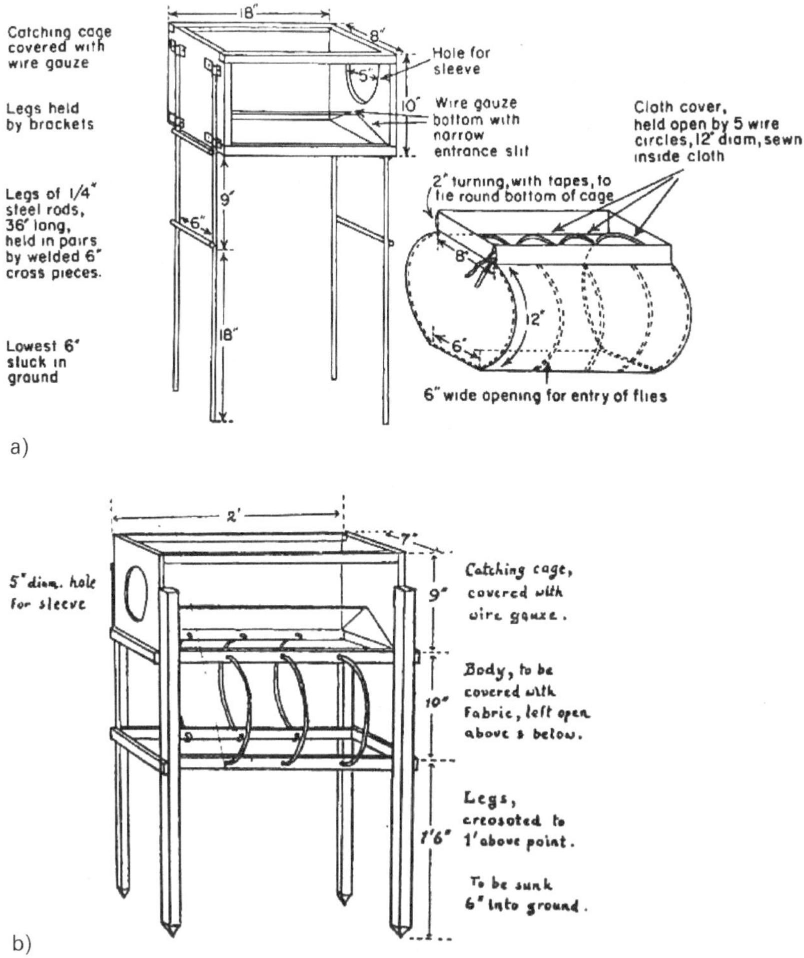

Figure 4.1a, b

The Morris folding trap, showing its removable parts for ease of carriage (top) and the fully assembled trap (bottom).

Sources: Morris 1961, 907 (top image) and Morris and Morris 1949, 495 (bottom image).

about OCP toxicity gathered pace, BTTC revisited the possibility of trapping *mhesvi*, especially in those districts where its threat was significantly reduced and where traps could act as surveillance apparatuses. Two mechanical traps were considered, based entirely on the artificial refuge model. One was the Morris trap, with a body of black cloth and a black screen just like Swynnerton's trap. The other was a Manitoba fly trap with a decoy composed of a black ten-gallon drum mounted fourteen inches above the ground (Thorsteinson, Bracken, and Hanec 1964).[6]

The idea of trapping *mhesvi* had seven selling points, enunciated in the BTTC's annual report of 1982. First, it integrated control and survey operations otherwise performed separately, thus saving costs. Second, it was a simple means for financial-resource-starved Africa. Third, it reduced the heavy demand placed on foreign currency, often associated with OCP spraying. Fourth, it placed only steady, constant demands on labor and equipment all year round, unlike OCP spraying's peaks. Fifth, it guaranteed a reasonably constant effect on *mhesvi* for the whole year, compared to the sharp but short pressure from spraying operations. Sixth, it allowed sufficient time to correct any errors without jeopardizing the entire program, unlike spraying cycles. Finally, compared to OCPs and "game elimination," it was an ecologically clean and friendly method.[7]

What follows is a discussion of the types of baits that were under serious scientific, technical, and economic consideration, each categorized according to the attractant or stimuli used and how it handled the *mhesvi*, the method of treating the attracted flies, and the costs and efficacy per each unit deployed. The underlying principle of a trapping system was that *mhesvi* had very small reproductive potential, such that a slight reduction in the *chipukanana*'s reproductive rate or increase in its mortality rate was enough to control its entire population. This could be done by attracting the *mhesvi* to artificial baits laced with killing or sterilizing agents.

Attracting the Fly

The role of *mhuka* as attractants had received attention prior to its escalation in late 1940s Southern Rhodesia; it was the key to why some traps succeeded for one type of *mhesvi* but not another and why a one-size-fits-all approach was deemed a terrible idea (Swynnerton 1921a, 1936; Harris 1938; Jack 1941; Vanderplank 1944).

From 1957 on, the most significant attractant research was performed at Kariangwe and then, from 1965 on, at Rekomichi Field Research Station. The research sought to establish the best attractants of *mhesvi*. The baseline for the investigation was quite clear: *Mhesvirupani* was not easily attracted to *vanhu* but was attracted to *mombe*, leading to the need to "compare catches made by various methods at the same time over the same territory."[8] The research focused on sight (visibility) and smell (olfactory) as attractants.

Visual Attractant Research at Rekomichi and Sengwa

The most interesting studies on visual attractants were conducted at Rekomichi (Rekomitjie) from 1966 on, with subsidiary research near the Ruyese River and at Sengwa. The experiments at Rekomichi began in August 1966 and ended in December 1967. After that, the research continued at Sengwa until October 1970, while the Ruyese study was conducted for only two weeks in September–October 1967 (Vale 1971, 331).

You may remember that *vatema* had long observed that *mhesvi* tended to be attracted to black forms, especially those resembling *mhuka*.[9] Old Mabuzana said as much to Swynnerton. The reason for moving at night was, in part, to neutralize blackness as a factor in visibility and render everything the same to *mhesvi*. The Rhodesians were working with the opposite dynamic; instead of trying to prevent *mhesvi* from biting, they deliberately deployed black forms during the day to flush them out.

For the Rekomichi experiment, five artificial refuges were found to attract *mhesvi*, all of them with a dominant black component. The *hut refuge* was composed of a hut measuring 6 × 6 × 6 in., with walls of sacking and clay mud covered by a thatched roof, all designed for *mhesvi* to come in and rest. The *trellis refuge* was a 10 × 8 in. wooden trellis, with a lean-to grass-thatch roof over a drum covered with soil for the same purpose. The *box refuge* was a forty-four-gallon drum with one end cut out, placed horizontally on the ground, and covered with soil to insulate against the sun's rays.[10] The *wigwam refuge* was a thatched, triangular structure offering shade over the entrance to a black insulated ten-gallon drum. The last refuge was a set of pipe refuges made of concrete pipe, 4 ft. long by 2 ft. in diameter, placed horizontally or vertically, and insulated with a 1.5 in. layer of thatch (Vale 1971, 333; see figure 4.2 and figure 4.3 for a closer view).

Each refuge was created within and applicable to three different vegetation types. Two variations were made—namely, the location and size of the drum and the materials used as cover (thatch, sacking, or black cloth). These changes had little effect on the number of flies attracted. More useful was the addition of a curtain that could be closed at a distance to keep out any flies that might have followed the observer, and a drum refuge site developed as a trap that allowed *mhesvi* to get in and rest, but kept them hostage.[11]

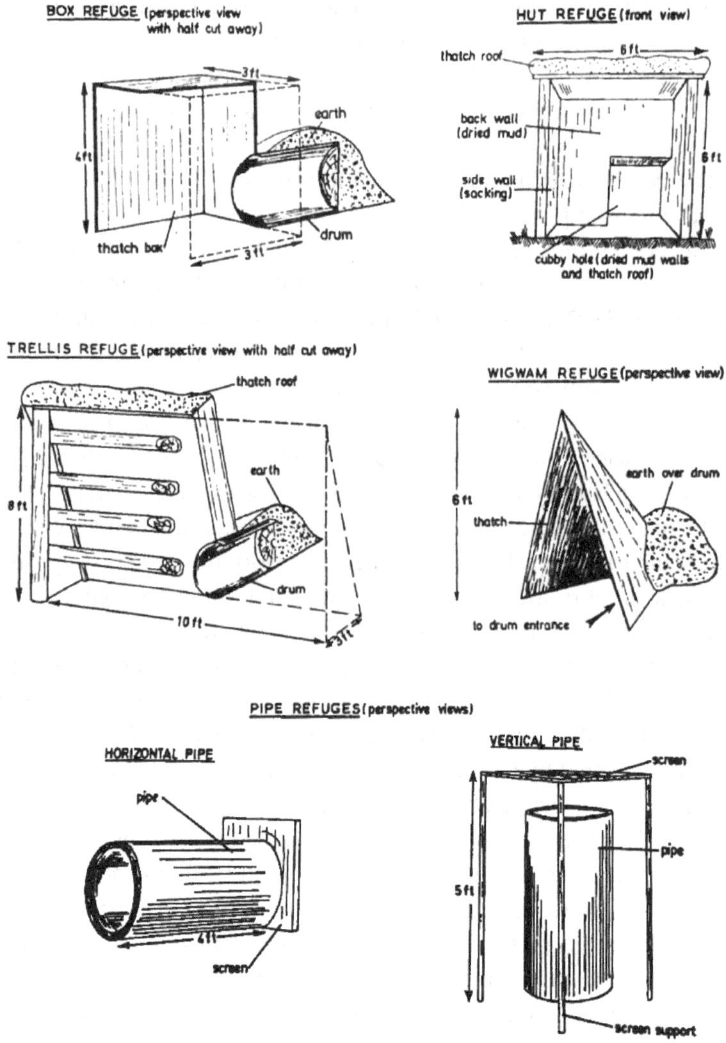

Figure 4.2

Five refuge models at Rekomichi and Sengwa in 1971.
Source: Vale 1971, 332.

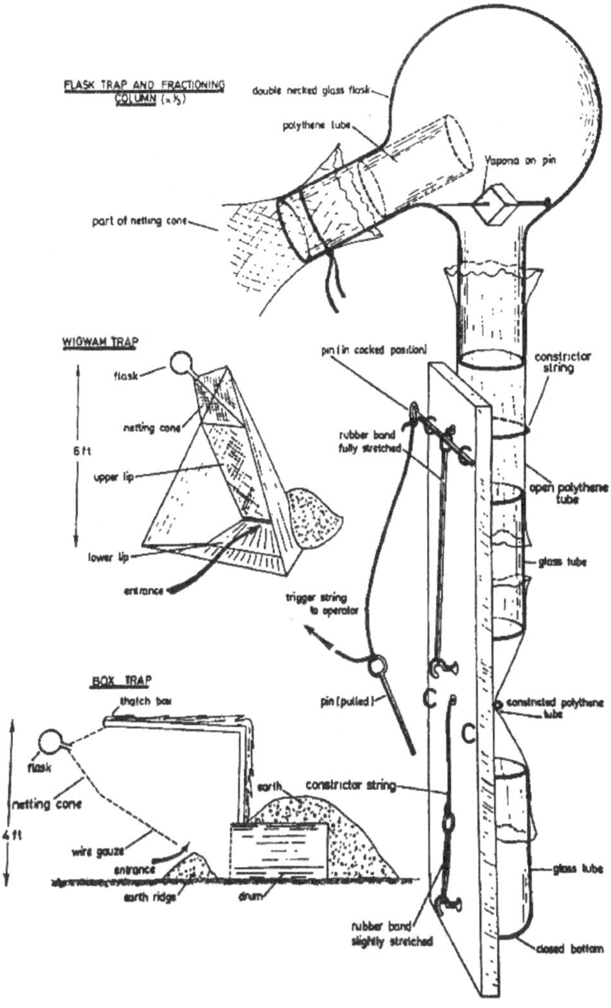

Figure 4.3

A closer view of the flask, wigwam, and box traps.

Source: Vale 1971, 334.

The "glossinologists"—as *vachena* specializing in studying *Glossina* began to call themselves—studied the composition of refuge site samples, focusing on the physiology of the flies caught. They compared these with samples from Morris traps, a tethered bait ox, and man-only bait.[12] Fifty percent of *mhesvi* caught in refuge sites had "incompletely digested blood visible through the abdominal wall."[13] The diurnal pattern of refuge behavior was of especial interest. Attractants like tethered oxen and traps were inspected at about 1400 hours, when the sun was blazing hot.[14] Seasonal patterns of refuge behavior were also studied, with catches peaking in late September and early October (the hottest period), gradually declining as the rains began, hitting their lowest in late December and early January, and staying generally low through to May.[15]

To assess the influence of refuge site positioning on the quantity of catches, over fifty drum sites were plotted within a one-square-mile area containing varieties of vegetation. Under observation from October to November 1966, one or two *mhesvi* were caught per day in the refuge sites in the open *mupembere* or *mubondoroko* (*vachena*'s *Combretum*) bushland, on *musasa*, *munhondo*, *muunze*, and *mupfuti* (all *vachena*'s *Brachystegia*) and other woodland on the Zambezi escarpment slopes, and *mupani* or *musharu* (mopane in *setswana*) woodland at its base. Moderate catches of up to fifty were obtained daily from areas with some shade and adjacent to steep banks, particularly gullies fringed with leafless thicket. Very good catches, frequently two hundred *mhesvi*, sometimes over four hundred, were obtained in deeply shaded areas, especially among leafy, riverine *musika* (*Tamarindus indica*) trees. How then could they explain "the great differences between the magnitude of catches from exposed situations and sheltered situations?"[16]

By the end of 1967, BTTC had successfully developed the *refuge trap* to catch the *chipukanana* as it sought sanctuary from prey or the elements or sought to rest. It was thus designed for resting sites and stationary objects. The centerpiece of the Rhodesian approach to refuge traps was the Morris trap described earlier.

The success of the refuge trap concept led researchers to believe that the use of mobile models of *mhuka* could help flush out *mhesvi* and expose the role the distribution and behavior of *mhuka* played in *mhesvi*'s host preference.[17] What the BTTC called *attraction studies* were composed of observations on numbers and behavior of *mhesvi* within the proximity of live *mhuka* and the development of mobile and stationery models as devices to explain these observations. This experiment was undertaken at Sengwa Research Station. The timider *mhuka* like *njiri* and *nguruve*, and *zvipfuyo* like *imbwa* and *mombe* were placed into one group, while the fickler ones like *mhara* and *dzoma* were placed in another. *Mhesvi* freely landed on the former, much less so on the latter.[18]

Smell

Many *mhesvi* landed near the eyes of the adult *njiri*, fewer on the adult *nguruve*'s, and none on those of the other *mhuka*. Large numbers engorged on *njiri* (warthogs), especially within two inches of its eyes. *Mhesvi* probed puppet of *njiri*, *nguruve* and *dzoma* stood next to live adult *njiri*, but not near the eyes. The conclusion: It was possible that olfactory stimuli played a role in *mhesvi*'s responses to *njiri*'s eyes.[19]

By 1972, it was now known throughout the tsetse research fraternity that *mhesvirutondo* usually fed upon hosts resting in fairly dense vegetation, where smell rather than sight seemed a likely attractant. Yet to be determined was the role of olfactory (odor) stimuli in host finding. Male *mhesvirutondo* proved more responsive to moving targets than females; the females preferred targets whose movement was interrupted (Gatehouse 1972). There was strong belief that olfactory attractants could improve *mhesvi* population survey techniques and give a new dimension to the means and ways of controlling *mhesvi* if deployed with pesticides or chemical sterilants (Turner 1972, 25).

In the 1972 field trials, the ox attracted the most flies, but large catches still were made with *mbudzi* (goat), *mubhemhe* (donkey), *njiri*, *hwayi* (sheep), *nguruve yemusango*, *nhoro* (kudu), *mhara*, and *nyati*.[20] Two hundred and twenty-five flies were caught in portions of cage and netting without bait, 442 in those with visual bait alone, 2,730 with just odor bait, and 3,868 with a combination of visual and olfactory bait. That translated to a ratio of 1:2:12:17. This compared with sixty-six for the small electric net without bait, 803 with just bait, 892 with odor bait only, and 2,902 with visual plus olfactory bait. The ratio was 1:12:14:44. These results led to two important observations regarding net size: First, odor alone was as good as visual stimuli alone when small nets were used, but "odour was six times more effective than visual stimuli when the large net was used." Second, adding visual stimuli to odor "increased catches only by half when the large net was used, but trebled catches when the small net was employed."[21]

Some of the nets were deployed to catch *mhesvi* on their own without any ox or person around. In its mounted position, the blue device would look like an ox; *mhesvi* would investigate this "animal," fall in, and get caught.[22] Without bait, catches were evenly distributed, but with the model as bait, catches were concentrated on the cage. With odor-only bait, the concentration on the cage was less marked than with the model, showing less precise orientation. Instead, catches were strongly concentrated on the downward faces, indicating an upwind flight to the baits. Finally, with both model and odor as bait, concentration was pronounced, both on the cage and on the downwind faces.[23] Two observations were made: First, few flies ranging meters off the bait could be attracted to a stationary visual stimulus; second, many flies were attracted by olfactory stimuli from afar, but needed visual stimuli at close range for precise orientation. The verdict: Odor alone had poor efficacy, even when used with small nets.

No significant difference was noticed between odors from a normal ox, one with a dampened coat to boost body odor, and yet another with its body thoroughly washed in water or a Teepol solution to decrease odor. To determine the importance or insignificance of body odor, a special pit was dug and split into two similar chambers using an airtight canvas sheet. The *mombe* was isolated in one chamber, with its head and upper neck passed through a laced slit in the sheet and isolated in the other chamber. The tests showed head odor yielded catches "several times greater" than body odor and indicated that head odor was responsible for *mhesvi*'s attraction to *mombe*.[24] Explaining the concept with a picture portraying an African field assistant pointing out the ventilator, Vale said: "To study attraction to stationary sources of host odour, an animal is hidden in a roofed pit and air from the pit is blown out at ground level through the opening to which the man is pointing. Flies attracted to the odour are caught by the electrocuting net suspended within the rectangular frame" (Vale and Phelps 1974, 1; see also figure 4.4). Air from pits was blown out at ground level close to an electric net to trap all attracted flies. Although the odor of ox was quite effective, large catches were still obtained using *mubhemhe, mbudzi, hwayi, dzoma, nguruve,* and *nyati*. *Munhuwi wemunhu* (human odor) proved ineffective and reduced the effectivness of ox's odor when deployed in combination (Vale 1973).[25]

In 1975, Vale determined that flies were attracted to a stationary visual target that released odors; the greater the odor, the greater the number of catches. The experiment deployed a ventilated pit like the 1974 one, but sited in semievergreen woodland on the Rekomichi floodplain. It measured 6 m long × 4 m wide × 2.1 m deep, with three 15 cm ventilation shafts equipped with an electric extraction fan 12.5 cm in diameter to ensure adequate evacuation of odor from up to six oxen. Finally, a black visual target (a cylinder 50 cm long by 37 cm in diameter) was placed 37 cm above the odor outlet. A 90 × 90 cm electric netting was placed downwind of the target to catch all attracted *mhesvi* (Vale and Hargrove 1975, 46; Vale and Phelps 1974).

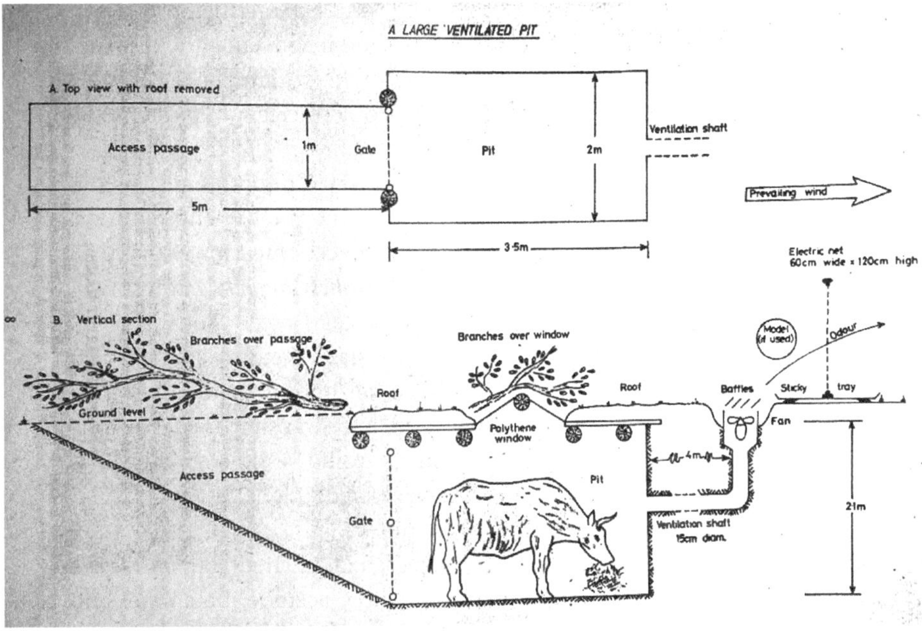

Figure 4.4
Roofed pit with a *jon'osi* inside to generate odor.
Source: Annual Report 1974. SACEMA/TA.

The researchers were expecting at least two things: First, they considered excluding visual targets because they were interested in odor targets, but used them anyway because they increased catches "by concentrating near the net those flies initially attracted by odour." Second, they expected that the electric net would not cope effectively with high numbers of flies attracted by six oxen (Vale and Hargrove 1975, 47–48).

There is no record of experiments for 1976–1981, because Rekomichi fell within a hotly contested warzone between *varwiri verusunungukolabalweli benkululeko* (freedom fighters) and the Rhodesian Security Forces, but we catch up with BTTC researchers after independence, while they were isolating the effective ingredients of the odor of *mombe* to dispense it artificially. Carbon dioxide and acetone had now been identified as the chemicals at the center of the smell. Applied in doses equal to the odor of *mombe*, they yielded half the number of flies the live bovine produced, suggesting more components in the odor of *mombe* beyond just carbon dioxide and acetone.

The race was on to isolate these components and explore their efficacy in lieu of carbon dioxide, which was "expensive and inconvenient to dispense in large-scale operations." That is how researchers found that "acetone alone, at economical doses of 5–500 mg/ha, could increase trap catches by 2–5 times."[26]

Several chemicals were thus identified and tested in both built lab and field, but just one—1-octen-3-ol—was subjected to intensive initial tests. When added to carbon dioxide and acetone, it accounted for virtually all the efficacy of natural ox odor (*hwema*) against *mhesvirutondo*, but just two-thirds of that against *mhesvirupani*. The chemical 1-octen-3-ol was effective alone or with acetone; the problem was that it became a repellent if dispensed in a high dose, unless diluted with air prior to exposure to flies. An air-mixing fan was used; alternatively, the odor was simply released one meter upwind of the trap so that the chemical would be diluted by the time it reached the trap. With these facts now mastered, the next step for BTTC was to establish tsetse behavior in odor plumes.[27]

By 1984, the odor-identification project had isolated four attractive elements found in host odors—carbon dioxide, acetone, butanone, and 1-octen-3-ol—each capable of boosting *mhesvi* catches several times over. Further attractants in ox odor remained unidentified; completing their identifcation was "likely to take a long time." The dung and urine of *nyati* seemed nonattractive, but catches from traps increased two to four times "by placing next to them sacks used as bedding by a bushpig." The attractive component was found to persist on the sacks for five to six weeks in a nonacidic form readily soluble in chloroform.[28]

The samples of the chloroform washings were sent to the Tropical Products Institute (TPI) in London and the Tsetse Research Laboratory (TRL) in Langford, Bristol. The process involved collecting host odor extract in Zimbabwe; fractionating and chemically identifying these extracts into their constituents at the TPI; lab studies at TRL to determine whether the chemicals the TPI had identified elicited a suitable behavioral response from *mhesvi*; and field studies in Zimbabwe to determine whether the fractions or known chemicals identified at the UK labs to produce suitable responses under built laboratory conditions could produce the same responses in the field. Thus, the trap activity on the ground was also geared toward the branch "meeting its obligations" to the collaborative project by refining its field methods for studying the response to odors.

In 1985, it emerged that there were at least three other as-yet unidentified groups of components—namely, an ox odor that was very volatile when passed through a charcoal filter; a rather involatile one from *nguruve* present on sacks brought into contact with these wild swine; and another involatile one, *mombe* urine. The ox odor was the distant attractant of *mhesvi* to hosts and expressed "marked increase" of catches with every dosage increase. *Nguruve* and *mombe* urine fractions were principally short-range attractants, especially responsible for boosting *zvipukanana*'s readiness to enter traps once visually attracted or odor-attracted. The fractions achieved a fivefold increase as attractants, but had a saturation point beyond which no further increases in catches could be obtained. In fact, sacks and urine were found to stimulate chemoreceptors of *mhesvi*, some of them repellents (small fatty acids at high dose) or neutral (cresols at various doses). Some were attractants, like a p-cresol sample the BTTC sent to Burkina Faso for comparisons.[29]

The discovery that carbon dioxide was necessary to interest *mhesvi* in the trap in large numbers significantly reduced the value of using catch-and-retain and sterilizing traps versus targets with OCP-treated netting. It was already known that only twenty cubic centimeters per minute (cc/min.) of carbon dioxide in place of 2 liters per minute (l/min.) present in ox odor was enough to entice *mhesvi* into traps, yet even such a low dose was too expensive. In lieu of carbon dioxide, BTTC resolved to modify the trap design or find a substitute for carbon dioxide while using targets with netting to control *mhesvi* in the meantime.[30]

All this research should bring to mind an earlier moment in history, when *vanhu vatema* burned dung indoors and underneath their *zvipfuyo* and smeared it on the skins of *mhuka* to ensure their safety against *mhesvi*.

Sticky Traps

I grew up trapping and catching *shiri*—lots of them—with birdlime. We made the birdlime ourselves. In Zimbabwe, boys tap rubber from at least three trees: *chitatarimbo* (which *vachena* now call *Euphorbia matabelensis*; see figure 4.5), *mukonde* (*Euphorbia ingens*), and *mutsamvi* (*Diplorhynchus condylocarpon*). In our village, we had plenty of *zvitatarimbo* (plural of *chitatarimbo*) growing wild in the forest or as a hedge in people's wetland gardens. *Kugumha hurimbo* (harvesting birdlime) is an old tradition, practiced for generations by our ancestors and continuing among youths today.

We would cut across the *chitatarimbo*'s trunk using knives to extract the white milk used to make *hurimbo*, the sticky rubber for trapping *shiri*. Two apparatuses were required: a knife and a small container. We would make a small, sharp knife by simply beating a three- to six-inch nail with a stone or claw hammer then sharpening (*kurodza*) the blade on a rough granite rock.

Figure 4.5
Chitatarimbo/Euphorbia matabelensis.
Source: Photo by Lewis Mavhunga, Mavhunga Village, May 2017.

The container was usually the base of a shoe polish tin or a disused canned jam or baked beans tin cut in half. We also carried a small amount of Vaseline Blue Seal petroleum jelly or Olivine cooking oil—an ingredient to make the rubber less stiff, more tactile. Each of us would collect his own *hurimbo* from the *chitatarimbo*. When enough was gathered, we added a little oil and put our small tins on a small *moto*. When the rubber showed signs of being well cooked, we took the tin off *moto* and stirred the mixture, either with the knife blade or a small stick. The substance now wound into a ball in the center of the tin, around the tip of the blade or stick. We then chewed it, spitting the saliva containing the excess juice, and stretched the birdlime in short pulls, bringing the two ends back together again, then stretching, until the *hurimbo* was well mixed.

Then, we were ready to trap. We wound the rubber around a specially selected, thin, and nontoxic stick called *mudziti*, which we placed atop the tree canopy where *shiri* rendezvoused en route to distant places or when helping themselves to *mhunga* (millet) fields, or we placed *midziti* (plural of *mudziti*) by the *zambuko* (drift), where *shiri* came to drink and bathe. Once a *shiri* landed, its claws wrapped firmly around the stick, it was stuck—but as it flailed its wings to make good its escape, the feathers also became stuck. As soon as it landed, we were already running to collect it. Two *dzimbahwe* proverbs capture the stickiness of *hurimbo* that we experienced as boys: First, "Poverty is birdlime, it sticks."[31] Second, "The bird stuck to the birdlime is the one that says *tsviri-tsviri*; that in the mouse trap says the sky has fallen" (Hamutyinei 1992, 40).[32] *Hurimbo* caught *shiri* alive.

One of the biggest surprises I encountered during research for this book was *vachena*'s appropriation of *hurimbo* to trap *mhesvi*. The ivory poacher Cecil *Bvekenya* Barnard, whose story is told in *Transient Workspaces*, built upon the local art of making traps laced with adhesive rubber from *zvitatarimbo* and other rubber trees to catch *shiri* in order to make his own brand of anti-*ndedzi* trap. He says he never lost a single *mubhemhe* to *ndedzi* in fifteen years of poaching in Hlengwe country between 1914 and 1929. His concoction was composed of animal fat, Cooper's dip, tree gum, and honey, which he smeared all over his donkeys. Any *mhesvi* that landed became stuck to the gum, and the arsenic poison would finish it off. "The donkeys certainly became pretty sticky-walking fly traps," but they survived and confirmed the concoction's efficacy (Bulpin 1954, 58).

The idea of turning *zvipfuyo* into mobile sticky traps did not catch on in government circles until the immediate post–World War II period. At that point, Southern Rhodesia's veterinarians and entomologists started applying DDT to cattle-dipping liquid chemicals to kill *mhesvi* as it landed and attempted to feed. The major concerns were that the animal sprayed might succumb to skin lacerations or hair loss or lick its skin and poison itself. The latter problem was resolved by concentrating the application around the neck area, the most olfactory part of the animal's body unreachable by its mouth.

Two experiments brought together OCPs and *hurimbo* as ingredients for the transformation of an ox into an automobile trap. One produced satisfactory kills of *mhesvirupani* when treating *mombe* with a solution of

9 percent DDT in oil twice a week—a very high concentration. A second mixed DDT in gum and ox serum, which was applied against another *mhesvirupani*. Three other studies involved mixing DDT in an adhesive made of coumarone resin. Yet another study used boiled linseed oil as an adhesive solvent, which, when dry, did not produce DDT crystals—thus resulting in low toxicity and therefore ineffective (Wilkinson 1948).

These researchers persisted with experiments because DDT-treated oxen had been successful against horn flies in the United States. Could *mhesvi* be exterminated by flooding their habitat with DDT-treated oxen that outnumbered *mhuka*? Could *mhesvi* be enticed to prefer *mombe*? Some researchers argued that because *mombe* were quite favored by *mhesvi*, they could, when injected with the trypanicidal drug phenidium compound, be kept in the *mhesvi*-infested bush longer, with the DDT preventing further bites (Vanderplank 1944). In Texas, Florida, and Kansas, ranch cattle dipped in or sprayed with 1 percent DDT aerosols, emulsions, and suspensions had kept cattle free of horn flies for weeks. Why, it could also work here!

Whiteside's experiment at Old Shinyanga in Tanganyika summarizes the method for treating oxen with DDT: "A solution containing 9 per cent. w/v pure DDT and 9 per cent. w/v resin in commercial groundnut oil was used. … The groundnut oil was heated to 105–110°C while dissolving the DDT and resin, and the resultant solution applied (cold) to oxen at the average rate of 110 cc. per ox, excluding wastage, corresponding to 9.5–10 grams pure DDT per ox, or about 450 mg. per square foot of body surface. The head alone was left unsprayed" (Whiteside 1949, 124). Yet, as Whiteside would discover in 1949, *mhesvi* in Africa was very different from horn fly in the United States: "Extermination—not merely 'control'—is required; the contact of tsetse with cattle is relatively fleeting—perhaps only 1 minute in four or five days—compared with that of horn fly; and whereas against the latter it is sufficient to treat the heads and backs of cattle, against tsetse the legs and belly are most important, and are difficult to spray efficiently" (123).

Could an efficient method of spraying be found? Not at the time. By 1949, none of these or other follow-up experiments had succeeded (Symes et al. 1948; Whiteside 1949; Woodcock 1949; Burnett 1954; Burnett, Robinson, and Leroux 1957). Simply because it had succeeded in the United States or Britain did not mean it would succeed anywhere else.

This was also what the Division of Entomology had found in Southern Rhodesia in 1947. The entomologists were adding DDT to dipping fluid so that as *mombe* plunged into dipping tanks and drenched their bodies, the emulsion would soak in and leave a residual effect on the skin as it dried. When a *mhesvi* landed upon it to feed, the OCP would kill the *chipukanana* through skin absorption. This chemical was applied to the dip mix at Muumbe dip in the southeast in 1947 to give an oral prophylactic to *mombe* in the face of a *mhesvi* onslaught from Portuguese territory. The view of the vet, entomologist, and Native Commissioner was that DDT could insulate *zvipfuyo* from *mhesvi*'s bite. However, the mix was too strong: "Certain of the ingredients of the emulsion caused an apparent scalding and a violent reaction resulting in temporary paralysis of the hind-quarters, and after two subsequent attempts, the project was abandoned."[33]

In any case, the rationale for using *chepfu* (poison) was to effect the mass destruction of *mhesvi*, rather than using single oxen moving around as bait, which would soon become clogged with trapped flies. That method was wasteful and extremely toxic to oxen. The idea was not abandonded, however. In 1965, BTTC and DNPWLM looked for a commercial insecticide that could "safely be applied to the coats of *mombe* in quantities which are toxic to tsetse flies"[34] and kill *mhesvi* within seventy-two hours, if the *chipukanana* was exposed for two minutes. The second objective was to find out how far *mombe* could tolerate the pesticide on their skins, so as to determine what concentrations "could be tolerated by them and yet lethal to tsetse."[35] The critical time to poison *mhesvi* was when it settled on the body of the animal to draw its blood meal, but the poison stayed on the body of the animal much longer.[36] The problem was that some of the OCPs, when applied to coats of *mombe*, "caused a depression in the production of cholinesterase that threatened to be lethal with further applications."[37] The holy grail would be to "safely incorporate an effective insecticide into the cattle dips used in routine tick control."[38]

Mhesvi became a bonanza for Big Pharma. By 1968, well over fifty insecticides had been tried as "screens" on the hides of *mombe*.[39] One of the manufacturing companies involved was Cooper Technical Bureau (CTB), a UK company based out of Berkhamsted and famous for making Cooper's Dip, an antitick dipping chemical for *hwayi* and *mombe*. CTB submitted to the Tsetse Department twelve insecticides for potential use in that role, including asuntol, neguvon, imidan, and 42/1/65. However, when these specific four were tested, they proved to be insoluble in water. Another company profiting was the petrochemical giant Shell Chemicals (UK), which supplied birlane (applied as a 24 percent solution). CTB changed its name and operated as Cooper, McDougall and Robertson (CM&R) until 1973, when it became Wellcome. As CM&R, it supplied sevkol (30 percent), ethion (30 percent), simuthion (60 percent), and DBM (38.63 percent). The compounds alodan, Dow E.T. 57, ruelene, bromophos, and bromodan were also mixed in a solution of eighty parts of petroleum ether and twenty parts of olive oil.[40] In 1969, twenty-two compounds were put on trial.[41]

We can now rejoin *hurimbo*. That same year, the branch announced that a 1:1 mixture of polybutene, manufactured by Klipfontein Organic Products Corporation of South Africa, and a *hurimbo* (gum) obtained from *chitatarimbo* could form an "effective sticky deposit" for trapping *mhesvi*. What was left now, it said, was "to find an effective sticky substance which could be mass produced from commercially available raw materials."[42]

It is significant that sticky techniques were now being integrated with attractant-research-based Morris traps to produce a new kind of trap. A synthetic rubber mixture of 9:1 polybutene to Adlatex (made locally by Glue and Chemical Products, Salisbury) had proved "simple to prepare" and maintained "a high level of effectiveness almost indefinitely."[43] Vale, the glossinologist in charge, felt that the color of the sticky substance was critical to "the efficacy of attractants." It had to match the color of materials upon which it was applied and that *mhesvi* favored (Vale 1969, 1).

Five attractants were deployed. The first was a Morris trap with a "body" of black cloth. The second had a black body plus a sticky screen on the inside of each "flank." The third had a black body and sticky screen hanging vertically inside the body of the trap. The fourth also had a black body and sticky screen arranged just like the third, but with a sticky screen on the outside of each flank. The fifth was a ten-gallon drum covered with a black cloth and mounted horizontally fourteen inches above the ground, resembling a cow. On each flank was a sticky screen. All sticky deposits were colored black. "It appears," Assistant Director Gerald Cockbill remarked in his 1969 report, "that many more tsetse visited Morris traps than were trapped by the cage recovery system. Since the sticky screens inside caught so few tsetse, it seemed that the cage recovery was inefficient because tsetse did not enter the body of the trap."[44]

Further experiments were undertaken to compare the recoveries from three scenarios: a sticky model alone; a sticky model plus *munhu mutema* (black man) walking three feet away; and a nonsticky model accompanied by *munhu mutema* with net in hand, catching flies landing on the model. The models were all pulled by hand. The unaccompanied sticky model yielded more male and female *mhesvi* than the nonsticky model. The entomologists drew one conclusion: "Since the model plus man combined with sticky recovery gave catches intermediate between those obtained from the sticky model alone and from the non-sticky model plus man incorporating hand-net recovery, it appears that the low recoveries by the hand-net technique are due to the presence of the catcher and inefficiency of the hand-net technique itself."[45]

Consistently, the problem was not to do with the sticky screen. The experiments with mobile attractants showed that the hand net system was a slow and inefficient way of studying *mhesvi*'s host preference; instead, the experimenters recommended the use of unaccompanied mobile models coated with sticky deposits and with visual stimuli resembling *mhuka*.[46] Conducted principally at Sengwa, the mobile research was mostly limited to a black drum with ten-gallon capacity mounted horizontally fourteen inches above ground on a little perambulator chassis. This model was moved on a straight run at the speed of 1.5 miles per hour using a long handle, with the operator pulling as he walked twenty-three feet ahead. Alternatively, a long rope was fixed to the chassis so that it could be operated by *munhu mutema* simply pulling while hiding a hundred yards away.

The experiments with the five attractants discussed previously—in mobile mode—were each conducted at the same time daily, for the same number of days, over a distance ranging from 200 to 1,400 yards. The *mhesvi* visiting the attractant were counted using the naked eye or by telescope, or those recovered with a drop net toward the end of the run or trapped by sticky deposits were counted manually. The sticky recovery model was covered with black sticky deposits, worn by a *mufrayi* carrying on his back a 3 ft. × 1 ft. screen with khaki sticky deposits.[47] The walking man generally depressed the numbers of *mhesvi* caught compared to unaccompanied models—the *mhesvi* was repelled, possibly, by visual stimuli.[48]

The question now was what to do with *mhesvi* after trapping it: Destroy it? Sterilize and release it? Here, we will limit ourselves to the first option—kill—and defer the second—sterilize and release—to the next chapter.

Insecticide Treatment

One method of killing *mhesvi* was the treatment of traps with insecticide, but as late as 1982, the biggest challenge remained the fact that most flies visiting such targets could not alight on insecticide-coated target surfaces. The solution was to place a sheet of fine netting close to the target and smear it with DDT wettable powder (WP). With this method, almost 100 percent of all *mhesvi* that collided with or landed on the netting died. There was another problem: Once heavy rains had fallen, the lethality of the netting was virtually washed away. Efforts to find a roof to protect the deposit failed, so BTTC persevered with a second option.[49]

The insecticide used in the field experiments was deltamethrin, a suspension concentrate that was moderately resistant to rainwash and which was also an ultraviolet light absorber. These two properties reduced the hydro- and photodegradation of deltamethrin considerably and paved the way for the design of a much cheaper but still efficient target without a roof.[50] The environmental pollution effects of pesticides that were hard to break down naturally were not considered. Deltamethrin had been proven to be effective when applied to netting at three-month intervals, and it now replaced the dieldrin wettable powder used previously. Deltamethrin also had two distinct advantages: First, unlike dieldrin, deltamethrin did not leave a white deposit on the netting and thus did not suffer the repellence associated with the former. Second, its deposits were more resistant to rainwash than those of dieldrin.[51]

The 1982 experiments resulted in the conclusion that pesticide-treated netting was more economical than traps baited with carbon dioxide and acetone. As the assistant director noted in his annual report, he hoped that the ongoing research would "lead to the development of targets that can be suitably cheap and convenient to compete favourably with conventional systems of tsetse control."[52] Armed with this knowledge, BTTC made several changes to its trap and target designs. The trap was the F^2, originally designed in a rigid form, but remastered into a portable, folding form for redeployment purposes. To increase catches, various flags were placed next to the trap.[53] In 1985, a new cage made of plastic bottles was added to the F^2, which killed *mhesvi* and funneled them into a large polythene bag underneath the trap to prevent clogging when the trap was not emptied frequently.[54]

Field trials of traps reached a milestone in 1984. First, targets baited with acetone and 1-octen-3-ol and coated with deltamethrin were deployed on the 5 km^2 Antelope Island on Lake Kariba at a density of four targets per square kilometer. A rapid decline and virtual extinction of *mhesvi* populations was achieved.[55] Meanwhile, an even larger field trial was under way from March to October 1984 near Chirundu, where 2,850 targets were placed in a 600 km^2 bush known as the Rifa Triangle. The target used—later called the R-Type (Rifa Type)—was a 0.8 m wide black panel in two planes equipped with a hooded downwind gauze panel swinging on metal support and wire frame. This was the "first offensive deployment" of targets following successful trials on Antelope Island on Lake Kariba. The *mhesvi* population declined rapidly, reaching near extinction by January 1985, or a 99.9 percent reduction in nine months.[56]

Given differences in terrain and in the behavior of different *mhesvi* therein, tactical and material changes in target systems were inevitable. One change related to the materials used in assembling the target. The results of a trial in *mhesvirutondo*-infested Rukute Valley were much better than previous results, but still much slower than the rapid decline of *mhesvirupani*. Two studies had shown similar results (Vale et al. 1986, 1988). The experimenters used the following four standard target types (the dates of the experiments conducted are included):

1. R-Type: 0.8 m wide black panel in two planes with downwind gauze panel, hooded, swinging on metal support with wire frame (Rifa, 1984)
2. S-Type: single plane, center black panel 0.7 m, with 2 m × 0.5 m gauze panels, swinging on metal and wire frame (1985–1992)
3. S-Type: all solid black, 1.7 m × 1 m on metal/wire frame (1992–1993)
4. S-Type: all solid, center panel black, two side panels blue (1994–1995)[57]

The first S-Type (Sengwa Type) targets proved to be twice effective as the R-Type. The switch to all solid materials was an attempt to prolong the life of the model's treated surfaces. The use of blue instead of black was meant to increase the target's visual attraction and thus reduce the amount of pesticide applied, and locally manufactured cotton cloth was the preferred fabric to lower costs. Reducing photodegradation of the color remained a challenge. The insecticide was applied using three methods: knapsack sprayer in situ (1984–1986), dipping and drying at the base camp (1986–1993), and painting on tables at headquarters (1994–1995). The concentration of pesticide varied from 0.1 percent to 0.52 percent.[58]

An expert report in 1995 cautioned, however, that "the general concern at the apparent lack of success is in fact based on the unrealistic expectation that eradication should be achieved in one year."[59] So long as experiments still used these four standard target types per square kilometer, each for a different type of *mhesvi*, it would require a significant outlay of material, transport, and labor, which were never always readily available, thus directly affecting the veracity of the results. The report's conclusion was rather grim: eliminating mixed populations per square kilometer would take four to five years.[60]

The attractant preferred, acetone plus octenol, was "as good as any other combination" then in use for *mhesvi*—and cheaper. Baits containing phenol and octenol were best for *mhesvirupani*, especially when combined with traps. Used alone in a sachet, octenol tended to evaporate faster, thus requiring constant reevaluation and replenishing.[61] A more perfected killing trap remained elusive.

Electric Traps

Using electric current to kill *mhesvi* was Swynnerton's idea, based on a request from the governor of Tanganyika in the early 1930s to equip all trains passing through the territory with fly traps. To stun *mhesvi*, voltages of 2–3 kV were required. Experiments to generate such voltage with motor engine–driven magnets and with combinations of a car radio vibrator and ignition coil were undertaken (Swynnerton 1933).

The use of electric current was designed to shock the *chipukanana* temporarily or kill it. These experiments were undertaken in 1971 and involved deployment of a large electric netting composed of a cage 1.35 m tall × 1.5 m long × 0.79 m wide and a net of 3.3 m long × 1.5 m high, stretching 0.3 m to 3.6 m from one side of the cage to the other. These two sections of cage and net were deployed in such a way that the maximum cross section would be exposed to prevailing winds. Then, a small black model of an animal was placed in the center of the cage, with olfactory stimulus (ox odor) in some experiments and zero bait in others.[62]

Two components of these electric capture systems were developed in 1971. One was an electric tray, designed to kill *mhesvi* alighting on the ground adjacent to bait, flying near the ground, or about to alight. The set was designed so that flies were electrocuted by a horizontal grid of fine electric wires and fell into a tray of netting, from which they were harvested as samples. The electric tray was pulled behind mobile baits.

The second innovation was an electric pen, a large, circular cage of electrified netting to electrocute all *mhesvi* flying toward a real *mhuka* (normally a black ox weighing 400 kg), shown in figure 4.6 with three "native assistants" standing with it inside the cage.[63] Three field assistants wearing khaki uniforms were "placed in the pen on six afternoons" for two hours and fifteen minutes from 1600 to 1815 hours. Their task was to observe and identify *mhesvi* colliding with the electric netting as they attempted to reach the ox inside (Vale and Phelps 1974, 2). The black men were at once a material part of the apparatus and observers of inbound *mhesvi*—a means of doing knowledge production.

The experiments with the electric pen also compared captures made using hand nets and the electric pen, with an ox and three black men as bait, and those using an electric pen with different baits: (a) an empty pen, (b) a pen with an ox, (c) a pen with an ox and three men, and (d) a pen with three men. Version c is shown in figure 4.6. The BTTC drew two conclusions: First, catches of female *mhesvirutondo* and *mhesvirupani* from one ox plus three men were greater when using the electric pen than with

Figure 4.6
An electric pen using as bait a *jon'osi* and three men.
Source: Annual Report 1972, 7. SACEMA/TA.

hand nets. Second, catches of male and female *mhesvirupani* using the electric pen increased when using the ox alone compared to the ox plus three men.[64] The conclusion was that models were more effective baits than men and that the presence of people reduced the model's catches.[65]

The electrocuting devices were designed to kill or stun *mhesvi* by passing high-voltage electric current and electrocuting it. The stunning and killing effect of a grid of fine wires running parallel, 0.8 cm apart, and forming grounded and charged poles of high-voltage current impressed experimenters. Five versions of the trap were experimented with at Sengwa (Vale 1972, 11):

1. *Electric surface:* To capture tsetse alighting on model animals or on the backs of men, an electrocuting grid was placed 0.8 cm above the bait surface.
2. *Electrified decoys:* Tsetse visiting baits often alight on or near tsetse already on the baits, perhaps for a sexual purpose. Decoy tsetse were placed on men or model animals, each mounted on a central charged wire and flanked by grounded wires. Decoys of male and female tsetse and even small rolls of cloth were highly effective. As many tsetse were taken from six decoys on a man's back as from the totally electrified surface of his back.
3. *Electric tray:* Many tsetse alight on the ground near baits. To capture such flies, a 90 × 90 cm tray of fine netting was mounted on small wheels 5 cm above

the ground and 2 cm below a horizontal electrocuting grid. Tsetse flying near the ground, and presumably about to alight on it, were electrocuted and fell, retained by the netting. The tray was pulled 23–113 cm behind mobile baits.

4. *Electric net:* To capture tsetse flying near bait, a vertical net, 90 × 90 cm, of fine nylon was flanked with an electrified grid 0.8 cm from the netting. The net was mounted on small wheels and pulled 23–113 cm behind mobile baits. The recovery chute at the base of the net was divided into front and rear portions to roughly separate flies caught near or far from the bait.
5. *Electric pen:* A large circular cage of electric netting, 3.4 m high by 6 m in diameter, was created to capture tsetse flying towards host animals tethered inside.

The system for these experiments was based on "the tsetse shorting a circuit between fine copper wires running parallel, 0.8 cm apart and charged with several thousand volts."[66] The electrocuted *chipukanana* would fall onto a sticky tray or be guided by a polythene chute to a nonreturn vessel. The electrocution experiments went even further. The ten-gallon black drum simulating an animal and the 3 ft. × 1 ft. screen that the black man in the experiment wore on his back were both electrified to capture alighting *zvipukanana*, usually hungry females feeding and "sexually appetitive" males chasing them. Ninety-five percent were captured by this method (Vale and Phelps 1974).[67]

The electric circuit was designed to work with absolute minimal power requirements (i.e., a small, sealed, two-volt accumulator, rechargeable with a solar cell array or car battery), because the trap was deployed in remote areas. The generator emitted energy only when a *mhesvi* was trapped. The circuit used an inverter with a step-up transformer and voltage-multiplying rectifier (Rogers and Smith 1977, 155; Vale 1974a, 1974b).

Conclusion: Converging Forms of Knowledge and Practices

The design of traps has been examined as a process of making means and finding ways of controlling the mobilities of *mhesvi*. Different types of *mhesvi*, different hosts, and different vegetation types required specific trapping devices; the trap that worked superbly against *mhesvirupani*, for example, did not work against *mhesvirutondo*. The trap was at once a control and a research device, via which *mhesvi* were caught and killed or captured alive for research. Its ease of construction, cost-efficiency, and self-operated killing capacity once set made it an attractive alternative to other methods.

The key factor in trapping was attracting the mobile *mhesvi* to the trap. As we discussed in chapters 1 and 2, the mobilities of *zvipukanana* and their predators, what attracted them (smell and vision), and how to manipulate them for the human good were already well-known. This chapter has shown how *vachena* translated and tapped into this *ruzivo* and practices based upon it. The appropriation was not a straightforward mobility of *ruzivo* from *vatema* to *vachena*, but the subjection of *ruzivo* to *vachena*'s ways of knowing and knowledge. The infrastructure of controlling *mhesvi* outlined in the previous chapter was brought to bear upon this *chipukanana*, its blood meals, and the habitats under experiment (not just experience) and forms of peer review acceptable to *vachena*'s knowledge production practices. For *vatema*, movement by night was a way of avoiding detection and molestation by *mhesvi* active during the day. They used darkness as a way of managing *mhesvi* through avoidance. The Rhodesians inverted these workings of visibility and mobility to encourage, not prevent, *mhesvi* to bite (bait) cattle, which now became instruments for detecting *mhesvi* presence. *Vatema* had deployed visibility to avoid the detection of the mobilities of *vanhu* and *mombe*, *vachena* to enhance it.

Even when unseen, an animal could not escape detection by *mhesvi*. Olfactory stimuli were particularly strong in *njiri*, *nguruve*, and *dzoma*. *Mhuka* hiding or moving in dense vegetation led *vachena* to the conclusion that odor was as good a stimulus as sight. When put to experiment, the two stimuli were found to complement each other well as elements of a trapping system. The active elements in the odor were found to be carbon dioxide, acetone, butanone, and 1-octen-3-ol. The discovery of carbon dioxide as the key ingredient in odor attractants should bring to mind the earlier and *vatema*'s continuing practices of burning dry dung indoors and underneath the bellies of *mombe* or smearing their skins with fresh dung to repel *mhesvi*. I have not found written evidence of direct mobility of that *ruzivo* into *vachena*'s experiments.

The same cannot be said of the direct movement of *ruzivo rwezvehurimbo* (birdlime knowledge) into the production of sticky traps. Here, I have drawn on experience as archive, detailing the tradition of making birdlime from *chitatarimbo*, *mukonde*, and other rubber trees in rural Africa. *Hurimbo* production was a maker space where the *ruzivo* of turning liquid into sticky substance happened. This is not tinkering or trial and error; this is the application of specific ingredients according to a set formula peer-reviewed communally. The space around *chitatarimbo* becomes a laboratory for that fleeting moment when the boys are out there cutting and cooking—then they are gone, the tree continuing its life as a plant in the wild, albeit bearing the scars of its domestic role in the upbringing of the boy child. So too stand the twisted or broken branches in the tree where *mudziti* once stood—the meeting point of *vanhu*'s means and ways of fulfilling culinary desires, on one hand, and *shiri* rendezvousing, sojourning, or brooding on the other. Two mobilities, one meeting point: the tree.

What we had made and strategically deployed for *shiri*, BTTC extended to *mhesvi*. Here, the mobility of *ruzivo rwevatema* to the core of white tsetse knowledge is very direct: from the poacher *Bvekenya* and his *mubhemhe* to experiments with *hurimbo* (*chitatarimbo* specifically) and OCPs to produce an automobile trap. As in previous chapters and sections, we have illustrated the value-adding elements, the redirection and adaptation to *mhesvi*, and the mobility of *ruzivo* from overseas. Again, there is no *cut* from Europe and North America and *paste* into Southern Rhodesia; *mhesvi* with its pestiferous mobilities demanded recognition of its specific qualities. By the time the reader reaches the discussion of the addition of electricity to trapping devices, I hope it becomes clear that *mhesvi* was becoming a mobile site where many forms of (chemical, electrical, mechanical, and biological) knowledge and engineering converged—all because of *mhesvi* and its pestiferous mobilities.

5 Attacking the Fly from Within: Parasitization and Sterilization

In 2007, the journal *Science* carried an intriguing article. Ethiopia was developing a sophisticated weapon against *mhesvi*. The idea was to produce one million male flies per week, blast them with radiation for a few seconds to render them sterile, then set them free in *mhesvi*-infested areas at a ratio of 10:1 (sterile to wild). The releases would be made several times, the repeated mating without offspring resulting in the annihilation of the *mhesvi* population. This "birth control for insects" was lauded as "an elegant and environmentally friendly method" (Enserink 2007, 310).

The article noted the success of the sterile insect technique (SIT) in eradicating the screwworm in North America; saving apples in Chile, onions in the Netherlands, and melons in Japan from different kinds of *zvipukanana*; and, of course, eradicating *mhesvi* on Zanzibar island. Now Ethiopia! Soon, *mhesvi* would be history in over thirty-five African countries and, with it, *gopé* and *n'gana*.

The most biting critique came from none other than Glyn Vale, one of the leading figures in the Rhodesia tsetse control research project. The factory for engineering these insects alone cost $12 million, for a technique that might be effective against one of over twenty-three kinds of *mhesvi*—and even then, without assurance that reinfestation would not occur. "I hate to see a poor country waste so much money," Vale said. Edinburgh University's Ian Maudlin went even further: this was a giant waste of money by a poor African country succumbing to the seduction of the International Atomic Energy Agency (IAEA; Enserink 2007).

Glyn Vale would know: Attempts to destroy *mhesvi* from within had been made for almost the entire lifespan of the (Southern) Rhodesia project and found to be more expensive and less effective than other methods, like OCPs and traps specific to one type of *mhesvi*. This chapter considers the local history of attacking pests from within their bodies, focusing on two methods. The first, *parasitization*, involved strategically (re)deploying *mhesvi*'s parasites to kill it or render it inhospitable to microorganisms deadly to people or their *zvipfuyo*. The second, the *sterile insect technique*, involved using chemicals and gamma radiation to sterilize male *mhesvi*, then unleashing them to mate with the wild females, driving their race into extinction.

The argument made is that attacking *mhesvi* from within represents the applied value of the knowledge of *mhesvi*'s bionomics and internal mobilities, both internally (*nyongororo* moving within its body) and in situations of intimacy (nuptial flights). In the first instance, the weaponizable element was the *nyongororo*—specifically, the points of contact between this micromobile organism and (potentially) vulnerable parts and systems within the *chipukanana*'s body. In the second, the weaponizable element was the sexual act, meant to result in procreation, but now genetically engineered to accomplish the ultimate genocide—the extinction of the gene line. Both required a meticulous understanding of the micromobilities of *hutachiwana* (which *vachena* called "protozoa"), developmentally and physically, and of the sperm within the small body of *mhesvi*.

Parasitization: *Nyongororo* as Weapon

The earliest record of *nyongororo* (parasites) of *mhesvirupani* in Africa is Leiper's report from 1910 (Thomson 1947). Catches during the wet season in 1912 and 1913 revealed *mhesvirutondo* to have similar *nyongororo* within them. They had been recovered from the Mpika area of Northern Rhodesia (Lloyd 1912) and from *mhesvirupani* on Lake Victoria in Uganda (Carpenter 1912) and Katanga (Rodhain et al. 1913).

In 1923, entomologist J. K. Chorley became interested in researching the parasitization of *mhesvi* with just that in mind. His superior, Rupert Jack, said of the idea at the time: "The study of the natural parasites of the fly has yielded results of great interest and shows that a very high death rate occurs at certain seasons of the year from this cause."[1] He did not specify what the *nyongororo* were.

However, Dr. William Lamborn, a medical entomologist then working on *mhesvi*'s parasitization in Nyasaland (*vachena*'s name for a country *vatema* called "Malawi"), sent Chorley two consignments of flesh fly *zvikukwa* that had been parasitized with a *nyongororo* that *vachena* called *Syntomosphyrum glossinae*. He then established a *nyongororo* strain on locally bred flesh fly *zvikukwa*. The same *nyongororo* was also bred from *zvikukwa* that Chorley had collected on the Munyati. The successful breeding of the *nyongororo* was a first step toward "ascertaining whether we could induce an artificial increase of parasitization with this species in the Tsetse fly's natural haunts." That is why "parasitised Flesh fly puparia were forwarded in regular succession" to Chorley on the Munyati River in 1923. As Jack noted, the experiment met with unforeseen challenges: "Unfortunately owing to various causes including the depredation of ants at his [Munyati River] camp and difficulty in breeding parasites in large numbers during the hot dry season, Mr. Chorley was unable to release great numbers of the insects in the tsetse haunts. This combined with the fact that a natural increase of the parasites occurred in the late dry season rendered conclusions impossible."[2]

Jack decided to continue breeding and studying the *nyongororo* at Salisbury laboratory to understand more about its life history and factors affecting its breeding, "so as to be in a better position to test its capabilities in the field during the coming year." Meanwhile, attempts were under way to breed another *mhesvi* parasite *vachena* called *Mutilla glossinae* in flesh fly *zvikukwa* (pupa) at Salisbury "with a view to artificial increase." The entomological section was also investigating the *nyongororo* of flesh- and dung-breeding flies *vachena* called *Sarcophagidae* and *Musidae* at Salisbury to determine whether their *nyongororo* might also breed in *zvikukwa*.[3] Between 1915 and 1935, researchers in Northern Rhodesia found and took interest in yet another parasite found in a *chikukwa* the investigators called *Anastatus viridiceps*. The *chikukwa* was associated with *mhesvirutondo* (Waterston 1915, 1916, 1917; Lloyd 1912; Ferrière 1935).

More findings followed in the post–World War II period. In 1946, while researchers were dissecting 1,500 *mhesvirutondo*, three "*Mermis*"-type worms, each 79 mm long were found, all of them during the wet season in Tanganyika. The worms were bigger this time and occupied "so much of the abdominal cavity as seriously to incommode the tsetse." Thus, in one specimen there were samples of two different bloods, each with recognizable corpuscles, as if the *mhesvi* had been unable to take up enough blood at one time to satisfy its needs and had been forced to take two meals within an unusually short time. In 1955, what *vachena* classified as the *Hymenopteran* family *Eupelmidae* was found in *mhesvi* in Southern Rhodesia and Nyasaland (Buxton 1955; Heaversedge 1968). Then, on February 27, 1969, a *Hymenopterous nyongororo* emerged from *mhesvirupani zvikukwa* taken near Izom in Northern Nigeria. Under examination at the Commonwealth Institute of Entomology in London, the *nyongororo* was confirmed to be *Anastatus sp.*, the first time these *vachena* had found this *nyongororo* parasitizing *mhesvi* (Baldry 1969).

In 1971, a male *mhesvirutondo* of *vachena* classes as *G. brevipalpis* was discovered infected with two *zviguraura* (larvae) of the *Mermithidae* family. All these *nyongororo* infections occurred during the wet season. As Moloo (1972, 159) concluded, "the infective juveniles hatch during this season and penetrate into newly emerged *Glossina* through thin places in the body wall. ... Since the incidence of infection is exceedingly low, the transmission to *Glossina* is almost certainly accidental."

Sterilizing the Male Tsetse

The second method of attacking *mhesvi* from within was to render its males sterile, thus preventing new insects from being conceived. Two methods were experimented with extensively, as discussed ahead.

Chemosterilants

Sterilize-and-release traps caught *mhesvi* that also could be retained as samples. The problem was that such traps were far more expensive and complex to assemble than catch-and-retain traps and required well-trained *vatema* to set and check them. Overall, these baits were generally affordable and effective when deployed in large numbers, but more expensive when deployed in smaller numbers. Cost-efficiency depended on the number of baits needed per area to produce optimal rates of decline in *mhesvi* populations.[4]

In a paper in 1966, R. J. Phelps is clear that "the sterile-male technique is a practical application" of H. J. Müller's 1946 Nobel Prize–winning work on the mutagenic effect of X-rays to induce mutations that would result in a sterile male. "In this context," Phelps says, "it does not mean a castrated male, but one which is normal in all respects except that damage has been done to the genetic material. Spermatozoa are produced by such males, and they are able to fertilize the female. However, no progeny are produced due to failure of proper chromosome pairing in the early embryonic divisions" (Phelps 1966, 29). Phelps had read North American research on releasing sterilized male screwworm flies, with field testing conducted in the West Indies (Baumhover et al. 1955) and, after its success, deployed in Florida and Texas at scale. It was due to these results that J. K. Chorley asked the British Colonial Office's Tsetse and Trypanosomiasis Committee to commission a study on the possibility of extending the technique to *mhesvi*. The resultant report was affirmative, with one study emphasizing dosage strength and the age of *zvikukwa* as critical determinants of success (Potts 1958). Another stressed the ratio of sterile males to females, urging that the technique could only work if the female *mhesvi* population was first reduced through the application of chemical pesticides (Simpson 1958). The third study suggested the use of attractants to draw flies into areas where they could be treated (Knipling 1963). There was one problem: The chemicals for sterilization were "dangerous to handle" (Phelps 1966, 31) and the technique could only be conducted when safer chemicals were on the market.

The result was a collaborative project between the United States Department of Agriculture (USDA) and the Agricultural Research Council of Rhodesia and Nyasaland (ARC), financed by the United States Agency for International Development (USAID). USDA would supply the personnel and sterilizing agents (mostly chemicals); ARC would mass-produce *zvikukwa* and *mhesvi*. The agreement signed in 1963 by H. C. Periera of ARC and J. H. Starkey, an administrator with USDA's Agricultural Research Service, committed the United States to supply $84,000 for the collaborative research. In exchange, it stipulated that "any invention resulting from this cooperative work and made jointly by an employee or employees of the United States Department of Agriculture and the Cooperator, or an employee or employees of the Cooperator, shall be fully disclosed, either by publication or by patenting in the United States, and any such patent shall either be dedicated to the free use of the people of the territory of the United States or be assigned to the United States of America."[5]

The Participating Agency Service Agreement (PASA 3–8) took effect on June 14, 1963. Then, from June 22 to July 5, Dr. Paul Oman (assistant to the director of USDA's Entomological Research Division) and Dr. David A. Dame (ARC's principal investigator, based on the collaborative agreement) visited Salisbury to review plans for executing the research in phases with local representatives of USAID and ARC. They also checked facilities, personnel, equipment, and other logistical matters before flying to Lagos for consultations with USAID and the Commission for Technical Cooperation in Africa (CTCA). The trip was intended to situate the Rhodesia research within the larger continental control of *mhesvi* and *gopé/n'gana*.[6] From September 25 to 28, Dr. Carroll N. Smith, lead investigator of the "Investigations on Insects Affecting Man" project at Gainesville, Florida, attended the CTCA's Meeting of Experts on Trypanosomiasis in Lagos, along with Dame. The meeting's purpose was to revise the International Scientific Committee for Trypanosomiasis Research (ISCTR) to create "a council to aid and organize trypanosomiasis research in Africa."[7]

By this time, R. J. Phelps's sterile male–release technique was showing that sex and reproduction could be turned into not just a point of intervention, but a means of effective *mhesvi* self-destruction, at the moment that *mhesvi* was engaged in its most intimate act. Laboratory-bred sterile males would be released in large numbers to mate with the females. By 1966, however, the challenge was that "the tsetse [was] refractory to laboratory maintenance; the most important symptom of this [was] the failure of laboratory-maintained flies to produce viable offspring at regular intervals," possibly due to neuroendocrine failure (Bursell 1967, 34).

The sterile male technique required the breeding of *mhesvi* in quantity—by stabilizing food supply, keeping them in mosquito-gauze cages with a tethered food supply inside, and using controlled environment rooms in cages large enough to contain an ox. Lab-kept *mhesvi* colonies were a recent phenomenon (Phelps 1966, 32). Dame and Schmidt (1970) found that mass sterilization depended on mass rearing of *mhesvi* in the absence of living host animals. Feeding *mhesvi* on different animals through natural and synthetic membranes revealed that the insect's mouthparts were adapted for piercing; it could not be induced to ingest from liquid surfaces (Langley 1972).

The record is again thin for the war period; by 1982, however, the BTTC was conducting field trials on several sterilization devices to replace retaining cages used in the catch-and-retain traps. These were designed to detain *mhesvi* in a chamber for half an hour so that more could enter. At that point, the chamber's position changed in a way that simultaneously swung the entrance shut and sprayed the flies with metepa aerosol. Once the *zvipukanana* were sufficiently sterilized, the chamber moved into another position again, opening the door and allowing them to escape. The chamber was one component of a three-chamber cylinder that rotated automatically at intervals, set at collect, spray, and release positions. Automation was electrically controlled using eight flashlight batteries that lasted for several months. A University of Zimbabwe (formerly University College of Rhodesia and Nyasaland) Zoology Department study had shown a wild *mhesvi* population exposed to odor-baited traps fitted with sterilizing devices to contain females with degenerated ovaries, while their male companions had "a high degree of sterility without a reduced longevity."[8] By then, research had shown that *mhesvi* could be permanently sterilized using chemicals. The sterilants were applied using injection, by dipping the *zvikukwa*, via wind tunnel spray treatment, and by contact with treated surfaces (Phelps 1966, 31; Dame and Ford 1966, 1968; Dame and Mackenzie 1968).

Gamma Irradiation

Parallel to the chemosterilants project, the International Atomic Energy Agency (IAEA) commissioned a study on using gamma irradiation (exposure to radiation) to sterilize *mhesvi*. Experiments were also conducted on young *zvikukwa* under laboratory and small-cage conditions. By 1965, two preliminary studies had concluded that gamma irradiation and chemosterilants could reduce the reproductive capacity of *mhesvirutondo* (Potts 1958; Chadwick et al. 1964).

ARC deployed *vatema* as *mafrayi* on the Kariba islands to locate and collect *zvikukwa*, which were then flown to Salisbury (Phelps 1964). Inside the lab, they were floated in methylated spirit to remove dead and parasitized *zvikukwa* and maintained at 25°C, plus or –2°C, and around 70 percent relative humidity. The *zvikukwa* were stored in 8 × 8 × 11 in. wooden cages with transparent plastic walling and mutton cloth at one end, and male and female flies were separated each morning into similarly sized cages, but with fine wire mesh walls, a cotton mesh top, a mutton cloth sleeve, and a wire mesh floor. A guinea pig was placed in the middle for one hour to provide a blood meal for the *mhesvi*. The gamma irradiation used the Eldorado G Cobalt 60 Teletherapy Unit at Salisbury Central Hospital, as follows: "The pupae were held in fine cotton mesh bags and exposed for varying periods on a 15-sq. cm surface at a distance of 55 cm beneath the source. The adult flies were enclosed in a 4 × 12 × 12 cm wooden-framed box covered with fine cotton mesh. The Cobalt room was maintained at about 26°C, a fan was used to assist aeration during the irradiation, and the material for treatment was conveyed in a Kaylite box to and from the laboratory" (Dean and Wortham 1969, 506). The effectiveness of radiation on reproduction was deduced from the number of *zvikukwa* the treated and untreated *mhesvi* produced. Mortality increased with dosage. The experiments revealed that irradiation did not affect the male's ability to inseminate females; in fact, sperm from irradiated males was "mobile and apparently behaved normally" (518).

In 1972, studies were conducted in the use of nitrogen and chilling to produce radiation-induced sterility in *mhesvirutondo*. Several studies found gamma irradiation "safer, more convenient and reliable than chemosterilants" (Curtis 1968, 1969, 1970, 1972; Curtis and Jordan 1970; Curtis and Langley 1972, 360; Curtis, Curtis, and Hamann 1973; Dean and Clements 1969; Dean, Clements, and Paget 1969; Dean, Dame, and Birkenmeyer 1969; Dean and Wortham 1969; Potts 1958; Itard 1968, 1970). In their field experiment at the Salisbury lab in 1968, Dean, Phelps, and Williamson gamma-irradiated *zvikukwa* of unknown age with 8,000–15,000 rad. They recorded a 95 percent reduction in reproduction in male flies one week old that mated with untreated females. Applying 4,000–9,000 rad to males emerging from *zvikukwa* three or four weeks after treatment, they achieved complete sterilization. In adult males treated with 8,000–16,000 rad, they recorded 95 percent complete sterility for the forty-five-day trial period (Dean, Phelps, and Williamson 1968). *Mhesvi* was extremely resilient to sterilants; it required far higher doses than any other problem insects (Curtis 1970; LaChance, Schmidt, and Bushland 1967). By using a nitrogen atmosphere (irradiation without oxygen), sterility could be achieved with less induced biological (somatic and genetic) damage to *mhesvi*, thus enabling it to feed, chase, and mate (Langley and Maly 1971; Baldwin and Chant 1971). It also was known at this point that radiation treatment in the air later in *mhesvi*'s life cycle yielded less reliably sterile insects. In the field, sterilized adults failed to reduce the population; *zvikukwa*, by contrast, succeeded, owing to "delayed flight muscle development" in lab-confined adults (Dame and Schmidt 1970; Dame and Ford 1968; Langley 1970). Hence, Curtis and Langley (1972) studied sterilization in the late stage of a *chikukwa*.

Meanwhile, the Tsetse Research Laboratory at the University of Bristol School of Veterinary Science was experimenting with introducing "chromosome translocations" into *mhesvi* to depress fertility through semisterile "heterozygotes." However, this strategy also depended on mass production of the translocations in the form of fertile "heterozygotes." The mutations were obtained from the *mhesvi* that *vachena* called *G. austeni* through radiation and careful selection for semisterility, with the offspring of these mutant individuals inbred to produce "homozygotes." The translocation homozygotes were mated with "close relatives" and produced numerous "translocation homozygotes" and "heterozygotes," one inexplicable wild type, and "some fully fertile matings which are expected to be founders of pure translocation homozygote families" (Curtis 1971, 425).

Hybrid *Mhesvi*

A third method was to produce a hybrid *mhesvi*. In 1972, Curtis reinvestigated hybrids of *mhesvirutondo* using contemporary rearing techniques. He returned to Vanderplank's earlier work, which sought to determine the effects of releasing alien types into *mhesvi* populations (Vanderplank 1947). This deployment of *mhesvi* as a weapon of mass destruction against fellow *mhesvi* without recourse to chemicals or gamma irradiation had a distinct advantage, not least because it avoided "the reduced viability or other abnormalities often associated with sterilizing doses of radiation … or with translocation homozygosity." As Curtis saw it, "the use of genetic incompatibility will only be effective where behavioural barriers to cross mating are weak or absent" (Curtis 1972, 250). The idea, therefore, was to simply cross one type of *mhesvi* with another and produce a sterile critical mass to release into the environment. Again, the problem boiled down to producing this critical mass. Rhodesia did not have that capacity yet.

Conclusion: Rhodesia's Tsetse Research in the Global Context

Our grandchildren may never see a cockroach, a Japanese beetle or a corn earworm. The pests may all be wiped out by new scientific weapons, deadly to insects, safe for humans.

Man is plotting to abolish some of his ancient insect enemies from the face of the earth. Insect experts, called entomologists, are fighting a research war on six fronts. Only one front uses pesticides that have come under so much fire recently. These are the fronts:

1. *Sterilization*—mass application of chemicals or radioactive materials can make insects incapable of having offspring.
2. *Traps*—insect traps will be baited with food or female "perfume" to lure thousands of unsuspecting insects, some that would be killed and others that would be chemically treated, then released to carry sterility and disease.
3. *Predators*—hordes of creatures, harmless to man, are being sought by scientists and released in infected areas to prey on and destroy harmful pests.
4. *Disease*—plagues will wipe out huge "cities" of harmful insects as diseases are mass produced and sprayed over large areas.
5. *Starvation*—insect food supplies will be cut off by planting crops that are immune to insects, taste bad to them, or grow at the wrong time of year for them to eat.
6. *Poisons*—new chemical poisons will be used in different ways. ("War against Insects" 1964, 74)

This *Science News* article was published in 1964. At the time, the USDA's Agricultural Research Center in Beltsville, Maryland, had performed experiments on the deployment of birth control for beetles, flies, caterpillars, *hutunga*, boll weevils, cockroaches, and screwworm flies—also called blowflies. The eggs of the latter, when deposited through a bite into the flesh of *vanhu* or *mhuka*, "hatched into maggots which burrow[ed] into the flesh" ("War against Insects" 1964, 74). Houseflies were being lured to feed on sugar solutions packed with chemosterilants, later hatching sterile eggs. Scientists were "meddling with the love life of the cockroach," manipulating the "perfume" the female emitted to lure the male. The perfume chemical was now identified, extracted, and produced synthetically for use as trap bait, with chemical pesticides performing the roach's last rite of passage. Light was being used to attract moths and other flying *zvipukanana* to their death or impotence—also at the hand of chemical pesticides. "Diseases for insects only" were being manufactured. Airplanes flew low, spraying

"Bacillus thuringiensis," a special chemical disease to kill gypsy moths, cabbage moths, and other moths. They also sprayed a hormone designed to prevent caterpillars from sexual maturity and oviposition. Several viruses were being developed to kill sawflies on forest trees, cabbage loopers on cabbage leaves, and other pests. Infected caterpillars were ground to obtain the virus, the resulting mash then mixed with a solution and sprayed over the land. *Zvipukanana*'s habitat and plants were being modified to expose and starve them—by way of more Hessian fly-resistant wheat, corn resistant to maize borers, earworms, and other worms, and destruction of tobacco stalks to deny the tobacco hornworm its winter housing. The *"Bacillus,"* the hormones, the viruses, the environmental destabilization—all these were described to the public as "harmless to man and other animals" ("War against Insects" 1964, 75; "Insect Chemical Warfare" 1962; Fleschner 1959).

This was not confined to science in the United States or Europe or built lab science; Chinese citrus growers also had for centuries deployed colonies of predator huang jin yi (yellow fear ants, also called huang gan yi [yellow citrus ant]) in orchards to protect *kumquat* fruit trees against pestiferous *zvipukanana,* especially black ants. In 1708, the writer Pei Wan Chai said that people were purchasing these yellow (weaver) ants and putting them in trees to attack the black ants and kill them (Fleschner 1959, 539). Enterprising farmers also grew mulberry trees, upon which they raised silkworms—not to make silk, but to feed them to the yellow ants, which they then sold to orange growers for a stipulated amount per nest (Huang and Yang 1987, 665). In addition to Chai's text, other records of the use of some *zvipukanana* to control pestiferous ones are found in the Ching dynasty encyclopedia *Gu Jin Tu Shu Ji Cheng* (1726) and *Nan Fang Cao Mu Zhuang* (Plants and Trees of the Southern Regions, 304 AD; Huang and Yang 1987, 666).

In the twentieth century, scientific laboratories seriously considered the role of insects as engineers. By 1963, the US Air Force Office of Scientific Research had taken strong interest in *zvipukanana,* blind fish, octopuses, and mice to learn from them—just as we discussed *vedzimbahwe* doing (Lener 1963, 27). Other scientists took interest in what they called the *tenebrionid* beetle of the *Stenocara* genus found in the Namib desert in southern Africa, which tilts its body into the wind to gather water. Droplets then form above its wings and roll down the *chipukanana*'s surface straight to its mouth. Material engineers were examining this process to create films capable of providing drinking water in dryland areas. Meanwhile, at the Institute of Neuroinformatics in Zurich, Rodney Douglas was examining the eyes of *chipukanana* to design artificial retinas. At Caltech, a laboratory was building a robot with movement based on a fly's visual system, and a computer mouse with an optical rather than a ball sensor, based on the way a fly's brain worked (Flannery 2002, 377).

The roles of *zvipukanana* as organic chemists were also recognized, especially as producers of pesticides. For instance, when attacked, the scorpion emitted a precisely aimed spray later found to contain 85 percent acetic acid, 10 percent water, and 5 percent octanoic acid. Among other pheromones (the chemical transmitters of information between members of a species) was what *vachena* called *bombykol*, a sex attractant of the female silk moth they called *Bombyx mori*. The "calling" female emitted the pheromone in small pulses at intervals. From 1968 to 1969, three glossinologists in Rhodesia investigated the possibility of *mhesvi* determining the presence of the opposite sex using smell sensors on the antenna or arista. They did so "by removing, with fine scissors, these appendages at their basal segment from flies," and then determining the success of mating "by dissection of the female flies and microscopic examination of the *spermathecae*" (Dean, Clements, and Paget 1969, 356).

In contrast, butterflies courted in broad daylight using vision, but males of certain kinds were found to have odorous organs *vachena* called *hairpencils* or *danaids*, which contained cetyl acetate, cis-vaccenyl acetate, and a heterocyclic ketone. When rubbed against the female's antennae during "hairpencilling," the pheromone (or *danaidone*) transmitted a sexual message that males could not refuse (Meinwald 1990, 30). However, as research on the *shayishayi* (monarch butterfly) also found, the pyrrolizidine alkaloid material for constituting the *danaidone* comes from the *shayishayi's* visit to a plant the researchers called *Heliotropium steudneri* (Meinwald 1990).

As an example of the weaponization or strategic deployment of nature, this chapter addressed the creation of *nyongororo*-resistant *mhesvi* through bioengineering of the insect so that, upon mating, no fertilization occurred. Rhodesia must not be treated in isolation, however: An experiment conducted on *hutunga* in Sao Tomé in 1946 sought to understand this insect carrier's natural resistance to *nyongororo*. The objective was to "build a better mosquito [that would] someday neutralize the deadly threat of malaria by making mosquitoes healthier, leaving the victims of Anopheles bites at risk of nothing worse than an itchy bump" (Levy 2007, 817). Research at Johns Hopkins University was seeking to manipulate the genome of *hutunga* (*Anopheles gambiae*) to produce malaria resistance by inserting an extra gene into the *Anopheles stephensi* that transmitted malaria in India. The question was whether lab *hutunga* would cope in the wild. Meanwhile, researchers at Caltech aimed to force transgenes into a wild insect population at rates much faster than those produced by normal

Mendelian inheritance. However, these bioengineers were attempting to cause a resistance to malaria that *hutunga* of Africa had long achieved naturally (818–819).

By the 1980s, the debate had shifted to the implications of releasing genetically engineered *zvipukanana* into the environment. Environmentalists and ecologists sued the US government to compel it to stop molecular biologists from releasing the recombinant DNA–carrying *zvipukanana*, which the investigators argued were "nothing more than well-known organisms with well-defined and predictable alterations" (Tangley 1985, 470). One proposal—from the National Institutes of Health's Recombinant DNA Advisory Committee in 1983—sought to release a genetically modified *"Pseudomonas syringae"* bacteria to make potato plants frost-resistant. A second—from the seed and pesticide giant Monsanto in 1984—sought to release on trial another *nyongororo* these *vachena* called the *Pseudomonas fluorescens* carrying a gene that produced a toxin lethal to *zvipukanana* that attacked the roots of maize plants. Ecologists rejected both because of possible ecosystemic effects on plants, *mhuka*, and energy and nutrient cycling. The biologists argued that because they were engineering single-gene changes, these organisms were basically the same as the original ones—but ecologist Frances Sharples of Oak Ridge National Laboratory disagreed. Past experiences had shown such single-gene changes to cause antibiotic resistance in bacteria and pesticide resistance in *zvipukanana*. "It's not the number of new genes but what their functions are that is ecologically important" (Tangley 1985, 472).

It wasn't only the smallest members of the animal kingdom that were turned into pesticides or even the animals for that matter. By 1940, bench scientists were also vigorously searching for ways of making plants function as pesticides. The objective was to improve microbial pathogens lethal to *zvipukanana*, cause defects in the pests, and transfer *chipukanana*-resistant genes to plants. There was one potential problem: the development of *zvipukanana* resistant to the *chipukanana*-resistant plants (Raffa 1989, 524). New phosphorous compounds equipped plants to "bite" *zvipukanana* that bit them—by carrying chemical pesticides through their sap streams so that their leaves, stems, flowers, and roots in effect became *mishonga* (poisons). Pyrethrum (produced in Kenya) extracted from dried, daisy-like flowers and nicotine from tobacco paralyzed beetles, flies, cockroaches, and other pestiferous *zvipukanana*. Plant extracts were used to stretch the efficacy of existing pesticides (hence the name *stretchers*). The goal of these systemic pesticides—developed in Germany during World War II—was to induce the plant's sap stream to kill pestiferous insects without harming harmless ones and by spraying leaves or the soil for root uptake (Morrow 1952, 330).

Meanwhile, in China, no effort was spared in the promotion and development of plant materials as pesticides, to reduce reliance on synthetic chemicals from the West. Building on a long history of herbal medicine, China embarked on a program of using indigenous plants with insecticidal properties in the 1950s, isolating, identifying, and chemically synthesizing them into pesticides. For example, cottonseed oil, juniper oil, mustard, and tobacco extracts were used to control wide varieties of agricultural pests, like rice borers and leafhoppers. The extensive industrial development and public use of herbal pesticides—indeed, herbal medicine in general—owed much to China's isolation from the Western world and the need to reduce dependence on imported magic. The question was how to develop and sustain a public health system under difficult conditions; this was the background to the "Eliminating the Four Pests" campaign against *mhesvi*, *hutunga*, rats, and fleas in the 1950s. Out of this campaign emerged a movement in communes, colleges, and research centers dedicated to indigenous pesticides (Yang and Tang 1988).

6 Exposing the Fly to Its Enemies

> The Umzila results, already described, show clearly ... that settlement properly planned will protect itself. ... Umzila's results even suggest that someday in the very far distant future the question will be settled by the natural increase of the now protected native population. "Properly planned" settlement in fly [country] will not consist in the giving out of isolated farms, scattered over the face of the country. ...
>
> There must be a definitely planned settlement scheme, affecting a large block of country together, on some sound agricultural basis. The closer the settlement can feasibly be the better, and first and foremost amongst the conditions of occupation must stand the effective clearing of the less freely deciduous types of woodland. ...
>
> Secondly, no ingress of large game must be allowed from areas still under fly. The best barrier ... might be a strong, patrolled fence; but if the fringes of the area are sufficiently closely settled it is likely, even from our present imperfect success on the lightly settled border, that this fact alone, with shooting, will suffice to keep off the elephants, buffalos and elands. It is these three animals that probably chiefly matter.
>
> Umzila's principle—the settling and clearing of the low-lying guard-area only, the enclosed hill mass then taking care of itself and being perhaps disposed of later at an enhanced value—would be well worth consideration and investigation. Under such a scheme, carried out with thoroughness, it seems at present fairly certain that cattle could, after a few years, be kept safely and in numbers on the dolerite.
>
> If, on the other hand, the settlement should have to be a gradual growth from small beginnings, its safest base would be the deciduous part of the British border, a block at a time being settled and special measures being taken against buffalos.
>
> The settlement of the two permanent fly-areas themselves would mean the end of the menace. (Swynnerton 1921a, 375–376)

This passage is taken from Charles Swynnerton's study of *ndedzi* in Muzvirizvi. He was laying out the various methods that the Gaza king Mzila had used and recommending them to the governments of southern and eastern Africa.

Among other stratagems, Swynnerton proposed five basic methods that the Gaza king used and which were deployed throughout southern Africa. The first involved controlled *ndzilo* (fire) to destroy not just the *xifufunhunhu* (insect) but also its habitat and hides, thereby exposing it to its predators. The second was forest clearance, intended for a similar purpose, as well as to separate *sviharhi* (animals) and the *ndedzi* that subsisted on them, on one side, from people and their *tihomu* (cattle) threatened by the flies, on the other. This was called *barrier clearance*. The third measure was prophylactic settlement, which involved the strategic deployment of people between *vachena*'s ranches and an advancing *ndedzi* front. It involved deliberately overcrowding people to cause deforestation and overgrazing, thus denying *ndedzi* its perfect living conditions. The fourth, albeit not unconnected, stratagem was to use fences to create permanent and patrolled game-free, cattle-free buffer zones that blocked and redirected the movement of *vanhu* and *tihomu*. The fifth was the use of *magocha* to shoot *sviharhi* and create buffer zones between *ndedzi*-infested areas and *tiko* (villages). The fourth method is the subject of the next chapter. This one, meanwhile, discusses the first two stratagems, with all places referenced shown on the map (see figure 6.1). The keywords in *xitsonga* and *chidzimbahwe* are listed in the glossary for easy reference.

Figure 6.1
The fronts against *mhesvi*, 1909–1970.
Source: Author.

In a sense, this chapter builds on and contributes to literature on arboricides or herbicides, which currently is strongest in the United States, where some of the chemicals and equipment used to deny *ndedzi* shelter originated. The richest literature focuses on the farm—specifically, on *mishonga yesora* (herbicides) deployed against weeds. This category included, by 1945, herbicides such as 2,4 dichlorophenoxyacetic acid (2,4-D), a growth regulator, which mimicked a plant's own hormones and caused the plant to literally grow itself to death (Anderson 2001, 2005; Daniel 2002). The role of institutions and scientists has received attention, with farmers fundamentally defining how industrial chemicals were adopted and deployed. This is how herbicides, along with the tractor, displaced cultural techniques between 1890 and 1940 (Fitzgerald 1990, 2003; Kline 2000; Williams 1987).

The chapter extends the concept of *herbicide* (better yet, *arboricide*) beyond the substances used to kill "problem plants" and toward the theory and practice of killing them (Mavhunga 2011, 152). Where the US literature limits herbicides to chemicals and their agricultural targets (weeds), I extend arboricide to cover all kinds of techniques used to manage or eliminate trees or whole forests and the multiplicities of mobilities within them that facilitate the spread of trypanosome-carrying *ndedzi*. Here, we see very direct mobility of *ruzivo rwavatema/vutivi bya vantima* into Rhodesia's program of forest clearance as an anti-*ndedzi* strategy.

Fire and Late Season Burning

In chapter 1, we discussed the widespread use of *ndzilo* (fire, fires) in pest-control management in the community. With *valungu*'s aggressive occupation and their establishment of Southern Rhodesia as white-ruled territory, Swynnerton (1921a, 325) observed that "under the white man everyone burns when he pleases." The first *ndzilo* were now being set while the grass was still "half ready to burn," even when there was hardly any leaf litter on the ground or enough wind to swirl the flames up into the treetops above and penetrate the thickets. As a result, the effects of *ndzilo* were drastically reduced, allowing flies driven from one burning place to fly above the flames into the burnt area and find refuge in the unburnt spots and leaves. The purpose of using *ndzilo* late was to destroy young shoots and germinating plants while sparing the more resilient, thick-stemmed, and taller trees. At the same time, discontinuing the use of *ndzilo* encouraged the roots of pyrophytic (*ndzilo*-tolerant) plants to shoot out and become undergrowth (383).

Swynnerton recommended that the *Companhia de Moçambique,* the authority in charge of Muzvirizvi at the time and which had commissioned his research, adopt the seasonal firing techniques that Mzila in particular had used well to control *ndedzi* in Muzvirizvi. Indeed, his son, Ngungunyana, might have continued his father's practice had the circling British and Portuguese not forced his strategic but doomed withdrawal to Bilene. Swynnerton advised authorities to delegate "the kraal natives themselves [to do] the actual burning." Even though certain trees protected *svimun'wana* (springs) and kept them wet, he was adamant: "Better an occasional spring lost than a continuance of the tsetse" (Swynnerton 1921a, 383). Here, he cited *Umzila's principle*: "The Zulu clearings, sufficient to remove *ndedzi,* do not seem to have caused any shortage of water. ... There should be no discontent over it amongst the natives," he said, "as late burning represents their own old custom, and, whatever their infringements, they still speak of it as the correct method" (384). People usually burned the grass around their huts and granaries early to protect against increased late winter valley *ndzilo,* which usually started as children burned grass to flush out mice and as adults cleared new land to plant crops. The plowed or hoed strips provided inadequate fireguard. Contrary to Swynnerton, however, the practice was not Mzila's "invention" but standard practice throughout Africa; in fact, to say Swynnerton "saw" or "found" *ndzilo* to be effective in controlling pests like *ndedzi* is to say that *vatema* were the ones he saw, found, and learned from.

As he had learned while still the manager at Gungunyana Farm, late fire burning could be "so postponed and regulated as to serve most useful purposes" against *ndedzi.* The grass would be mature and very dry, yielding "a really fierce fire" that destroyed *mahlomalavisi* (chrysalises) and breeding places. The scorching *ndzilo* also brought about "the probable hardship to the *ndedzi* itself of a widespread and thorough removal of shade, concealment and food at a hot, dry time," rather than "burning by small installments, [which] makes escape easy" (Swynnerton 1921b [1960], 31). Late burning (see figure 6.2) also caused indirect effects. It found the grass and fallen leaves in their driest and most abundant state, the weather hot and the winds great, the flames therefore achieving maximal destruction of low woody growth and temporary elimination of high shade.

There was no need to burn every year, but only every second year; even then, it was not just "the best method" but "the right time to burn" that mattered. It was best to burn late one year and skip burning altogether the next to ensure "an extra accumulation of ... extraordinarily dry, effective fuel." Whether burning annually or every other year, the hottest and driest

Figure 6.2
Late burning in Southern Rhodesia.
Source: Proceedings and Transactions of the Rhodesia Scientific Association 1960.

months (September–October) were best, especially just before the rains (Swynnerton 1921a, 1921b [1960], 32).

As Swynnerton had learned from Mzila, late burning provided rich ash for "the excellent grasses left us here by the Zulus [Gaza], who burned late and regularly" (Swynnerton 1921a, 384). Regular firing suppressed the dense wooding that otherwise kept good plants down. The flames killed or maimed flying *ndedzi* and scared away *sviharhi*. Fire denied *ndedzi* shelter, destabilized it into flight, and rendered it visible to swooping *svinyenyana* (birds). It also destroyed insectivorous *xinyenyana* (singular of *svinyenyana*) populations, whose breeding season started in October, the perfect time for late burning. However, Swynnerton concluded that "useless fires" were destroying forests anyway, so *svinyenyana* "may as well therefore be destroyed by useful ones" (385).

To be clear, Swynnerton's work in Muzvirizvi was commissioned by the Portuguese via the *Companhia de Moçambique*. However, his recommendations of late burning every other year, with fines for unplanned *ndzilo* and guards deployed to monitor, was not confined to Mozambique. This recommendation was also implemented in Southern Rhodesia following the Nemakonde *n'gana* outbreak of 1927.[1] These organized or controlled grass *ndzilo* were "fierce and complete" but did not eliminate the pest, which took refuge in pockets inaccessible to the flames—especially in swampy breeding areas, where *mahlomalavisi* remained virtually untouched. The 1928 flames were even less ferocious and incomplete and caused only a "negligible" effect on *mahlomalavisi* and the *ndedzi* population.[2]

The *ndedzi* surge continued. By 1933, the early optimism surrounding late burning had evaporated. Elsewhere that year, Tanganyika had a "successful" burning experience, Uganda "promising," and Nigeria "not so favourable." As Chief Entomologist Rupert Jack of Southern Rhodesia noted in his annual report, the problem of late burning lay in its dependence on "extensive areas of heavy grass," whereas most of Southern Rhodesia's *ndedzi* areas were in mopane country of very thin and scanty grass. Test after test had given "very discouraging results" and it was "obvious that general application of this measure [was] out of the question."[3]

Moreover, by the early 1940s, some tree types were exhibiting resilience to *ndzilo*, which seemed to stimulate rather than suppress increased regrowth from root suckers.[4] Indeed, the role of *ndzilo* in the maintenance of open savannah and the prevention of thicket formation presented "a complex ecological problem of great importance in applying anti-tsetse measures."[5] This was particularly so in southeastern Rhodesia's mixed thicket (bushland), which, when left undisturbed and unprotected from *ndzilo*, became "a tangled mass of almost impenetrable thicket, known locally as 'Jessie bush.'"[6] It was the perfect haunt for *ndlopfu* (elephant) and *mhelembe* (rhinoceros). Whereas open-type savannah woodland could be maintained through recurrent *ndzilo*, soil erosion was a problem; however, no soil-disturbing agents such as *tihomu* were allowed in.[7]

Up until 1955, emphasis within Africa's entire *mhesvi*-control fraternity had been on the positive effects of *moto* (fire) on vegetation composition. New research that year called such bullishness into question. Fifty percent of *mhesvirutondo* blood meals were now known to come from the pig family of animals that *vachena* called Suidae, in areas in which *nguruve* constituted just 10 percent of game population (Weitz and Jackson 1955). In the Fort Johnston area of Nyasaland (Malawi), where the Muslims regarded *nguruve* as unclean, *njiri* (warthog) thrived well after locals hunted all other *mhuka* for the pot (Mitchell and Steele 1956). Foliage constituted almost all of the animal's diet, but after seeds ripened and food translocation to roots was advanced, roots—not just grass—became an important dietary factor. In the dry July–October period, *njiri* survived almost entirely on roots, and the *nguruve* had "adapted their breeding season to coincide with the period at which the grass roots contain[ed] maximum food value; that is before the food reserves have been drawn upon for the growth of new spring foliage" (Mitchell 1963, 27). Most grass types, especially *zengeni* or *shengezhu* (what *vachena* called *Hyparrhenia spp.*), continued to produce some foliage fed from food reserves stored in the roots. Other grasses, like *tsangadzi* (what *vachena* called *Loudetia superba*), stored food in large rhizomes and deferred shoots until mid-October, regardless of *moto*.

This evidence showed that late burning achieved the opposite effect of what was intended. As Mitchell demonstrated:

> Frequent late burning has resulted in the displacement of woody vegetation by coarse grasses *Hyparrhenia* spp. and *Loudetia superba* in particular. Protection from fire or very early patch burning result in thickening up of woody vegetation and a suppression of the coarse grasses. Frequent late burning therefore, by increasing *Loudetia superba* and *Hyparrhenia* spp. dominance, renders conditions favourable for an increase in the population of warthog and a consequent increase in the density of *Glossina morsitans*. It is thought probable that the intensification of burning which has taken place over the last sixty or seventy years in Northern Rhodesia has been in some measure responsible for the spread of tsetse which has occurred over the same period. (Mitchell 1963, 28, citing Trapnell 1959)

Based on this damning analysis, enthusiasm dampened and focus turned even more energetically toward other methods. One thing was clear: No method worked everywhere, because no two contexts were the same.

Mechanical Forest Clearance

Swynnerton undertook two experiments in 1918 in Chipinge, one year before the Rhodesian government started its own in Gwai and Shangani. His first was designed to establish the effects of clearing undergrowth in primary forest. He instructed his workers to clear out an 80 × 70 yd. area and to drive two black oxen through it the next day. Only three *ndedzi* accosted the oxen, whereas swarms had attacked them before. In the uncleared areas, fifteen flies were caught. He concluded that clearing undergrowth was effective in "banishing" *ndedzi* (Swynnerton 1921a, 374).

The second experiment sought to ascertain the width of undergrowth clearing necessary "to protect a strip of road from attacks by *G. brevipalpis* in sunny weather" (Swynnerton 1921a, 374). Swynnerton proved that only minimal clearing was needed to proof a piece of road so long as *tihomu* moved along it in sunshine, when *ndedzi* was sheltering from the sun and predators. By contrast, "a considerably wider clearing would be needed to render it safe at all hours and in all weathers" when *ndedzi* was out and active (374).

The Division of Entomology's experiments from 1918 on were intended to test the efficacy of barrier clearing and must be regarded not merely as experiments but as the only measures the government had to contain *mhesvi*. The first was discriminate and targeted a dry-season concentration area for *mhesvi* in the Sepani vlei of Sebungwe. Discriminative clearing (targeting of specific types of trees or forests) was based on the idea that *mhesvi* did not just move or live randomly throughout *sango*.[8] The experimental objective was to determine the effect on *mhesvirutondo* of eliminating evergreen trees along the riverbank. The experiment was abandoned due to two continuous seasons of abnormally high rainfall that discouraged the normal concentrations of *mhesvi* and *mhuka*.[9]

The second experiment was a barrier-clearing exercise that involved cutting down all trees, eliminating undergrowth, and late burning, with the aim of preventing the spread of *impukane zegangeni* (*mhesvirutondo*) from the Shangani to the Gwai Rivers. The work was abandoned following the outbreak of influenza in Matabeleland North. Further experiments in discriminate elimination of evergreens followed in Gwai in 1927, with the objective of determining whether depleted shade might discourage *mpukane*. However, the experiment was terminated in the next year to focus solely on "game destruction."[10] Settling *abazingeli* (hunters) along the Gwai was deemed a far cheaper way of clearing the "true habitat" and keeping *mpukane* at bay.[11] Jack found discriminate clearance "almost as distasteful as destruction of game" but was prepared to give it a chance if it could eliminate *mpukane*.[12] In 1933, preliminary experiments started in poisoning indigenous trees with the objective of "furnish[ing] some information on methods of treating cleared barriers."[13]

Experiments later conducted in Hurungwe in 1957 were promising, not least because the settlement of *vatema* from Bikita and Chivhu followed right after discriminate clearing. *Mhesvi* disappeared within six months. The advantage of cleared areas was their inhospitableness to *mhesvi* breeding long after, before regeneration occurred—sometimes as much as ten years later. In areas where *shiri* (birds) swarmed, tree destruction exposed *mhesvi* that might take refuge in the trunks' greyish, camouflaged bark containing cracks inaccessible by beaks.

Although selective or discriminate clearance suited evergreen river lines, block clearance was preferred in expansive areas that had to be rendered unsuitable to *mhesvi*, thus forming a wide barrier to *mpukane* crossing from infected to uninfected areas. As Jack noted, there had to be a considerable population of *vatema* suffering from a shortage of safe grazing for their *izinkomo* for such block clearance to provide sufficient incentive for them to settle there and thereby offer themselves as free labor to government. Second, there had to be "good tribal discipline"—that is, loyal *vatema* that could be counted upon to supervise fellow *vatema* in the absence of a white man, whose health was susceptible to fever in these margins between European-occupied territories along the Zambezi and southeastern borderlands.[14]

Barrier clearing consisted of clearing lines through *sango* wide enough to prevent *mhesvi* from crossing. The clearings had to be between one and ten miles wide. It would have been easier to maintain the cleared strips of land by settling *vatema* in it, but few were prepared to move to these unsuitable areas.[15] The Division of Entomology also considered planting conifers or eucalyptus as a thicket barrier against *mhesvi* encroachment from Mozambique. Indeed, experiments in Tanganyika had shown *mhesvi* to dislike the interior of extensive thickets. However, the division found the method "doubtful to say the least."[16] The experiment was tried in a smaller area in eastern Chipinge, but went no further.

In any case, there was doubt whether *mhuka* were even responsible for bringing *mhesvi* across the border or any certainty about whether a border clearing would help. In 1935, Jack gave three reasons for why a strategy focused on controlling the traffic of *mhuka* was a bad idea: (1) The *mhesvirutondo* followed *vanhu* for up to ten miles, or even more; (2) it also caught rides on *vafambi* (travelers) and didn't need *mhuka*; and (3) it could always fly, stop, fly to cross on its own.[17] Jack had advised the government in 1932 that a much narrower clearing was more reasonable for *mhesvirupani*, which did not catch rides on *mhuka* as much. Hence, the clearance axis could go alongside open grasslands (natural clearing) and high ridges (altitude inhospitable to *mhesvi*), thus minimizing the work of chopping trees down. Using such a method, the Tsetse Branch had employed local *vatema* to clear about thirty miles of the border at a cost of about £1,500 over three years. Between 1932 and 1934, the length of the border clearing in Chipinge was extended to thirty-five miles.[18] Again, barrier clearance was front and center in *Umzila's principle*.

However, as the 1940s show, unlike the Gaza king, the Rhodesians failed to control the movement of *vanhu* and *mhuka* to any commanding degree. In 1939, a *ndedzi* invasion from Mozambique breeched this barrier, burst forth into Chipinge, and infected one thousand *tihomu* with *n'gana*, leaving four hundred dead at thirty-two farms. In response, the government embarked on an extensive program to double the width of the clearing, slash regrowth, and burn the cut grass and wood. It recruited a large workforce composed of locals under two white supervisors and cleared the Budzi to Cheredza River frontage near Mt. Chirinda. After a drastic decline of *n'gana* to just 132 cases producing twenty-four deaths at nine farms in 1940,[19] the cases rose again in 1941 on a far wider scale than before.

This time, *mafrayi* were dispatched to extend the clearings in the Nyamadzi valley, which Jack's successor Chorley identified as "the main channel along which the two types of *mhesvi* involved enter the Colony."[20] In 1943, several farms—Wolverhampton, Helvetia, Chibudzana, Sherwood, East Leigh, Southdown, and Grampians—were subjected to "total clearing" on the outward (border-facing) side and "partial clearing" on the inner side facing the farms.[21] The following year, the border clearings on the Mount Selinda, Farfell, Pendragon, and Bayswater farms were widened, slashed, and burned.[22] Meanwhile, the coniferous and eucalyptus barrier proposal was seriously considered, but shelved due to Chipinge's distance from the Mutare-Beira railway line.[23] By 1952, the total area of barrier clearing had reached 59,188 acres, and the government was putting some of this land to tea crops (private growers) and conifers (Forestry Department).[24]

By 1954, (white) public objections to game destruction had forced the government to appoint a Commission of Inquiry to find other *mhesvi* and *n'gana* control methods. It recommended only discriminative bush clearing combined with close settlement and OCPs, with strictly supervised discriminative game elimination continuing as a temporary strategy until the former methods were perfected. More funds were allocated for mechanical bush clearing and OCPs spraying, with hunting gradually restricted to controlled hunting areas (CHAs) under the Department of National Parks and Wild Life Management (DNPWLM; Cockbill 1967).

It became clear in 1955 that *mhesvi* was reestablishing a stranglehold on the border clearing, with control operations being confined to eliminating regrowth.[25] *Vatema* were flocking to the better wages and working conditions that the white farmers and tea and wattle companies offered, shunning the backbreaking work of felling huge trees.[26] Some local farmers were overgrazing their lands and failing to organize *moto* fierce enough to destroy the regrowth, thus promoting overgrown bushes habitable to *mhesvi* and grass too short to sustain a raging, *mhesvi*-killing fire. The operations were assuming "more and more the character of pasture improvement ... than anti-tsetse operations."[27] Meanwhile, senior officials were dismissing discriminate clearing as a "spectacular and very expensive failure" and a wild goose chase based on an unproven assumption that whole *mhesvi* populations concentrated in specific areas when only small fractions did—and, even then, because specific *mhuka* were in the locale.[28]

This is what made organochlorine pesticides attractive—the element of mass destruction.

These developments in Southern Rhodesia must be placed in the broader African context. In 1952, Glasgow claimed quite boldly: "Of all the various methods of control which have been devised for various types associated with particular vegetation, the only method which can be guaranteed to succeed with any types of tsetse in any situation is sheer clearing" (Glasgow 1960, 86)—that is, the complete removal of all woody vegetation and its replacement by pasture or fields. Some were not so sure about sheer clearing, however.

Chemical and Mechanized Phytocides

Here, we can make very direct connections to the US literature we began the chapter with. Arboricides were coming from the United States, Canada, and the United Kingdom. In 1958, investigations were launched into the practicability of killing trees with organic arboricides or phytocides in the control of regrowth in clearings. The research continued in the 1960–1961 operational year (Cockbill 1961). Two soluble chemicals were selected: 2,4-Dichlorophenoxyacetic acid, a common pesticide/herbicide for controlling broad-leaf plants, and 2,4,5-Trichlorophenoxyacetic acid, a chlorophenoxy acetic acid herbicide used to defoliate broadleaf plants.[29]

The 2,4-D chemical is an organic compound (chemical formula: $C_8H_6Cl_2O_3$) that kills broadleaf weeds by inducing them to grow uncontrollably while sparing other plants around them. It was first published as a selective herbicide in 1944. The next year, American Chemical Paint Company started selling a 2,4-D herbicide it called *Weedone*. It became the first such compound to selectively destroy broadleaf plants while leaving narrow-leaved ones alone, and thus it "replaced the hoe" (Hamner and Tukey 1944). Later, Dow Chemicals became the biggest manufacturer of the herbicide.

The World Health Organization (WHO) International Agency for Research on Cancer (IARC) has listed 2,4-D as "possibly carcinogenic to humans," though it admits this classification is based on "inadequate evidence in humans and limited evidence in experimental animals" (Loomis 2015; IARC 2015). 2,4-D is placed alongside coffee and red meat, in a category of carcinogenicity called Group 2B, which is much milder than Group 1, but it is still extremely toxic ("Agents Classified" 2015). The "2,4-D General Fact Sheet" (2015) notes, significantly, that the chemical was an active ingredient of the notorious Agent Orange that US troops used extensively in the Vietnam War, even though 2,4,5-T was responsible for the health effects associated with the bombing. 2,4-D caused fertility problems in men (abnormally shaped sperm; NPIC 2015).

The chemical 2,4,5-T (chemical formula: $C_8H_5Cl_3O_3$) is a synthetic chlorophenoxy acetic acid herbicide also designed to defoliate broadleaf vegetation. Like 2,4-D, it was developed in the 1940s and used extensively as an agropesticide before being gradually discontinued in the 1970s. It also gained notoriety as an ingredient of Agent Orange (composed of 50 percent 2,4,5-T and 50 percent 2,4-D). Today, it is associated with US carpet bombing in Vietnam, but this belies its earlier devastating deployment by the British in Malaya (Sodhy 1991) and, as we will discuss later, similar deadly chemicals in Rhodesia against *varwiri verusununguko* (freedom fighters), whom *vachena* called "terrorists." During manufacturing, traces of 2,3,7,8-tetrachlorodibenzo-p-dioxin (TCDD), a persistent, carcinogenic organic pollutant with long-term effects on the environment, contaminates 2,4,5-T. Although such contaminants can be reduced with good temperature control, pre-1970s manufacturing did not have such controls. Consequently, the US Department of Agriculture terminated all 2,4,5-T use in food crop production in 1985. Internationally, dicamba and triclopyr have since replaced 2,4,5-T because of confirmed evidence from intentional overdoses and unintentional high-dose occupational exposures that caused weakness, headache, dizziness, nausea, abdominal pain, myotonia, hypertension, renal and hepatic injury, and delayed neuropathy (CDC 2016).

Both 2,4-D and 2,4,5-T were used extensively in Southern Rhodesia's arboricide against *mhesvi*. They were diluted to 3 percent and 6 percent mixtures in lighting paraffin, according to a method borrowed from similar experiments in Mozambique. Lighting paraffin was more expensive than diesel, but it was preferred because it caused less rapid leaf fall and permitted much more of the applied toxin to penetrate into plant tissue. Application to the foliage was made with pneumatic knapsack sprayers at 75 lb. per square inch. About fifty tree types exhibiting coppicing were treated, each with 3 percent and 6 percent mixtures applied to the foliage in January, February, March, and April. Twenty-three types of trees were cut at about two feet from ground level and the cut ends of their stumps treated with the 6 percent mixture over the same period. As a control, corresponding trees were cut and left untreated to indicate the extent of regrowth under the prevailing conditions. During February, March, and April, deep cuts were made into standing trees and a 6 percent mixture applied to bark and sapwood. Two growing seasons were needed before researchers could offer opinions on the exact effects, though early results indicated that some trees were "fairly resistant to 2,4,5-T and 2,4 D applied in this way," whereas others were more susceptible. Already there was fear that using these arboricides on some types within a mixed association might spread chemical resistance instead of killing the trees, as was happening with trypanicides.[30]

On the advice of giant British agrochemical company Fisons Pest Control Ltd., a follow-up trial was conducted in 1966–1967. This time, the arboricidal agent was no longer a butyl ester concoction but the weed-killing compound Tordon 22K, manufactured by Dow AgroSciences Canada Inc. of Calgary. Tordon 22K was applied to freshly cut stems of regenerated *misasa* (*B. spiciformis*) at Lettie Swan Farm in Chipinge District. The trees had grown to a height of eight to ten feet and their stems up to three inches in diameter. Before the experiment, the vegetation had been suppressed through slashing and burning, in line with the anti-tsetse program to create a barrier to *ndedzi* incursions from Mozambique. In 1958, Lettie Swan Farm was excluded from annual treatment, leading to rapid regeneration of woods. At the same time, the tree growth stunted the grass cover to a level at which annual seasonal fires ceased to occur. The coppice thickened, and when the experiment team moved in the regenerating shoots were so dense that operators found it impossible to move with pneumatic knapsack sprayers through them to apply a foliar spray.[31]

The only solution was to cut each shoot clean through and apply arboricide using paintbrushes to the freshly cut ends. This application was carried out on February 1, 1966. Then, on December 9, the arboricide was applied as a foliar spray to half of the surviving leaf-bearing shoots, with the rest used as controls. The shoots that had been cut and left untreated as controls on February 1 were divided on December 9 so that half of them became controls and the rest were treated with foliar-sprayed arboricide at a concentration of 1:100. The results suggested that Tordon 22K in water could prevent regrowth. Applied as a 1 percent foliar spray to untreated cut stems, the pesticide "killed or prevented regrowth of about the same proportion as the 4% foliar application following application to the freshly cut stems."[32] Tordon 22K cost £22.10 per gallon; added to the cost in man hours (labor), the cost of spraying was £12 per acre to remove *misasa* that had regenerated on Lettie Swan farm.[33]

Bulldozers, Chainsaws, and Preexisting Methods

South of Chipinge, the thinking within Tsetse Branch until 1943 was that while the Runde and Savé Rivers offered a natural barrier to the movement of *vanhu* and *sviharhi* when in flood, during the dry season they carried very little water. In fact, the Runde was even completely dry. In either case, there was quite considerable pedestrian traffic across both rivers. At Chivirira Falls on the Savé Gorge, the river was only about a hundred yards wide. There was nothing to stop the *ndedzi* advance from Mozambique continuing quickly across the Savé into the Gonarezhou Forest in the Ndanga and Chivi Districts. Chorley therefore advised the government to take "very drastic measures" to avert a sure disaster. There were no objections.[34]

Border-clearing work commenced on the Rupembi River in the Savé Valley in 1955. The primary tool of mechanized tree destruction was the mechanical chainsaw, which still required significant labor from *vatema*, albeit working in pairs and in far fewer numbers than the axe-felling "gangs" required.[35] The operations were a kind of experiment in comparing costs and efficiency of mechanical chainsaws to that of hand labor with axes. The area selected was composed of thickets in deciduous woodland impenetrable to *xindedzi xa nhoveni* (*mhesvi* of the forest) and *xindedzi xa nkova* (*mhesvi* of the valley) during the dry season. The project's success would be judged according to the effect on traffic control figures and by random catches. Further, the future incidence of *n'gana* or the presence of the *xitsongwatsongwana* (microorganism) that caused it in *tihomu* in Musikavanhu African Reserve would indicate the efficiency of the clearing as an anti-*mhesvi* measure.[36]

Two years later, the methods were extended to four main forest clearing operations: in the Savé West and Savé East Game Destruction Scheme, the Rupembi and Honde River Discriminative Clearing Scheme, and the Bandai Clearing Scheme. By 1958, the two latter schemes were finished and the first abandoned; the Savé East scheme was the only remaining one, in which *mafrayi* were busy with drainage line bush clearing. The Tsetse Branch hoped to put the area to close settlement.[37] It was also during that year that "the very large task of modifying the vegetation of the black basalt zone so as to form an adequate barrier to further advance of tsetse" began.[38] Teams of *mafrayi* were tasked with felling trees in the alluvium of the drainage systems of the Rupembi and Honde rivers running southeast across the international boundary between Beacons 105 and 108. The lower reaches of these rivers inside Mozambique were thick with *ndedzi*, leading to the importance of creating a barrier clearing to curb any likely westward movement.

Meanwhile, another team of *mafrayi* with axes was deployed to hand-clear an area in Bandai of approximately four thousand acres lying at the southern end of the Ndanga East Reserve on the Savé's west bank. The objective of the clearing was to cut off any likely cross-border invasion into the Peri dip and Rupangwana dip area. The Tsetse Branch hoped that after the initial clearing, the Native Affairs Department would take over the slashing and late burning necessary to suppress regeneration and promote grass growth.[39] However, owing to lack of locals wanting to perform *mafrayi* work, the work was suspended after just a thousand acres had been cleared. It was only in September 1958 that work to clear the outstanding acreage resumed.[40]

In July 1958, the government approved a budget for three Caterpillar D7 *mabhurudhoza* (corruption of bulldozers), to be operated by the Conservation and Extension Department. The D7 was a medium *bhurudhoza* made by the Caterpillar Tractor Company of Peoria, Illinois, and Stockton, California.

The first D7 was made in 1938, setting the stage for many versions with different horsepower ratings in subsequent decades (see figure 6.3). The first machine arrived on the Mkwasine River on December 3, 1958, followed by the second on January 20, 1959. Before they moved in, District Entomologist John Farrell had mapped the vegetation communities to guide the machine teams—composed of drivers, axmen, and mop-up personnel. The two machines soon proved inadequate.[41]

While the D7 *mabhurudhoza* were busy at work on the Mkwasine, *ndedzi* was captured on the Chiredzi in April 1959, thus providing "a more precise indication of the direction which operations should take."[42] After completing the Mkwasine drainage clearing up to the Sangwe Reserve boundary fence, the machines rolled into the Chiredzi river valley. Riverine drainage was cleared for eight miles along the Mkwasine on the western side of the Sangwe boundary fence. Thirty miles of tributaries entering the Mkwasine in this vicinity were also cleared. The width of the clearing was 50–70 yd. on either bank. On the Madela tributary, the removal of *chinanga* (the hooked-thorn tree *vachena* now called *Acacia nigrescens*) woodland created a clearing half a mile wide, inside which occasional trees were spared for shade. The *mabhurudhoza*'s progress was 0.8 acres per hour per machine; all told, they cleared 2,165 acres at a cost per acre of just under £7.[43] In August 1959, the *bhurudhoza* teams moved into the Nyamasikana tributary of the Mkwasine close to Chidhumo Clinic, clearing 690 acres by the

Figure 6.3
The Caterpillar D7 *bhurudhoza*.
Source: https://www.youtube.com/watch?v=My1fzbOxwi8.

end of September. Inside the first three hundred acres, the working teams piled all felled bush at the white landowner's request. It was an expensive undertaking; the clearing itself cost £7. 8s. per acre, but the piling cost £5 per acre.[44]

By the early 1960s therefore, the *bhurudhoza* had entered the anti-*ndedzi* operations. In addition to the 2,900-acre clearance at Mwangazi, these monsters also blazed through 2,700 acres at Lusongo. Both motorized destructions of trees were aimed at creating a barrier to the path of *xindedzi xa nhoveni* advancing from Mozambique. The bulldozing operation cost per acre was £6.10s.; the fairly flat terrain and limited density of forest depressed the costs somewhat. By contrast, the clearing of eighty acres in Musikavanhu African Reserve had cost £14 per acre on account of the heavy riverine bush near the Savé River. These *mabhurudhoza* were driven by *vatema* working in teams to not only fell or uproot trees, but also construct and repair roads or tracks. The road works cost £18. 18s. per mile to carve out.[45]

That was not the only job the *bhurudhoza* teams were charged with. The land was very rocky, and complete clearance using machinery was impossible, meaning that trees left standing had to be cut or treated by hand. Instead of chopping whole trees down with axes, particularly for the tough-as-rock hardwoods like *musimbiri* (ironwoods), the *bhurudhoza* team members on foot ringbarked the trunks to expose the sapwood, then applied Tordon 22K. In Mwangazi alone, these teams treated 362 acres using a combination of sheer felling and ringbarking.[46]

Machines did not replace axes in difficult terrain; 206 acres of riverine bush could only be removed by felling with axes in a section where the Mkwasine passed a ravine, right near the Sangwe Reserve boundary. The road would meander through bush, trying to find maneuverable space between the big trees. Where there was a corner (and the line had to be straight), Farrell instructed people to stump the trees so that the tracks could be straighter. Stumping was also performed when there were tough roots and big trees and the grader found it hard to clear.[47]

There were finer details that hand clearance could perform that machines could not; thus, the axmen covered all stumps with piles of brushwood "in the hope that fire would kill them in due course." The financial cost aside, locals (both *vachena* and *vatema*) raised concerns that riverbank clearing—especially the uprooting effects of *mabhurudhoza*—would result in massive soil erosion. Therefore, during the Chiredzi clearance, the decision was made that actual riverbanks would only be hand-cleared at a cost of £10 per acre. In other areas, with narrow bands of riverine thicket, axmen were deployed to cut and stump.

The *bhurudhoza* work was completed in 1961 and resulted in a marked improvement in the *n'gana* situation, but there was "a marked a deterioration with the whole picture" similar to "the worst times of 1957 and 1960."[48]

Questioning Discriminative Clearance

Two constituencies questioned vegetation clearance as a technique for controlling *mhesvi*. The first was *vachena* concerned about environmental consequences without any tangible evidence of the effect. The writings of Edward Bursell, then lecturing at the Division of Biological Sciences at the University College of Rhodesia, exemplify this white critique. In a 1967 paper, for instance, Bursell described the method as "a dismal failure"; the Pilsons had shown the method to affect "only a certain fraction of the total (tsetse) population" (Bursell 1967, 33–34). In other words, discriminative clearing was based on a misreading of the mobilities of a peripatetic insect.

As he noted two years later, the "theory" of discriminative clearing was based on the concept of "stereotyped behavior"—namely, the observation that tsetse populations seemed unevenly distributed in the general environment, higher in some "concentration areas" than in others. Such concentration areas were conspicuous by their vegetational features, "sometimes as a contact between two vegetation types, sometimes by the presence of a double-storeyed canopy, and so on." *Mhesvi* was always present in high densities in such areas, presumably as a sign of "some innate behaviour pattern, an attraction for the tsetse of that particular concatenation of physical features."[49] Whatever the reason, concentration areas played a "special part in the economy of the tsetse population," and the weaponizable aspect was that if the trees could be cut down, the *mhesvi* population would be eliminated.

Instead, the theory of discriminate clearance defined the major functions of *mhesvi*'s everyday life—sex, feeding, and sheltering—according to vegetational differences. Therefore, it was useless to massacre every tree in sight. Just targeting the "true habitat" of *mhesvi* within a larger forest ecosystem could lead to a rapid decline in the *chipukanana*'s population density and drive it to eventual extinction. A large-scale operation of that nature had been undertaken in Abercorn in Northern Rhodesia and had exterminated *mhesvi* over three hundred square miles after felling trees covering just 2.2 percent of the entire area. Another scheme, somewhat smaller in scale, was undertaken in Hurungwe in 1957 with very promising results, even if the speed of elimination could not be predicted. In the small Hurungwe pilot scheme, the settlement of *vatema* followed right after discriminate clearing, and *mhesvi* disappeared within six months.[50]

In areas *shiri* (birds) occupied in droves, Bursell urged that tree destruction was an ineffective form of indirect assistance to this predator for a different reason: Bark offered *mhesvi* camouflage and physical protection. The bark's color was gray and somewhat charred, and it had cracks into which the *chipukanana* retreated. For *mhesvirupani* and *mhesvirutondo*, arboricide was ineffective short of wholesale clearance, because *muunze*, *musasa*, *munhondo*, and *mupfuti* all had suitable bark. Thus, Bursell considered the rationale of discriminative clearing "essentially apocryphal," having found in its support the theory of discriminative clearing. His ideas came from discussions between glossinologists at the Central Tsetse Research Laboratory "at about the time when the concept of discriminative clearing was born."[51] Tremendous goodwill had accompanied the Abercorn success, and it was subsequently extended to the whole south-central-east Africa region. Substantial reductions in *mhesvi* population density were recorded in some areas, whereas in others only negligible reductions were reported. Instead of admitting the limitations of discriminate clearing as an anti-*mhesvi* strategy and reexamining exactly why Abercorn had succeeded, glossinologists sought excuses. Bursell lamented the cost of such stubbornness:

> Had we done so we should not have lost sight of the fact, that there had been an outbreak of rinderpest in the area just prior to the clearing operations, and that the consequent reduction in the density of game animals might have had something to do with the spectacular results achieved. But we decided to soldier on, even though evidence for the plasticity of tsetse in relation to vegetation became more formidable with every new situation that was investigated, even though the empirical basis of the method was eroding and a theoretical basis was all but lacking. In some areas the tsetse population would appear to be associated with the evergreen vegetation of major drainage lines; in others such vegetation was deserted in favour of an ecotone between savannah woodland and the open grassland of *mbugas* or vleis; in yet others the fly appeared to be associated with a sparse acacia woodland, with the trees widely scattered in rolling grasslands, and so on. There seemed to be progressively less in the way of a common denominator between the conditions favoured by the same species of tsetse in different regions of the country, certainly as far as vegetation was concerned. Or perhaps the common denominator was so common as to be useless for practical purposes, namely, shade.[52]

Bursell anticipated criticism that he was taking an extremist position from those of his peers who saw discriminative clearing as an imperfect yet still necessary method. Since 1959, he had dismissed the method as a "spectacular and very expensive failure"; each time, supporters of the scheme had given examples of success that had not been reported in scientific publications. In recent times, several researchers had shown that discriminative clearing may well have been a wild goose chase: "The apparent concentration of tsetse populations in specific parts of the general environment, on which the concept of discriminative clearing was based, may represent little more than an artifact of sampling. That it is not the whole population which is so concentrated, but only a small fraction of it, comprising males in a particular stage of the hunger cycle, when they happen to be particularly susceptible to sampling by traditional techniques. This discovery completely destroys the empirical basis of a control method which has no theoretical basis and which doesn't work."[53] Like the late burning and sheer clearing, therefore, the manner in which discriminative clearing was deployed appears to have been a misreading of *mhesvi*'s mobilities.

In directing its attack against the physical environment, the Tsetse Branch was attacking *mhesvi* at its strongest point, not its weakest. Instead of discriminative clearing, what was necessary, according to Bursell, was sheer clearing, the effectiveness of which was known and well documented. It deprived the *chipukanana* "completely of its requirement for shade and so cause[d] exposure of all stages of the life cycle to lethal levels of direct insolation." Bursell was not necessarily suggesting wholesale clearance of vegetation, but saying that "unless one does this, one may do nothing."[54] Any strategy had to be built on the vulnerable part of the life history of *mhesvi*, where the *chipukanana* was most closely dependent on its environment.

In other words, the attack on *mhesvi* was to be made not on *mhesvi* itself, but the environment; the attack therefore needed to focus not on the habitat of *mhesvi*, per se, but on that of its host. *Mhesvi*'s feeding habits, "its blood-sucking mode of life," involved "as many striking morphological and physiological specializations as any other function."[55] For example, these included the structure of its mouth parts and their efficiency in piercing the epidermis of vertebrates; the salivary glands and their secretion that contained an anticoagulant to prevent blood from clotting during its passage through the fine tracts of the alimentary system; the midgut, where proteases dominated the enzymes that digested blood proteins; the excretory system, which played a key role in the disposal of certain nitrogenous constituents of the blood meal; and the metabolic system, with its biochemically adaptive propensity toward a rich protein diet.[56]

The second critique of vegetation clearance came from *vatema* in *ndedzi*-infested areas. For them, trees were not merely "flora" and *sviharhi* no mere "fauna," but social institutions. Even today, stumps of the big, evergreen hardwoods of the *khaya* (*munanga*) cut down during the 1960s can still be seen along the Muchingwidzi and Runde rivers.[57] Figure 6.4 shows one of the trees cut down by the *ndedzi* people.[58] Some trees, like the *muchakata* and *marula*, were sacred; underneath them, *vanhu* held rainmaking and other ceremonies to commune with their ancestral spirits, and other riverine groves were burial grounds of their ancestors. Because the trees were protected, they grew densely, thus making them reliable refuge for *ndedzi*—especially *xindedzi xa nkova*.

Watson Machiukele was born in 1933 and was in his thirties when *vanhu vetsetse* (the tsetse people) arrived with their axes, chainsaws, *mabhurudhoza*, and chemical phytocides in the Chivonja area of Chipinda Pools on the Runde: "The people of Mafanele who were staying there had to be removed by *vanhu vetsetse*. They were cutting down the big trees, because they said

Figure 6.4
The stump and entire trunk of a very hard tree felled during the 1960s, still intact on the banks of the Muchingwidzi.
Source: Black Bvekenya Project 2011.

that is where tsetse lays its eggs. We were not happy about the tree cutting because the big trees that held the land together were being removed. The wind coming from the east was strong, and the trees were like mountains shielding us from it. Some of these trees were sacred; it was where we communed with our ancestral spirits. Tsetse destroyed the resources we used to communicate with our ancestors."[59]

With a sense of guilt, Machiukele continued: "I was one of the people involved in cutting the trees for *vanhu vetsetse*."[60] Machiukele's dilemma was that he could not stand up to *hurumende*. Without the money, he was certain to default on paying taxes, in a district administered by a man named Allan Wright, whose reputation for cruelty had earned from *vatema* the name *Chibwechitedza* (one whose heart is as hard as a rock and slippery as a pebble).

Conclusion: The Machine in the Garden, The Bulldozer in the Countryside

The concept of arboricide has been extended to account for the theory, practice, and instruments of destroying forests and rendering geophysical space into a means and way of controlling *mhesvi*. This is an instance in which means and ways are no longer just so-called expertise or physical, human-made artifacts, unless we are prepared to consider the land so cleared an artifact. There is no question now that *hurumende* relied on local *ruzivo* for its strategy to control *mhesvi* using arboricidal methods. Local people assisting Swynnerton, and Swynnerton himself, were the bridge that enabled the mobility of that *ruzivo* into *hurumende*'s official practices.

The chapter started to touch on the important theme of chemicals deployed against *mhesvi* that have since been classified as *almost certainly* carcinogenic. I emphasize *almost certainly* because of the way in which bench scientists are careful to avoid making definitive pronouncements based on built lab-generated or circumstantial evidence, absent actual field observations, fearing legal consequences. As will become even clearer in later discussions on OCPs, that reticence is due to fear of court battles, especially for scholars in the United States and Europe. The toxicity of 2,4-D and 2,4,5-T also raises two important questions in the context of *vatema*'s history. One relates to the health effects of these two substances when washed into water bodies; the other marks a shift in pest control from *vatema*'s organic means and ways and into synthetic industrial pesticides, with negative health and environmental consequences.

Industrial chemicals are one dimension of mass destruction; the other is "the bulldozer in the countryside" (Rome 2001) or "the machine in the garden" (Marx 1964). In the first instance, this narrative differs from Romer in its location—deep in the forest, not in urban sprawl. In the second, where Marx was concerned with technology's violent interruption of pastoral scenery as the United States industrialized, the focus of this chapter is located far from the whistling sound of steam locomotives that animated Henry David Thoreau's *Walden* (1854). Concern about the machine tearing through *sango* is inspired by Leo Marx, no doubt. However, whereas his machine was creating industrial space, my *bhurudhoza* is opening the land to motor vehicles carrying DDT and other OCPs; to fences demarcating *mhesvi*-infested from *mhesvi*-free land; and to destroy evergreen hardwoods that harbored *mhesvi* and which in some cases were sacred to *vatema*. Therein lies a shared interest in what Marx calls "the landscape of the psyche" (Marx 1964, 28) and the rationale for extending arboricide beyond instruments to theory and practice, to an ideology of *mhesvi* management by destruction rather than strategic deployment within a shared environment with other species.

7 *Cordon Sanitaire*: Prophylactic Settlement

"The alternative, which is generally to be preferred, is to use man to hold the land, or to advance into a fly belt," wrote Director of Tsetse Fly Operations J. K. Chorley in a 1953 article. "If the land is suitable the villager will clear bush and plant crops, cut wood for fuel and burn thicket for grazing; he will harry game and his goats and later his cattle will continue to hold down the thicket. In this way an area free of such a fly as *G. morsitans* can be established, indeed has been established."[1]

In the article, Chorley acknowledged Mzila's model of vegetation clearance and prophylactic resettlement as the future of operations to combat *mhesvi* and *n'gana* in Rhodesia: "This historical account is given not to indicate methods we should use, although some of our administrative officers may wish they possessed the powers of life and death enjoyed by a Zulu Chief, but as a challenge. What was done by a Zulu Chief 100 years ago can be done today by us with our infinitely greater knowledge of the tsetse's biology, of the cause and cure of the disease it carries and with our modern mechanical resources."[2]

The deliberate re-placement and overcrowding of *vatema* as forest-clearing agents and shields against *ndedzi* was handled in ways that cynically twisted Mzila's methods. The argument was made in scientized terms. On one hand, overcrowding *wantima* (blacks) with their *tihomu* (cattle) would overburden the soils and *svidvelo* (pastures) and "lead to widespread erosion, poverty and other ills," as Chorley had seen in parts of Tanganyika. On the other, too light a population density would cause minimal effect on vegetation and create ideal conditions for *ndedzi*.[3]

This chapter throws light on the main elements of this method, focusing on fencing, resettlement, and the experiences of resettled people. The argument is that *vatema* and their *zvipfuyo* were deployed as human and animate means and ways of pest control and an outer ring of early warning systems to protect white settlements and *zvipfuyo*. At the same time, fences were installed to channel the movement of *vanhu* and *mhuka* and sanitize it of carried *mhesvi*. The keywords in *chidzimbahwe* and other indigenous languages are listed in the glossary for easy reference.

Cordons Sanitaire: Fencing as a Prophylactic Infrastructure

Today, when visiting Nembudziya, people still talk of a road called *Eight Wire*, named in reference to a notorious eight-strand fence, the hardwood poles of which still stand between Chota, Nembudziya, and Gumunyu.[4] The name bears testimony to the enduring mark of the barbed wire fence as a means and way of controlling *mhesvi* and a marker of boundaries between infected and clean spaces. More importantly, it is a site where *hutsiny'e hwemabhunu* (the cruelty of the Boer) or *hudzvanyiriri hwevachena* (the downpression of white people) was felt in those moments when *vanhu* or *mombe* transgressed the wire. People were beaten up mercilessly, sometimes even shot—and not just here, but anywhere the fence of the *purazi rebhunu* (the Boer's farm) or *waya yehurumende* (government fence) existed.

By 1970, a principle of *cordon sanitaire* had emerged that was simple in its logic. Hunters went in first to clear game. Once shooting teams had cleared an area and *mafrayi* had certified it clear of *mhesvi* and erected a fence bordering the uncleared side (a game fence, locally called *fenzi yemhuka*), they proceeded to set up another fence on the side of *misha* (a cattle fence, locally called *fenzi yemombe*). This area would be clear of *mhesvi*. The space in between became the cattle free, game free corridor.[5] This corridor was the sanitized lane; the two fences on either side were the cordons or lines.

Sometimes, just to be safe, a third fence called the *middle fence* was erected as extra security in the event of game or *mombe* breaching the first lines.[6] Otherwise, the game fences were also deployed as "flanking" mechanisms to prevent game from escaping from killing fields,[7] and they were shifted further into new areas as hunters moved systematically forward.[8]

Flanking fences were erected to counter the advance of *mhesvi* in a set area by placing a formidable barrier. Many such fences had been erected in Gokwe since the 1930s, not least the one along the Mupfure River in 1930. However, it was common for *mhesvi* to also outflank the flanking game fences, as it did along the Munyati fence line in 1946, in the Ngondoma area.[9]

To be an effective barrier, the game-free, cattle-free corridor had to be of sufficient width. Already by 1932, experience in Nemakonde had demonstrated that "a 10-mile wide game-free cattle-free belt was insufficient to prevent all flies from crossing the area."[10] The game fence served the purpose of lineating (marking out lines to follow) and delineating (dividing up the land), thereby allowing the field teams to control vectors of *mhesvi*.[11] The object of fences was seen more as "defence rather than reclamation of tsetse infested country ... to put a greater distance between the fly limit and the occupied country."[12]

For these *vachena*, Mzila had achieved this aim very effectively through clearing a wide buffer zone, decreeing that his subjects draw near their king, and deploying armed patrols. Now, first Rupert Jack, then Chorley, and then John Ford all used wire fences. Materially, the game fences were "stout fences" made out of hardwood poles and eight strands of high-tensile steel wire. Steel corner posts, standards, and droppers came much later.[13] In the fenced area, all big *mhuka* could be held hostage and slaughtered; as they died out, *mhesvi* found no alternative food and also died.[14] Cleared areas could either act as buffer zones between infested and noninfested areas or as paddocks with cattle-dipping tanks constructed for veterinary disease control.[15]

Once constructed, these hundreds of miles of wire only stayed in place as long as needed. In 1941, the Public Works Department took down the game fence created in 1926 in the Kadoma area and used it for other purposes after the area was declared clear of *mhesvi*. The southern fence erected in Doma in 1925 was sold, and those established in Nemakonde Southwest in 1930 were dismantled at the end of 1941. Once *mhesvi* was conquered, new grid lines of wire fences were strung for the purposes of controlling stock movements and to prevent *mombe* from straying into or being deliberately grazed in *mhesvi*-prone areas.[16]

The department bought wire and nails, then either commandeered African convict labor or paid a pittance to dig holes, fell poles from the proximate *sango*, and erect the wire. Whether a private contractor or a government department was in charge of construction, *vatema* did the work.[17] In summer, a tractor with a hole-digging attachment was used, but as the season grew drier and soils rockier in places, *mafrayi* got down on the ground to dig manually with steel jumpers.[18]

The relationship between *sviharhi* and the (Savé West) fence is quite interesting. The fences initially suffered damage as *magocha* harried *sviharhi* toward them, but after a few months *sviharhi* were found "to move up to the fences, inspect them, and then move away."[19] Large herds of *mangwa* (zebra) and some *nyarhi* (buffalo) going to the Mkwasine to drink were "held up in their eastward movement for about two weeks, during which time some 300 zebra, six buffalo and one eland died near the fence."[20] Other *sviharhi*, like *ndlopfu* the elephant and *timhala* (impala), simply turned back or trekked southward toward the Chiredzi or Runde rivers. Still, many *sviharhi* that followed the fence eventually outflanked it, moving through or around the incomplete section. To address the problem of *sviharhi* following the maintenance roads leading to the game fence and thus getting stranded at it, the department began cutting its maintenance roads to run parallel to the fences. *Vantima* employed as "orderlies" (messengers) were also deployed on bicycles to patrol twenty-mile stretches of fence daily, taking note of breaks and the numbers and kinds of *sviharhi* involved. To increase the visibility of the fence and reduce damage due to animal movements, large, white-painted metal disks were suspended at intervals on the wires.[21] The appearance and noise of the discs startled *sviharhi*, and they subsequently steered clear of the fences.

Not all fences were erected to stop or canalize the movement of *sviharhi* or *tihomu*, but most were. For instance, in Ndanga East Reserve, a cattle fence was constructed running from the Savé westward along the Murondonzi River to meet with another fence running north to south. The fence was designed to prevent the movement of *tihomu* from Ndanga East into known *ndedzi*-infested areas in the south and to "restrict the wandering of the native population who could, and undoubtedly did, serve as *vehicles for the carriage of the tsetse*" (my italics).[22] With the fence in place, all *vanhu* and *tihomu* traffic to and from the reserve was now inspected for *ndedzi* at specific surveillance points in the fence, such as Ndari Gate. The Native Department also constructed another fence along the western boundary of the Ndanga East and Sangwe Reserves to further protect *tihomu* from straying into *ndedzi*-infested areas. Minor fences were also set up to direct pedestrian—and, to a lesser extent, *tihomu*—traffic toward the inspection gates.[23] The border fence was also erected to channel pedestrian and cyclist traffic in and out of Southern Rhodesia through "deflying" points.[24]

Local people paid £3 per month were recruited to erect *ndedzi* and cattle fences under the supervision and direction of a white man—as in the case of the *ndedzi* and cattle fences in Chibwedziva. First, teams cut trees to clear the path along which poles would be erected. The fence erected was *waya yemakurundundu yetsetse* (wire nailed to "crude" poles, with the bark not removed) because it was only temporary. Out in front, surveying and pegging the line that the fence was going to follow, was another white man, named Donati.

The man who was leading the fence gangs was locally called *Ngomakulu*, whose title was *baasboy* (the boss's boy, or "African assistant"). His name was therefore apropos: Ngomakulu (*isizulu* for *Ngomahuru*) in *chidzimbahwe* means *big drum*, which in *dzimbahwe* traditions was the megaphone or talking drum of the king (*mambo*, or *nkosi* in *isindebele*). Now *muchena* insisted on being addressed and treated as *nkosi*. Like most *baasboys*, Ngomakulu also was a powerfully built man—an insurance against mutiny among the fence-cutting gangs and to mete out instant justice, including becoming angry on behalf of the *nkosi*. *Ngomakulu* was a *mundau* from across Savé, whereas Donati was an Italian national hired specifically for the purpose of installing fences. Donati was in front with "African assistants," a team cutting trees behind him, another following up and stumping (*kugobora*), and, further behind, a team digging *goji* (holes) and installing *ntsandza* (poles), and, finally, the team inserting and tightening the strands.[25] This was in 1962, and bulldozers were on the Chiredzi, but not yet that far east; people there were the human bulldozers. They were paid £3 per month.

The fence was complete by 1963, and BTTC turned to erecting the game fence from the railway line to the Runde. District Commissioner Allan Wright commandeered *vantima* convicted for failure to pay taxes and for frivolous offenses to cut the fence line.[26] The corridor between the two fences became a game-free, cattle-free zone to break contact between *sviharhi* from Gonarezhou Game Reserve and *tihomu* in Matibi II Tribal Trust Land.[27] By 1968, this five-mile wide Guvulweni-Chepfu Tsetse Corridor, lying between the Runde and Mabalauta, had been hunted out. No *tihomu* were allowed here. The corridor's sole purpose was to stop the spread of *ndedzi* westward and to prevent *nyarhi* carrying foot-and-mouth disease and *n'gana* from exiting the game reserve and infecting *vantima*'s *tihomu* in Matibi II and, after that, *valungu*'s ranches.

Previously, only an old brush-pole game fence had run along the international boundary with Mozambique; between the late 1960s and early 1970s, it was replaced with two parallel, all-steel game fences set a mile and a half apart. The fences were composed of a 7 ft. high railway line straining posts concreted into the ground and carrying multiple strands of high-tensile oval steel wire. Occasionally, bull *ndlopfu* broke through, but the fences were generally too strong for *sviharhi*. There were unfenceable places that were too steep and inaccessible, such as Chilojo Cliffs on the lower Runde, where it was impossible to cut down trees or build fences; here, the BTTC resorted to aerial spraying of DDT (Thomson 2001, 12, 20).[28] The fences had achieved their purpose; neither *sviharhi* nor *tihomu* could cross the corridor, and at last the pestiferous mobilities had been tamed.

Resettlement as a Prophylactic

What was the purpose of creating a *cordon sanitaire* if the trees were going to continue providing habitat for *mhesvi* and if fences fell into disuse because of *majuru* (termite) attacks on poles, breakouts by *mhuka*, or theft of wire by locals? What would be a better way to create a permanent buffer zone between *mapurazi* (white farms and ranches) on one hand and *mhesvi* on the other than to settle *vatema* in overcrowded conditions so that they cut every tree, grazed the grasses almost into the ground every summer, and hunted out every animal? What could be a more perfect way to deny *mhesvi* its bloodmeal and its shelter? (For orientation throughout this section, see figure 6.1.)

In 1928, the Southern Rhodesia government introduced the first "antitsetse resettlement" scheme, under which *abantu abamnyama* or simply *abantu* (*isindebele* for *vatema*) were forcibly settled in the dry, *mpukane*-prone Gwai Native Reserve, squeezed into tiny land holdings while being granted free title to land if they moved to and stayed in the resettlement area for a considerable length of time. The government was hoping that title deeds would entice the massive numbers of *abantu* needed for settlement to act as an effective barrier against *mpukane*. However, the scheme proceeded very slowly, and the objective of using settlement as a *mpukane*-clearing strategy was not successful.[29]

The reluctance of *izinkomo*-owning people to settle in *mpukane*-infested areas stemmed from a long experience with this *isibungu* (insect), going back long before the coming of *amakiwa* (whites). They knew what the *isibungu* could do and were not interested in the white man's silly experiment. Of course, those without *izinkomo* had nothing (else) to lose. Only a few that were "used to" an *izinkomo*-less existence, having been forced to the inhospitable margins by the more powerful Ndebele and Tswana, could settle in such areas willingly. In fact, because these borderlands were rich in *inyamazana*, *inkulumende* found that most of these *abazingeli* (hunters) "indigenous to the fly areas show[ed] no desire to leave the infested country and in fact tend[ed] to drift back into it, if officially removed."[30]

Farther east, the Hurungwe resettlement scheme is an example of *vachena*'s attempts to introduce *vatema* with their *mombe* to "deflyed" areas to screen *mapurazi* (*vachena*'s farms) (in Karoi) from *mhesvi*-infested areas (in this case, the Zambezi valley). The scheme started in 1928. The following year, the *chipukanana* invaded the native reserve and inflicted heavy losses of stock among resettled *vanhu*.[31]

Another early anti-*mhesvi* shield was the resettlement of *abantu* and their *izinkomo* in the Kana-Shangani River junction in 1939, during a period of seeming success against *mpukane*. However, in 1943 the *isibungu* returned with much virulence, decimating entire herds and forcing the withdrawal of those still alive.[32]

As the situation improved, the government made land available for tobacco farms in the southern Hurungwe and Karoi areas near Magunje for white veterans returning from World War II and for postwar immigrants from 1945 onwards. *Vatema* living there were all uprooted and forcibly resettled with their four thousand head of *mombe* in *mhesvi*-prone areas of Hurungwe Native Reserve.[33] By 1951, over eight thousand head of *mombe* had been brought in.[34] The Tsetse Branch and Native Department insisted on swelling herds and locating the *mombe* stockades close to each other to reduce the distance between homesteads, fields, and *hufuro* (pastures). However, the authorities left no room for sons becoming adults, marrying, and needing land to start their own *musha* (homestead) and *mhuri* (family). Already overcrowded at the time of settlement, the reserves could not sustain the rising population as it grew from two million to three million in the 1950s. The government started subdividing the reserves even further, to a point where urban-based men returning from work arrived at month's end to find their land holdings reduced (Palmer 1977, 243–244).

Prior to the Native Land Husbandry Act (NLHA) of 1951, the government designed two main types of land use: the block system and the unit system. The *blocks* were large areas of arable land hundreds or thousands of acres in size, surrounded by correspondingly larger grazing areas, with homesteads and *matanga emombe* (cattle stockades) along one or more edges of the block. The system was considered undesirable for two reasons: First, it was deemed inefficient in terms of manuring (fertilizing) the land in the middle of the block with dung that *mombe* excreted in the *hufuro*. Second, it was seen to canalize *mombe* traffic along the fence line and to water points, causing serious erosion.

By contrast, the *unit* system had much smaller patches of arable land separated by grazing veld, serving just a few families clustered in *misha*. This system enabled easier access to land and grazing and reduced heavy tracking and erosion. The government preferred it for Hurungwe Native Reserve not only because large tracts of arable land were at a premium, but also because it suited the close settlement essential for keeping out *mhesvi*. However, a density of six families per square mile was the heaviest concentration possible—inadequate for controlling *mhesvi*. In any case, the unit method was not applicable everywhere. Some areas with fertile soil not only could sustain more *vatema* with limited acreage, but also supported good natural vegetation ideal for heavy *mhesvi* concentrations. In such areas, block settlement, with dwellings, water points, and grazing corridors, was preferred to achieve intensive settlement with limited erosion and tracking.[35]

The new arrivals were immediately subject to agricultural extension work and the application of NLHA, which emphasized strict conservation methods and "good" farming practices, enforced stock-to-carrying capacity ratios, individualized grazing rights, and compact land units registered in every individual's name. To combat erosion while still achieving close settlement against *mhesvi*, care was taken to ensure against heavy trampling of paths by *vanhu* or *mhuka* traffic, especially on tracks leading to and from water supply points.[36] People remember two draconian aspects of *nhimura* ("the slashing," their term for NLHA): forced destocking or limits to their herds, and *makandiwa* or *madhunduru* (contours) designed to arrest soil erosion.

In 1950, disaster struck. After game destruction, *mhesvi* had adapted to a new and timid host and blood source: *mombe*. It thrived. The first *n'gana* cases were confirmed in February 1950 at a village in northern Hurungwe. Apparently, the beasts had strayed into *mhesvi*-infested country. Things seemed to be under control; chemotherapy was administered, and the strain was stamped out. Then, in May, more outbreaks—this time in the southwest—left five hundred *mombe* of *vatema* dead. The Veterinary Department and trained *vatema* working for the Native Department moved in with chemoprophylactics, but the respite was short-lived. By mid-1951, *n'gana* covered an even wider arc, killing more than 2,300 *mombe*, three hundred cases each month, and reaching the white farms of Karoi, east of Hurungwe Native Reserve.[37]

Only at that point, after *mapurazi* (white farms) recorded *only* eighty-three cases and two *mombe* deaths, did the implosion become a state of emergency. The government immediately resolved that the Hurungwe Native Reserve be evacuated of all *mombe*. The movement was planned for August and September 1952. In its aftermath, a multipronged strategy was put into operation. *Magocha* were deployed to intensify "game destruction," while TFOs and private hunters were given incentives for slaughtering *nzou*, including keeping its ivory. This effort to starve *mhesvi* occurred alongside an assault on *hutachiwana* with chemotherapeutic interventions throughout the affected areas.

Three fences for which construction began in September 1951 were completed in May 1952: one game fence along the Hurungwe-Gokwe boundary on the Sanyati, one strong farm fence along its eastern boundary, and one rough *mombe* fence north to south straight through the middle of the reserve. Other cattle fences were later erected along the northern and southern boundaries, thus completely enclosing the reserve and turning it into a vast game-free, cattle-free area.[38]

However, this effort was wasted. By July 1952, just 5,500 of the 8,000-strong herd were still alive. A new problem arose: *Hutachiwana* was becoming resistant to drugs, principally dimidium bromide, turning *mombe* into a vast reservoir of drug-resistant *hutachiwana* for *mhesvi*. The government decided to arbitrarily evacuate all *mombe* to the northern banks of the Mupfure River and force the owners to look after them there, leaving only *mbudzi*, *makwayi*, and donkeys, believed less susceptible to *n'gana*. The Provincial Native Commissioner (PNC) explained it thus to *vanhu vatema*: "These cattle will only be *lagisa*'ed ... on the north bank of the Umfuli, and owners must group together for herding arrangements."[39]

As we discussed in chapter 1, the Ndebele, Tswana, and *vedzimbahwe* deployed *ukulagisa/kuronzera* as a pest-management stratagem; here, we note *hurumende* doing exactly the same, directly referencing the herding arrangements as *ukulagisa*. We also discussed people keeping *mbudzi* and *imbwa* in the *mhesvi*-infested areas of the Zambezi. The difference in approach is the arbitrariness, with *hurumende*'s actions being not preventive but for damage control. *Mhuka* "unfit to be moved" were taken over by the government at dipping tanks or sale pens, the owner being paid "at compound grade and estimated live weight." The animal was slaughtered on site, the carcasses removed and buried or burnt as far as possible or converted into biltong (dried meat). The compensation rate for *n'gana*-related deaths was set at three pounds per beast, which the PNC deemed "a fair one."[40]

But what constituted "fair" when *hurumende* arbitrarily used only a monetary or property value for *zvipfuyo* that *vatema* also valued in spiritual, social, economic, or other ways? They felt *mombe* were priceless; to remove them from *vatema*'s lives was to disarm their owners of a critical spiritual and social armament, to rip off not just the flesh, but also that which conjoined mortal and ancestor. The PNC stated: "We are going to make special arrangements in regard to your agriculture and your ploughing during the time that you cannot have cattle and these are being worked out."[41] Yet, as Chorley admitted, "Many African people have a close attachment to their cattle and are unwilling to be separated from them. They prefer to stay with their cattle and see them die rather than be separated."[42]

Vatema's Experiences of Prophylactic Resettlement

During the 1960s, after deciding that Gokwe District was sufficiently cleared of *mhesvi*, the government embarked on a propaganda exercise to persuade *vatema* who felt overcrowded in the western, southern, and central provinces of Rhodesia to resettle in the district. This section focuses only on some of those who came from Bikita District (Fort Victoria District; now Masvingo) and Charter (now Chivhu) District to settle in *Ishé* (Chief) Nembudziya's area. The government provided lorries to ferry these families. To the north of Nembudziya along the Zambezi were the local *vechishangwe* (the *shangwe* people), so-called because of their tendency to farm in the valleys. Remembers Reuben Mavenge: "We called them *vechishangwe*, which they hated, saying 'We are *vakorekore*, don't call us *vechishangwe*, *shangwe* is a place,'" they would protest angrily[43] (see also Nyambara 2001, 2002; Worby 2002).

The first group of immigrants from the south arrived in 1963 from Bikita and the Chivhu-Sadza area. Others also came from Marozva in Bikita, where they had lived by the generosity of the Duma under *Ishé* Marozva. However, because most local land was seized by *vachena* and parceled out into *mapurazi*, Marozva now wanted land he had given to the Murozvi chief, *Ishé* Gumunyu, back. Therefore, Gumunyu, along with his two siblings, Jiri and Masuka, left to settle in Hurungwe and Gokwe, separated only by the River Sanyati (also called Munyati further upstream). Two other sons of Tohwechipi, Ushé and Makotore, remained.[44]

Others, such as Raymond Muzanenhamo, born in 1942, came from Chivhu and settled first in Chief Chireya's country, then in Mhondoro, then in *Ishé* Neuso's country in Sanyati, before finally arriving in Nembudziya.[45] Still other groups came much later, in the 1970s, after the initial groups—who became *vekupureya* (spraymen), *magocha*, and *mafrayi*—had long been settled.[46] When these immigrants arrived, there were no people living in Nembudziya—bar *vechishangwe*, Ishé Dandawa's people of Korekore lineage, who lived in the vicinity of the Gandavaroyi Hills, named after the sacred waterfall and pool into which those convicted of witchcraft were thrown alive (Mapara and Makaudze 2016). Most of these arrivals from Bikita were Rozvi people, descendants of Chirisamhuru and siblings of Riwanika (Lewanika), who had crossed into and settled in what became Barotseland (Varozviland).[47]

Just as in the adjacent areas of Lupane and Nkayi, the newcomers saw themselves as more "modern" because they had exposed themselves to large-scale farming and Western equipment and machinery; they were organized into cooperative societies, guaranteeing them capital and technical support; many held master-farmer certificates and grew cash crops like cotton and introduced their production to Gokwe; they were members of the nationalist political movements; and so on (Nyambara 1999; Alexander and Ranger 1998). *Vechishangwe* called these strangers from the south *madheruka magochamiti*. *Kudheruka* means "suddenly showing up without invitation or forewarning," which is what these *strangers from nowhere* did.

Magochamiti came from these *madheruka*'s practice of cutting and burning trees on the uplands to clear the land, kill pests, and produce ash fertilizer—all activities preceding the planting of crops. By contrast, *vechishangwe* had no need for these activities because they planted crops in the riverine valleys.[48] Here we have two identities, two encounters, based on where each farmed: the upland forest, favored haunt of the *mhesvirutondo*, and the riverbanks, preferred habitat of *mhesvirupani*. As land-clearing agents, *vechishangwe* and *madheruka* complemented the control of *mhesvi* perfectly—at least on paper. Some *madheruka* chose to settle in the fertile soils of the Sanyati Valley. They were warned that they would die of *nyong'o* (malaria), but went anyway; they lost all their children there.[49]

The new area was thick with *mhuka*, among them *nhéma* (rhinoceros), *nzou, nyati, dzoma, nhoro, njiri*, and *nguruve*.[50] However, strong in their faith in ancestral spirits, both *vechishangwe* and *madheruka* had no reason to fear these *mhuka*, particularly *nzou*. They say the animal did not bother anyone who meant no harm to others; it reserved its ire for murderers, prostitutes, philanderers, and those who dabbled in bad medicines or witchcraft to harm others.[51] Said one elderly woman in March 2016: "If you are an evildoer—then yes."[52] *Nzou* did not get into people's fields, but would go around the enclosures. The violence of these *mhuka* as the century progressed thus is not hard to explain: "We have followed those people, the white people, who destroyed our *hunhu*, we threw away our *chivanhu* [culture], our *vadzimu* [ancestors] have abandoned us."[53]

Inevitably, the presence of *mhuka* and movement back and forth between cleared and infested areas meant *madheruka* faced the problem of *mhesvi* and *hutunga*.[54] Because of the *mhesvi* presence, *hurumende* banned all *mombe* from Gokwe. Only donkeys were allowed, and only a few people, mostly among *vechishangwe*, had them. Otherwise, most *madheruka* relied on tilling with a hoe, or zero tillage. Without draft power, plows were not even necessary; the farming was thus limited to homesteads and small gardens near rivers.[55] *Ishé* Gumunyu later owned a tractor, but did not plow for everyone or always—just those who performed *magobo* (stumping) to clear the chief's fields in return for tillage. The equivalent of the land that one had stumped was plowed.[56]

On the other side of the Sanyati River, the district commissioner for Hurungwe had commandeered *vatema* to build a big kraal in which all *mombe* were kept and pastured safely from *mhesvi*, but no *mombe* were allowed in the homesteads—just the donkeys.[57] Those who had donkeys used them for plowing. A plow could be purchased for £2. 5s. at Gokwe Center, and the donkeys were bought in Makonde and Hurungwe. As we discussed, donkeys were more resistant to the bite of *mhesvi* than *mombe*.[58] It is the common understanding locally that "the donkey would survive when bitten; the blood of a donkey is stronger than that of cattle. That of goats is stronger than that of cattle."[59] Strength here is measured based on resistance to disease.

Madheruka and *vechishangwe* were only able to keep *mombe* after the civil disobedience campaign, when the Zimbabwe African National Union (ZANU) leader, Ndabaningi Sithole, toured Nembudziya.[60] At that time, nationalist leaders, who included Robert Mugabe, were detained at nearby Sikombela Detention Camp (see figure 7.1), with rights to visit the surrounding areas and conduct political activities. Sikombela itself was well within the *mhesvi* belt; it was, along with Gonakudzingwa in southeastern Zimbabwe (Mavhunga 2014), strategically designed to dump these "hotheads" in inhospitable, animal-infested forests to "cool off." This form of prophylactic settlement was designed to isolate the vocal elements of the nationalist movement from the cities, but they ended up subverting the entire countryside. In the end, they had to be moved to maximum security prisons further inland.

When Sithole arrived in Nembudziya, he found that all people had were *mbudzi* and donkeys. He said: 'Why do you only have *mibhemhe* and *mbudzi*? Why not *mombe*?' And the people said: 'There is *mhesvi*, and the government has said *mombe* can't enter because they will all die.' Ndabaningi said: 'No ways, let them die while you at least have the opportunity to eat meat. Find *mombe*.' That is how people started keeping *mombe*. The white veterinarian named Johnson was a thoroughly despised man, and people worried he would have them all thrown in jail—but *mombe* were now there to stay. People began plowing larger acreage.[61]

Madheruka had never known *mhesvi* in Bikita—at least in their lifetime. Thus they had no *ruzivo* on how to prevent it from biting them and, once it did bite, how to treat its effects. The locals relied on *mafrayi* at the tsetse gate to prevent *mhesvi* from coming in and escalating the situation. "It bit you until it was full then left you," one said.[62] The *mhesvi* in the area apparently did not transmit *gopé*—only the painful bite and *n'gana*. Says one elderly woman who arrived with the first emigrants from Bikita: "*Mhesvi* terrorized people. Do you know that if it bites you, you feel like you have been pierced by a needle? Yeah, it pierced like a needle, looking for your blood, to suck so that it fills its stomach."[63]

Figure 7.1
Inmates at Sikombela, including Robert Mugabe (foreground), reading books in the 1960s.
Source: The Sunday Mail (March 6, 2016).

There were many *mapere* (hyenas) in Nembudziya, and they preyed upon *mbudzi* and *mibhemhe*. "So," the same old woman continued, "to safeguard these *zvipfuyo* from *mapere*, you would sleep in one room, your *zvipfuyo*, and you. *Mhesvi* would follow *zvipfuyo* that had entered the house. Particularly the goat pen; that was the most tsetse-infested."[64] People had no toilets, and they relieved themselves in the bush—and when they "went to the bush" (*kuenda kusango*), *mhesvi* detected them and followed. Killing the *chipukanana* was impossible because it kept shifting places, each bite feeling like a razor cut. *Mhesvi* bit by day, *hutunga* at night.[65]

There is one known case of suspected *gopé* in 1968, involving Ambuya (Grandma) Misi, wife to *Ishé* (Headman) Misi. She says when she was bitten at Dandawa, she developed *mapundu* (boils), had a devastating fever, and a persistent sleep. She was admitted to a hospital, given two injections, and placed on intravenous fluids for three days. She only woke up the third day, finding a nurse at her bedside, who told her what had happened since she was rushed in by ambulance.[66] Otherwise cases of *gopé* in Nembudziya were rare. "You died of other things," said a neighbor. "What killed people was *nyong'o* (malaria) caused by *hutunga*, not *mhesvi*."[67] Yet that is true only for areas to the south. The further north people went, the nearer they came to the shores of Kariba, the source of several sleeping sickness cases—including fatal ones—throughout the 1960s. That is where Dandawa is located.[68]

Clinical medicines came to Nembudziya much later, in the mid-1970s. Up until then, people traveled all the way to Gokwe Center or Sanyati to be treated, which made traditional medicines very important. *Madheruka* arriving in *vechishangwe*'s country first knew about *mhesvi* when they left Bikita and Chivhu, which were much colder and more elevated, whereas Shangwe country (Gokwe) was very hot and at a low altitude.[69] When *madheruka* are asked what traditional medicines they used against *mhesvi*, the answer is standard: "We had no mechanism to prevent tsetse from biting us."[70] Evelyn Musengi expresses *madheruka*'s complete dependence on clinical medicine in this way: "Unlike *vechishangwe*, we knew absolutely nothing about *mhesvi* and therefore had no *ruzivo* of herbal medicines obtainable from the forests."[71]

Vechishangwe's intimate *ruzivo* of herbal medicines, strategic deployment within the environment, and inoculants was based on long residence in the *mhesvi*-infested areas.[72] As relations improved, *vechishangwe* taught *madheruka* the names of key herbs and medicines derived from them. One such plant was *zimunhuwenhuwe* (smelly plant), which looked like sweet potato and smelled like *tsvina* (human excrement). The medicine was fed to the patient through the rectum and acted as a purgative.[73] To protect against *hutunga* and *mhesvi* accompanying *mbudzi*, people placed *mbudzi* dung on top of burning charcoals so that the smoke would act as a repellant against the pests. Where *zumbani* (eucalyptus or mint) was available, people would stick it into the wall (where grass thatch-roof meets walls), or put it on burning charcoals to smoke *zvipukanana* out of the house or suffocate them.[74] The occupants returned a while later, after *zvipukanana* were dead or gone.[75]

Once bitten, *vechishangwe* had yet another therapy for *gopé*: eating a very hot pepper. They would crush it, put it in a cup, and drink it. It served as an emetic; when the patient vomited, relief would come. The same medicine was applied against *nyong'o*; the patient would vomit the offending yellow substance after which the fever was named.

Nyong'o must be understood within a larger (spi)ritual context. In *dzimbahwe*, land was not just a geophysical expression; *dzimbahwe* was a "supra-spiritual commonwealth" that fell under five territorial spirits, complete with subordinate structures: the Matopos *Mwari/Mlimo* (Ranger 1999; Werbner 1989; Daneel 1970), Mutota/Nehanda (Lan 1985; Mudenge 1988), Chugumbi/Dzivaguru (Mudenge 1988; Bourdillon 1978); Musikavanhu/Chapo (Rennie 1978), and Nevana (Alexander and Ranger 1998; Tapson 1944). The latter was in the Gokwe and Nkayi areas—which was Sebungwe under Rhodesian rule—home to *vaTonga*, *vaRozvi*, *vaNyai*, and *vechishangwe*. Big ceremonies commemorating the start and end of harvests were intended not just to thank the spirits, but also to ensure good health.

In the entire belt from Gokwe to Lupane, endemic seasonal fever was called *nyong'o* (*chidzimbahwe*) or *inyongo* (*isindebele*). *Nyong'o* was "a non-fatal disease of the rainy season attributed to gorging on the first fruits" (Alexander and Ranger 1998, 223). That is why the festival of the first fruits every year was held with offerings to the spirits, who—along with kings and chiefs—saw to the management of all pestilence within their territories. *Nyong'o* was blamed on eating "fresh, sugary and green foods such as watermelons, sweet reeds, greens and pumpkins," not *hutunga*, which were repelled by burning or rubbing "strong-smelling herbs and leaves," not least *msuzwan* or *mutandamsenya* (literally, "a very smelly log"; 224) (Lukwa 1994). *Hutunga* themselves were not killed nor malaria prevented. *Mombe* too suffered from *nyong'o* when changing from eating dry winter grass to fresh green grass as the rains began. *Nyong'o* was also found in the air, water, soil, and vegetation. It was treated with bitter herbs deployed as emetics and purgatives, to cleanse and revitalize the body. Two other medicines, *mukombehwa* and *murumanyama*, were taken when a person fell ill. The medicines would be put in water, and a big stone placed in *moto* (fire). This stone would then be put in a dish containing the water, and the patient would go on all fours over it, the whole body and the dish being covered with a blanket. The patient was supposed to open his or her mouth and inhale the medicated steam, almost to the point of passing out, before being taken out from under the blanket and placed in the shade to recover.[76]

There was no hospital in Nembudziya until one of these *madheruka*, Cleto Zharare, took the initiative to build one with his own savings from his psychiatric nursing job at another foundation started by *munhu mutema*, the priest named Jairos Jiri. The story of Zharare Clinic is outside the scope of this book, but it speaks to an overlooked theme in the history of knowledge, means and ways, and innovation under Rhodesia[77]—namely, that of *vatema* who built and ran educational, business, technology, public health, and scientific infrastructures such as clinics and grocery stores for their own communities. There was nothing political about what they were doing—just the imperative to take risks, make money, and improve the lot of their own people.

Conclusion

In the introduction, I signaled that *mhesvi* forced the Rhodesian state to deploy *kugarisika kwevanhu* (human settlement) as a prophylactic structure against it. In beginning this chapter, I highlighted that people removed to these *mhesvi*-infested margins considered themselves ejected to live like other *mhuka*—as *mhuka*. They felt like *dirt*, to use Mary Douglas's term. But Douglas was thinking of dirt from the eye of the beholder—namely, the perception of something or somebody as dirt. I am talking about the feeling of being treated as dirt (*tsvina*), what it felt like for *vatema* to be ejected from their ancestral lands by *vapambevhu* and forcibly resettled as *madheruka* in a place befitting dirt. This is called *kubatwa setsvina* (being treated like dirt). I have shown that contrary to Douglas and Wildavsky (1982, 102), the borderland asserts its presence to *hurumende* because of *mhesvi*. Once resettled on the margins, *vatema* cannot be left on their own with *mhesvi* lurking; they must be controlled just like the *mhesvi* so that the insect cannot breech their villages to reach *vachena*'s heartland.

The irony of prophylactic settlement is precisely that it was *vatema*'s idea, now deployed to displace them from their lands and turn them into a preventative means to fight the encroachment of *mhesvi*. This chapter has traced the direct mobility of prophylactic settlement (as an intellectual idea and a practice) that *vatema* practiced to the control of *mhesvi* under the regime of *vachena*. This is quite contrary to the work of Kjekshus ([1977] 1996), who sees the advent of *vachena*'s regime as destroying rather than appropriating *ruzivo rwevatema* to control the environment. This does not mean that *vachena* took all *ruzivo* at face value or that no *ruzivo* and practices were destroyed; instead, this is a call for more careful readings of moments of knowledge translation, which we will not see if we read too much into the civilizing mission narrative.

Appropriating the *ruzivo rwavatema* while turning them into surveillance equipment and land-clearing machines, then spreading the propaganda of the Rhodesia project as introducing knowledge and civilization from Europe, exposes Europe's imperial project in Africa as a fraud. It shows, yet again, how the settler project was built on *ruzivo rwevatema* and (not just) the labor of *vatema*—and that is one of the least explored secrets of Europe's occupation. The fraud was sadistic: taking ideas invented by *vatema* and using *vatema* as guinea pigs to ensure its success. No *vachena* were settled in these *mhesvi*-infested borderlands—only *vatema*, who under *hunhapwa* were designated, as we discussed in this book's introduction, as eugenically inferior. Their lives could be experimented with, and if they died, it would not be homicide; they would have succumbed to other *mhuka*. It was survival of the fittest out there. "I am fixed. ... I am laid bare," Fanon said ([1952] 1967, 115–116). Reduced to a contrivance, a device against pestiferous *zvipukanana*, the status of *vatema* as instruments was confirmed. Cabral (1974, 30) was right: "To co-exist [with *vachena*] one must first of all exist."

The role of *vatema* in prophylactic settlement was now that of "an instrument of production," what Aimé Césaire called *thingification*—the transformation of the black person into a thing—in this case, a machine or "an instrument of production" (Césaire [1955] 2000, 42–43). Robbed of the *ruzivo* now deployed to make him an instrument, the deintellectualization of the black person was complete.

And yet!

Always, in these moments of utter despair, I look for moments of creative resilience. Of "African nationalists" dumped at Sikombela to vegetate, only for them to fan out into the countryside and subvert it in defiance of *hurumende yehudzvanyiriri* (the oppressive state) to embark on *kuzvisunungura* (self-liberation). Of *vechishangwe* deploying their *ruzivo* of medicine and their spiritually anchored practices to deal with *nyong'o*. Of *madheruka* that extend the *ruzivo* they have appropriated from *vachena* on the central watershed to their new home, where they can be seen engaging in thriving cotton production, well-organized cooperatives—and building a clinic when *vapambepfumi* have left them to the mercies of *mhesvi* and *hutunga*.

Vatema at work, rehumanizing themselves, reintellectualizing themselves—turning extreme adversity into a future for themselves and their children.

8 Traffic Control: A Surveillance System for Unwanted Passengers

Ke nna lexokolodi le leso leputlelele la nkô ye nthso,	I am the black millipede, the rusher with a black nose
Se-nwa-meetse le 'dibeng tsa baloyi.	Drinker of water even in the fountains of the witches.
Wa re ke tla lôwa ke mang?	And who do you say will bewitch me?
Ke paletse le-ija-motho le 'fsifsing la nkata,	I triumphed over the one who eats a person (the sun) and over the pitch-black darkness,
Mo dinkatawana le dinyamatsana di bokollaxo madi bosexo le mosexare.	Where the carnivorous animals drink blood day and night.
Ke nna lexokolodi le lese lepopoduma le dumêla teng.	I am the millipede, the mighty roarer that roars within.
Baxêxo ba nhteile ba re ke nna Ke-sa-ya-Borwa.	My people have named me, they say I'm-still-going-south.
Ke hlanamile xa e se nna marwala-dithoto,	I have changed, I am no longer a carrier of
Namana ye nthso yaBorwa,	goods, the black calf of the South,
Ke nna moloyi-moso	I am the black witch
Moloyi wa bosexo le mosexare,	The witch of day and night,
Ke nna Ramaêtô setsubalala lesokeng, mohlôya-tsela,	I am a traveler the vigorous rapid one and hater of the road,
Ke nna lexoletsa mollô teng.	I am the one that kindles fire in the stomach.
Ke laditse pitsi kxang	I have won the horse,
Re ile re siana ya re ke lebelô, mohlaba wa re ke nabile,	When we raced I was the fastest, the sand filled the air
Ka feta nna namane e nthso.	And I passed, I the black calf.
Kei le ke fihla motse-molla-kôma,	I arrived at the place where the circumcision drum was beating.
Ba re ba mpotsisa, ka re ke tswa xa ntintilane,	When they asked me I told them that,
Ke tswa setsibye,	I come from a place which nobody knows,
Ke tswa naxeng tsa kxole.	I come from the unknown, from a far-away country.
Ba re mphaxo O tla tsea wa eng?	They asked me what kind of provision I would take?

Ka re xa ke tseye mphaxo ka etsa
mafsêxa a a xeno
Nna ke lalêla ka tlala mo-ja-o-sa-hlalle.
Kei la mathudi boxadi bya Ramaêsela,

Xe nka hwetswa mathuding
mokxolokwane ó ka lla
Wa etsa sebata-kxomo xe nkwê e swere
ya mosate.
Xaxeso ba nhloboxile.

Xa se nna ngwana-lapa,
Ke lexokolodi le tumisa khuiti,
Xa ke ditelwe ke tlala,
Ke ditelwa ke bana ba naxa;
Xa ke ditelwe ke maoto-bohloko,

Le xo loba ke loba xo bôna
Xa ke rate xo huêla dikôma
Dikôma xa se tsa bo motho;
Baxeso ba itaeletse xe ba ntesa k aba
lexwara-xwara.

Metse nkabe e se ya thopya

Nna sexakalala mohla motse ó eme ka
dinao,
Naxa e re: 'Ke tla ba khutisa kae
mafsexa a?
Ke dula ke le dihlako mo tseleng

Ke wêla-wêla mekoting
Ke etsa noka xe e êla,
Ke rwele motse wa monna yo moso
Ba ka ntirang benye-tsela

Nna lexokolodi
Le leso se polakêla-dinakô/
(Demetrius Segooa, in van Zyl 1941,
130–132)

And I said I do not take provision like
these cowards of yours.
I sleep without food, I, the omnivorous;
I shun the verandah where Ramaêsêla is
married,
If I be found on the verandah a
triumphal outcry will be heard
Like the great cry when the leopard has
victimized the royal animals.
At my home they have lost all hope of
ever finding me
I am not a house-child,
I am the millipede that praises the vlei,
I am not delayed by hunger,
Nor am I delayed by sore feet,
But I am delayed by the children of the
wild;
To pay tribute, I pay tribute to them.
I do not want to die for the sacred
The sacred belong to nobody;
My people have committed national
suicide by allowing me to become a
deserter
Villages would not have been taken
into captivity.
I the brave, when the village stands on
its feet (in danger)
And the country says: Where shall I
hide them these cowards?
I remain with my feet on the road
(always travelling)
And go falling-falling into the dongas,
I imitate a river which is in flood
Carrying the village of a black man.
What can they do to me, the owners of
the road
To me the black millipede
That rushes for scheduled times?

This is a *setswana* poem about the train, likened to a black millipede traveling in very difficult conditions, such as the heat of the sun and the thick, impenetrable darkness of the night, the hills, and the mountains. Its journeys are endless; surely, it also has to be tired like every other being. As something that conveys all kinds of things to their destinations, the train calls itself a being that consumes everything (van Zyl 1941, 153). It is vulnerable here to *vatema*'s poetic innovation, to *vatema*'s intellect. Why, its many wheels are feet—so many, like a *xekolodi*'s! From that perspective, the *setswana* poem would fit perfectly within emerging portraits of the locomotive and automobile: as cultural objects and spectacles. Trains and railroads, bicycles and cycling, and more recently airplanes and cell phones (Schivelbusch 1977; Mom 2004; Seiler 2008; McShane 1994; Sheller and Urry 2004 and 2006; Creswell 2006; Cwerner, Kesslring, and Urry 2009)—these narratives of means of transport and communication have dominated even the so-called new mobility paradigm (see Mom et al. 2011). In these emerging narratives, if it is not the means of transport or the physical infrastructure that carries it, it is the traveler—the human traveler.

Elsewhere, I show how ordinary people in Mozambique—and, indeed, Zimbabwe today—have turned the road into a *thriving, transient marketplace* (Mavhunga 2013, 2014). People can be seen bringing to the roadside all kinds of merchandise to sell: charcoal, chicken, vegetables—anything that might tempt the motorist to stop. The human dimensions and meanings of cars to *vachena* and *vatema* have only begun to be explored (Gewald, Lunning, and van Walraven 2009; Green-Simms 2009; Hart 2016). To be fair, the importance of the car as an historical element in *vatema*'s experience—or perhaps the experience of the car in Africa—had been signaled as early as 1986, but apparently the call was not followed (Kopytoff 1986).

Much of the recent transport scholarship on Africa focuses on human-fabricated and inbound modes of transport (cars, trains, airplanes; e.g., Gewald, Lunning, and van Walraven 2009; Pirie 2009). The railroad and road literature has dealt extensively with construction, with teams of African forced laborers cutting and digging through thick forests and *hutunga*-infested swamps to build roads, railroads, and later airstrips to host these incoming Western artifacts (Akurang-Parry 2002; Akintoye 1969). Because they were press-ganged into this arduous work (Heap 1990, 2000; Akurang-Parry 2000; Law 1989; Machin 2002), the majority exercised several stratagems to escape their conditions: migration to neighboring territories ruled by a different European country, temporary flight into the bush, and downright sabotage of bridges, roads, and railroads (often named after "important" *vachena*, for all the back-breaking slavery that went into building them by *vatema*; Likaka 2009).

Once built, the railroad tracks became material extensions of *vachena*'s territorial aggression, linking labor reservoirs to mines and farms, and these sites of production to coastal ports, from which minerals, cash crops, rubber, and timber were shipped to factories in Europe and the United States. The existing literature does not make explicit this point about the outward-facing nature of railroad infrastructure that *vachena* designed, and *vatema* built (Robinson 1991; Bekele 1982; Dubois 1997; Pirie 1993, 1997). By contrast, Tanzania and Zambia's ambitious TAZARA railway line built in 1970–1974 was aimed not only at connecting the two countries, but also at facilitating the shipment of freedom fighters and matériel for the liberation of Zimbabwe, Namibia, and South Africa (Monson 2009). Today, the view of *vatema*'s heads of state is that China is helping Africa build infrastructure to link African countries to each other, where *vachena* were only interested in extractive infrastructures facing toward Europe (Foster et al. 2008).

The research on cars is still only beginning, but substantial literature exists on roads. Like the train, the coming of the automobile led to massive conscription of *vanhu vatema* as "road-cutting gangs" (Zhao 1994; Chiteji 1979; Sunseri 2002). Contrary to earlier research however, such road work affected and involved not just the men conscripted but the women left at home who fed them, who Kathleen Sheldon (2002) calls the "pounders of grain." At the construction site, Landeg White (1993) has superbly captured the drama of *vachena*'s bridge construction. The negative impact of roads (displacement—making it easier for *vadzvanyiriri* to downpress *vatema* even more) and their benefits (easier transportation) have received some attention from historians and policy studies (Stephens 1994; Chilundo 1995). Historical studies of automobiles themselves only began in the last decade, with a focus on the car, bus, lorry, and motorcycle as environmentally, economically, culturally, and politically transformative means and ways (technology) (Gewald, Lunning, and van Walraven 2009; Pirie 1993, 2011; and several other articles).

This chapter takes the mobility discussion in a totally different direction—away from trains, from *vanhu* (humans) and means and ways as the central actors, to *mhesvi* subverting the transport systems that *vanhu* contrived. This is to further the thesis of this book—the idea of *mhesvi* as mobile workshop, this time as a passenger taking a ride on pedestrians, disabling ox wagon transport, riding on automobiles and on bicycles, and forcing *vanhu* to institute mechanisms and infrastructures of traffic control. The glossary at the back of the book should help the reader understand *chidzimbahwe* and other regional keywords.

How *Mhesvi* Determined if a Person Rode or Walked

Prophylactic settlement could only be effective with good control of foot, bicycle, and automobile traffic in and out of *mhesvi*-infested areas. Herein lie profound connections between mobilities of *vanhu* and mobilities of *mhesvi*.

The tapestry of footpaths illustrates the role of foot transport as a mode of conveyance from place to place and of haulage transport, especially for trade, migration, and military expeditions. It shows that footpaths were the first roads of Africa, ox wagons and palanquins (*machila*, or hammocks) among the first "cars" (i.e., if we take *car* to be shorthand for *carriage*). *Vatema* not only physically carried the white man's burden; they also carried the white man himself, as a burden reclining and dozing off in *machila*. Later, the *machila* was improved into a *gareta* (bush cart), which was basically a chair with two long handles at front and back, with one *mutema* pulling in front and another pushing from behind the chair (Gewald, Lunning, and van Walraven 2009, 25).

Bulls, donkeys, mules, and horses were ridden and used as pack animals or to draw wagons, sledges, carts, and plows. Rivers were crossed via drifts or wooden bridges. *Magwa* (canoes; singular *igwa*) and *zvikepe* (boats; singular *chikepe*) were made and deployed as freight and passenger craft across and along rivers and from one coastal settlement to another. Many of the *magwa* that incoming *vachena* used in the hinterland, starting with the Portuguese (since 1498) and then the British and many other itinerant *vachena* (subsequently), were and are still locally made (Sheriff 2010).

Well into the 1950s, ox wagons still plied the beaten track, with two black men on foot—one an outrider (conductor), the other the driver—and in between them at least eight spans of oxen towing heavy loads, often including the white client who paid for their labor services. Since the nineteenth century, South African men played an enabling role as foot transport vehicles in *vachena*'s encroachments of their land and those of others: by missionaries, traders, explorers, concession-seekers, and hunters. These mobilities prepared the way for *vachena*'s partition of the region. By their micromobilities inside the horse, ox, or donkey's body, *hutachiwana* that caused *n'gana*, and horse sickness, rinderpest, and African Coast fever, immobilized *zvipfuyo* as means of transport for *vachena*, forcing them to walk while exclusively relying on *vatema* to shoulder and head-port their burden (trade goods and supplies). This can be seen clearly in practically every travel writer's account (as referred to in chapters 1 and 2).

In the northern, northeastern, and southeastern areas, the biggest problem was movement of *vatema* across *vachena*'s boundaries to and from mine, farm, and emerging urban workspaces since the beginning of gold mining in South Africa in 1886. During famine years, mine agents scoured the countryside for *vanhu vatema* prepared to trade their labor for grain (van Onselen 1974, 276), providing free transport as far as the roads and footpaths into the countryside allowed. A pattern emerged in which *vatema* from Northern Rhodesia (colloquially called *mabwidi* in *chidzimbahwe*) preferred to work in the mines, while those from Nyasaland (*Manyasarandi* in *chidzimbahwe*) took up farmwork (Scott 1954). Figure 8.1a shows men carrying *misengwa* (luggage) embarking on their journey to collection points, at which Southern Rhodesian government lorries awaited them (figure 8.1b).

By 1950, non–Southern Rhodesian black employees made up 50 percent of the total workforce; of these, 56 percent were in Mashonaland (maize and tobacco farms around Salisbury and Umtali) and 40 percent in Matabeleland (at Wankie colliery and the Nyamandhlovu sawmills; and the rest miscellaneous; Scott 1954, 45–46). Most of these figures must be read as *vanhu* passing through *mhesvi*-infested areas separating Northern Rhodesia, Nyasaland, and Mozambique from Southern Rhodesia, potentially carrying *mhesvi* on their bodies.

The pedestrian and ox wagon background detailed previously clears space for consideration of two incoming things that local actors strategically deployed as means and ways of moving around in southern Africa: the automobile and the bicycle. They matter to this discussion because of the way *mhesvi* subverted them into means of short- and long-distance transport. The train is excluded because there is no evidence of any such subversion by the *chipukanana*. This chapter emphasizes the intellectual agency of *vatema* in seeking means and ways of earning a living in the wake

a) b)

Figure 8.1a, b
Embarking on *rwendo* (journey) to the mines and farms on foot (left), and boarding lorries from the mines (right).
Source: Scott 1954.

of increasingly restricted access to land and the biting effects of taxation under *vachena*'s rule. The resultant mobilities from *misha* (villages) to *migodhi* (mines), *maguta* (towns), and *mapurazi* (white-owned farms), including across borders via undesignated crossing points, inadvertently offered *mhesvi* ready means of transport. Herein lies an interesting history of mobilities through which means and ways (vehicles), people (migrants to and from work), and *zvipukanana* (*mhesvi*) became vehicles for *hutachiwana*.

The Automobile and *Mhesvi*

Starting as surreptitious affairs from individual homesteads, and winding through neighboring *misha* and *dondo* or *sango* (forest; plural *masango*), footpaths merged into beaten tracks across and along the borders, staying that way until they reached railheads and, from the mid-1920s on, automobile roads. By the mid-1930s, there were over three thousand miles of road networks available to the Free Migrant Labour Transport Service for labor recruits who signed up to go to designated Southern Rhodesian mines and farms registered with the Rhodesia Native Labor Bureau (RNLB). The Nyanja-speaking recruits from Nyasaland called this transport *ulere* or "free" (Scott 1954, 36). *Vedzimbahwe* called it *urere* (free-bee; see figure 8.2). These bus and truck routes followed the Zambezi inside Northern Rhodesia (now Zambia) to Kalabo; another followed the same river into Mozambican territory to tap into Nyasaland (now Malawi).

Two *ulere* routes are of interest to *mhesvi*-related traffic control: one from Luangwa and Kafue (Northern Rhodesia) through Chirundu to Sinoia (now Chinhoyi), the other from Msusa and Misale (Nyasaland) through Darwin (Dande) to Mutoko. Later, the Misale-Mutoko route expanded to Chikwizo in 1947 while the Zobwe-Tete and Honde-Umtali (Mutare) ones were also absorbed into the *ulere* system (Scott 1954, 40). These two routes were connected to an older tapestry of paths to the Rand, which later followed the Mozambique-Southern Rhodesia border to Pafuri and thence to the Rand. Another followed the Savé from Vilankulo to Masenjeni and the Shabanie (Zvishavane) Mine recruiting depot at Marumbini and thence either to Shabanie or to Pafuri and the Rand. One of the collection depots for road transport to the Rand was located at Pafuri, in the armpits of the Limpopo River.

The *ulere* lorries were part of a larger automobile presence in post-1920s southern Africa. These vehicles came from three US companies (Ford, Chrysler, and General Motors) and three British automakers (Morris, Austin, and Land Rover), which had a virtual monopoly in the southern African market

Figure 8.2

Regional migrant labor routes from the 1890s to the 1950s.
Source: Scott 1954.

until the 1960s. From that point on, German, Italian, Dutch, Swedish, Japanese, and South Korean brands took over. The shipment of preassembled automobiles to southern African ports increasingly began to be replaced, from the late 1920s on, with the development of regional car assembly plants and retail branches of the big automakers. Cars automatically necessitated the development of roads, stemming from three factors: first, the railroad companies' desire for feeder roads to link farms and mines without having to extend the railroad system; second, the need for mine owners and farmers to recruit and transport *vanhu vatema* from the countryside to the mines, farms, and towns; and third, the emerging tourism industry's quest to link newly established game reserves, historical monuments, and "wonders of nature" like Victoria Falls as one vast trans-Zambezi product exclusively and discriminatively for the enjoyment of *vachena*. Government statistics show that 1,722 private motorcars were registered in 1934. Of these, 1,407 were from the United States and 308 British-made ("More Motor-Cars in Southern Rhodesia" 1936, 21).

For *vapambepfumi*, the road motor vehicle was proving to be "a powerful supplement to the railways in the development of Rhodesia" ("The Civilizing Influence of Roads" 1929–1930, 144). The most obvious reason was flexibility of access. The road could reach "immense blocks of settlement, far removed from the main line of the railways." Such areas had every ingredient needed for agricultural success, but experience the world over had shown that "the building of branch railway lines to assist the development of agricultural areas [was] not economically sound." Prior to the adaptable materiality of the road motor vehicle, there was no alternative to the branch railway "when a district had outgrown the transportation limits of the ox wagon." Yet Rhodesia avoided the "uncommercial risks" of the branch line. The construction, operation, and maintenance costs were too heavy in relation to the value of agricultural traffic (144; "The New Pioneers" 1929–1930, 102).

The Iron Tortoise and the Inciter

In the same period (1920s onwards), the coming of the bicycle added a new dimension to the speed of the traveler on foot, while retaining the element of flexibility. Pedestrian and cycle traffic were as much if not more of a challenge to the control of *mhesvi* (the inciter) as motor vehicles. In fact, by 1960, the director of Tsetse and Trypanosomiasis Control observed that bicycles had become "the most efficient carriers of tsetse to the control points. Flies carried per 100 cars were 0.7; per 100 cycles 10.7 and per 100 pedestrians 1.9."[1]

Without ascribing to it any malicious intentions, *mhesvi* was subverting means and ways of transport such as cars and *hambautare* ("iron tortoises," as *vedzimbahwe* called bicycles) and organic vehicles such as people and draft cattle into its own means of transport. If US automakers completely dominated southern and central Africa's roads, British manufacturers had a virtual monopoly on the iron tortoise—the cars of *vatema*. Bicycle makers exporting to Rhodesia included Norman Cycles (Kent; the Norman); British Salmson Ltd. of London (the Cyclaid bicycle); the Hercules Cycle and Motor Company Limited of Aston (the Hercules bicycle); Phillips Cycles Ltd. of Sethwick (the Phillips); New Hudson (the New Hudson Tourist Roadster); Rudge-Whitworth Ltd. of Coventry (the Rudge); the Birmingham Small

Arms Company of Birmingham (the BSA); Sunbeam Cycles Ltd., Birmingham (the Sunbeam); Armstrong Cycles of Birmingham (the Armstrong), and Raleigh, Nottingham (the Raleigh).

Vatema deployed these *hambautare* as their favored mode of transportation between workplaces in *maguta* and *kumusha* (villages), not least because automobiles were, throughout the Rhodesia period, the preserve of *vachena* and a very few *vatema* who could afford secondhand cars. The bicycle was in demand not solely for conveying its rider from one point to the other; *vatema* also remodeled it into a transient workspace for performing all kinds of work (Mavhunga 2014), as a platform for staging their own modes of everyday innovation (Mavhunga 2013).

Two stories often told about bicycles in *vatema*'s experience of *hudzvanyiriri* have a bearing on the *mhesvi* theme of this book. The first relates to World War II. When the war ended, combat veterans from Burma and Malaya, where Japanese enemy fire had not discriminated between *vatema* and *vachena*, returned to a segregated Southern Rhodesia. While *vachena* were awarded farms, *vatema* were rewarded for their service merely with bicycles. To add insult to injury, *vatema* were forcibly removed from their lands to make way for these new landowners and were resettled in *mhesvi*-prone areas. Thus positioned in the buffer zone between the *mhesvi*-infested areas and the white-owned farms, *vatema* acted as a human shield against veterinary disease and as vegetation-clearing agents to suppress *mhesvi* habitat or incursions. Many returned to find their families removed to the *mhesvi*-prone areas to make way for white officers and white soldiers who opened ranches and new farms under the Land Tenure Act. They were still required to carry a *stupa* or *chitupa* (an identity document *vatema* were to carry always or face arrest) and to follow the Native Registration Act, which mandated that all *vanhu vatema* must carry an extra pass in addition to *chitupa* (Matibe 2009, 5).

Hambautare (*xikanyakanya* in *xitsonga*, after the sound of pedaling, *kanya-kanya-kanya*) must be located within a larger economy of *vatema*'s importations and strategic deployments of Western-made goods. Like possessing a musket in the late nineteenth century, ownership of a bicycle meant that someone was a real man. The bicycle was one of many consumer goods produced either in Europe or locally in the factories that *vachena* had established. On white-owned farms, people learned to operate farming equipment such as plows, cultivators, ridgers, and motor vehicles. When returning to their *misha*, they went into the "blacks only" sections of cities to buy clothes, shoes, blankets, *hambautare*, floor polish, shoe polish, petroleum jelly, beauty cream, metal cooking pots, hunting knives, sugar, soaps, matches, cigarettes, soft drinks, sewing machines, wrist watches, radios, gramophones, cameras, furniture, and other goods to take home. They loaded these goods into the carriers of the "chicken buses" or the "long chase" (long-chassis omnibuses) at *misika yemabhazi* (marketplaces for buses; bus stations) to begin the long, dusty, and bumpy journey home (Mavhunga 2014, 71-98).

Zvechirungu (*chidzimbahwe*) or *svexilungwini* in *xitsonga* (meaning "the things of the white people") were also imported in the form of ideas carried out of the industry or city in the head and transplanted into *musha* to express new modernities. This is how iron or asbestos-roofed brick houses, cement-plastered and painted walls, grocery stores, table manners, and the four o'clock tea traveled from *vachena*'s suburban house in the *guta* (singular of *maguta*) to *kumusha*, sometimes via their lodgings in the crowded black quarters called *marukisheni* (locations). *Vatema* purchased the goods not just for their own use, but also as resources for resale and as equipment for business. Through a combination of thrift, risk-taking, and innovation, some of these men later bought cars, amassed sizeable herds of *mombe*, built "modern" houses in their rural homesteads, built grocery stores at the local shopping center, and even started bus companies and hotels (Mavhunga 2014, 136–140). Black entrepreneurs like Mwaera and Machipisa in Highfield Township, Moses Chikuhwa of Glen Norah, and George Tawengwa of Mushandirapamwe Hotels and Buses fame all began humbly, riding on their retrofitted bicycles selling tomatoes or exchanging grain for *huku* (chicken). Bus operators like Isaac Maziveyi, owner of two buses under the Maziveyi Omnibus Service stable, were in business by the early 1950s. The likes of Mverechena, Matambanadzo, Chinaka, and Mucheche became brand names of buses and hotels, but the bearers of these names arose from very humble origins (Chikuhwa 2006, 106). Others (men as well as women) distinguished themselves as owners of tailors' shops, often run as family businesses that sprang up at shopping centers in urban locations and rural areas, sewing cutoffs collected from urban textile or garment-making factories into *hembe dzemapisi* (clothes from pieces).

A dearth of new bicycles or repairs in the *varungu*'s workshops spawned the development of bicycle-repair shops, mobile (bicycle-borne) and under-the-tree welding workshops, and tire-repair workshops in the countryside. The remittance of overseas and locally manufactured things to *kumusha* depended on the existence of *tsika* (culture) and facilities for thrift and retirement packages that allowed some *vanhu vatema* to buy and install grind (or hammer) mills, to build *magirosa* or *zvitoro* (grocery stores), or to establish a bus company plying rural routes. The feedback loops between *guta* and *musha* that made such savings and investments possible were the very same ones that transformed *munhu mutema* traveling back and forth by bicycle or on foot into vehicles for carried *mhesvi* (Mavhunga 2014, 138–140).

Carried Fly: The *Mhesvi* Passenger and Traffic Cleansing

The development of *migwagwa* (roads) was considered paramount to monitoring the movements of *mhesvi* and its passenger *hutachiwana* and the potential vehicles for both: *vanhu* and their *mombe*. The dilemma facing the government's use of rural development as a strategy of controlling *mhesvi* was how to utilize roads for surveillance against the *chipukanana* while preventing it from catching a ride on *ngorodzemoto* (carriages of fire; or *motokari*, motor cars) plying these roads. That is why from the 1920s on the government set up *cleansing chambers* (see figure 8.3) and *tsetse gates* to monitor and cleanse cars, cyclists, and pedestrians of *carried fly*. Almost all

Figure 8.3
The fly chamber through which all *motokari* on busy roads passed.
Source: Proceedings and Transactions of the Rhodesia Scientific Association 1960.

ngorodzemoto were owned by *vachena* prior to 1950, with very few *vatema* who could afford them. By contrast, all cyclists, pedestrians, and rural commuters (lorry and, later, bus passengers) were *vatema*.

The term *traffic control* is first mentioned in the chief entomologist's annual report in 1928, expressing alarm at the increasing danger of "motor vehicles being used more freely for prospecting, etc." and carrying *mhesvi* with them (Jack 1930).[2] In May 1929, a bill was passed in the legislative assembly to "secure the necessary powers for the control of traffic from fly areas." By the end of the year, however, the *mutemo* (law) had not been implemented, in part because "effective treatment of motor vehicles, without having recourse to the use of deadly poison, constitute[d] by no means a simple problem."[3] The first comprehensive, practical steps to control traffic coming out of *mhesvi*-infested areas, right on their edge of such belts, began in 1930 and proceeded well into the 1970s.

In what became known as the "Zambezi Front West" and the "Zambezi Front Central," the road (*mugwagwa*) that cut through *mhesvi*-infested and noninfested areas enabled motor vehicle traffic to pass through that had to be cleansed of carried *mhesvi*, leading to the assignment of cleansing chambers. The Kariba Dam also placed further barriers to cross-border mobilities that were already difficult—except by boat for *vanhu* and *mhuka*. There were two major roads. One was the Salisbury-Lusaka highway via Chirundu Border Post, which passed through Hurungwe Native Reserve and the *mhesvi*-infested areas of Makuti and Chirundu. The other was the Bulawayo-Livingstone route passing through Gwai and Shangani and the *mhesvi*-infested Mapfungautsi plateau.[4] There were other (minor) roads going to tin and tungsten (Sebungwe District) and mica mines (Hurungwe), both in *mhesvi* territory.[5]

From 1939 to 1945, the Chirundu highway became an important route for moving black troops and supplies traveling to join the Allied War effort in Burma. *Mhesvi* lurched onto the truck convoys, providing a headache for the guards manning the chambers at Chirundu, Makuti, and Vuti.[6] In the postwar era, the massive drought of 1948 dispersed *mhuka* in all directions in search of water and grazing, carrying *mhesvi* on them into Hurungwe and triggering the catastrophe covered earlier. Then, from 1953 to 1963, during the construction of the Kariba Dam, more dispersals occurred due to the displacement of the Tonga people from the Gwembe Valley into Hurungwe and other areas.[7]

The river barriers of the Zambezi were entirely absent on the southeastern border with Portuguese-ruled Mozambique. The most critical mobilities remained those by foot, hoof, or paw. That is why the fences were necessary: not only to create buffer zones, but also to channel *kufamba kwevanhu nemhuka* (human and animal traffic) to tsetse gates for inspection. Here, the transborder movements on the Zambezi Front East (Rushinga) and the South East Front (Savé and Runde regions) illustrate how *vachena*'s arbitrary borders had simply cut straight through *misha* (villages) organized along kinship lines and set up transgressions by other kinds of animals and plants. As people now visited their relatives, they carried *mhesvi* back and forth (Mavhunga and Spierenburg 2007).

A major problem from the onset of fences was that of roads passing through both *mhesvi*-infested *matondo* (plural of *dondo*, forest) and *misha*. Some led to several active mines, farms, emerging towns, and neighboring countries. Others were maintenance and patrol roads for tsetse control work that soon became the only public roads available. Either way, all roads in Hurungwe Native Reserve were fairly busy and had to be manned.[8]

To ensure that vehicles, pedestrians, and cyclists using these roads did not carry *mhesvi* out of Hurungwe, there was only one exit point located on the eastern boundary of the reserve. It was here that a cleansing chamber was installed in July 1952. All traffic was barred from crossing anywhere other than at the designated gates, with the exception of a few stiles erected over the eastern fence to enable *vatema* on foot passing between the reserve and the farms to cross.[9]

Pedestrians, cyclists, and motor vehicles were controlled lest they become vehicles that carried *mhesvi* from infected to clean areas—a process *vachena* called *mechanical transmission* or the problem of "carried fly."[10] This *traffic cleansing* took place at deflying chambers and tsetse gates (see figure 8.4), where "carried fly" catching rides on *ngorodzemoto* (motorcars), *hambautare*, and *vafambi* was apprehended. Administrative centers like the chief's court, Native Commissioner's offices, dipping tanks, cattle sale pens, and shopping centers pulled human traffic toward them, thus acting as magnets for the movement of *mhesvi*.[11]

The "cleansing" or "deflying" chamber was established on roads and the "tsetse control gate" on footpaths *vatema* used. At each cleansing chamber was a gate guard (*mufrayi*) dressed in uniform. In his hand was a fly net and hand spray pump. The traffic arriving was supposed to stop at the control point, where *mufrayi* first examined it for *mhesvi*. Any clinging on were caught in the net. The guard also sprayed the motor vehicle around and underneath to unsettle any *mhesvi* that might be relaxing or hiding there. Just in case the critters made for the shade of the open-sided, grass-roofed huts, their undersides were liberally sprayed with persistent OCP. This is something that came later in the 1950s; prior to that, arsenic was used.

Figure 8.4
Fenzi yetsetse and *gedhi retsetse*: a typical tsetse fence and gate, with *mufrayi* standing behind the gate to inspect traffic.
Source: Proceedings and Transactions of the Rhodesia Scientific Association 1960.

Attracted by the shade, the flies flew in and landed literally on their own deaths; OCPs killed them through skin contact.[12] The issue was not whether the guards manning these chambers caught flies every month, but whether motorists, cyclists, and pedestrians passing through them were *mhesvi*-free going into uninfected or deflyed areas.[13]

Conclusion: Transport Systems and Dangerous Insects

Before takeoff, planes from several airlines flying out of Africa are sprayed with a pesticide aerosol. The New York–bound South African Airways flight that makes a stopover in Dakar, Senegal, is sprayed, as is the Emirates Airline flight that stops in Lusaka, Zambia, en route to Dubai. Apart from causing eye irritation for people wearing contact lenses, the aerosol is "completely harmless"—or so we are told. It kills *hutunga* and other *zvipukanana* that might be hiding under our seats or clothes. It is good for us. What can you do—get off the *ndege*?

This chapter has shown that the connections between human-fabricated transport systems and portable, tiny *zvipukanana* carrying deadly viruses is not new. We see it today with *hutunga* carrying Zika and ticks carrying multiple viruses. The significance of measures we see at airports or at the checkpoints on roads as we leave game reserves is the link between microbial mobilities inside *zvipukanana*, *zvipukanana* riding on our cars, buses, and planes, and these latter transport systems becoming conveyors of people, *zvipukanana*, and *hutachiwana*.

Seen from *musha*, the sites where *vachena* had designated cleansing chambers and tsetse gates become workplaces. The *mugwagwa* (road) and *nzira* (footpath) that *vatema* used in their everyday itineraries on bicycles or on foot and which later *vapambepfumi* passed through in their automobiles were a site of knowledge production where the mobilities of cars and pedestrians at once become (potential) mobilities of *mhesvi* and the *hutachiwana* they carried. It was because of *mhesvi* that the gate and chamber were established; they were an infrastructure of the mobile workshop: the *mhesvi* on the move, forcing *vachena* to keep it under surveillance, providing *vatema* with work. Without *mhesvi*, the control of the movement of *vanhu* and *mhuka* would not be necessary—which was another way of saying that the traffic being controlled ultimately is not that of cars or bicycles, but the traffic of *mhesvi* itself, because of the deadly passenger it carried inside it.

9 Starving the Fly

My name is Mugocha Mavasa, son of Mubhulachi Mavasa, son of Marhule, son of Nyambiti, son of Makulani, son of Hlati, son of Mugwangwani, son of Malanzela, son of Hlati, son of Xinyori xaHumba. This is my ancestry.

I used to be called Julius Mavasa, but many call me *Mugocha* (the one who is always barbecuing), because I once worked *mutsetse* [in the Tsetse Department]. So now my name is Mugocha, my birth name is no longer heard of anymore. I do not recall when I was born. It is a long time ago (possibly in the 1940s) at Chitalahimbera near Bhaule, here in Chibwedziva. Chitalahimbera means "that which fills up even from drizzle," because every time it rained, even briefly, the pool filled up quickly.

I joined *vanhu vetsetse* in the early 1960s as a teenager. Gillett, who was called *Ngungunyana*, was in charge. ... We were all summoned to the camp so that they could select who could be *magocha*. So we went there. I was still young and I was wondering to myself whether I would be able to use a gun. When we arrived at the tsetse camp, we went to sleep. In the morning *Ngungunyana* started selecting, simply by pointing: "You, come here, you come." Some of us were too young, and we stood behind the elders, the three of us: me, Koko, and the son of Zhuwawo, or "Fifteen," my father's old friend. We were resigned to an old pattern where *valungu* [*xitsonga* for *varungu*, white people] usually selected the old-timers already skilled in guns they had learnt under instruction from their own fathers. ...

Then he pulled me out of the line. I was shocked because I knew nothing about hunting with guns. I was convinced that if I fired a rifle, the *mbumburu* [bullet] would strike at me and I would instantly be dead. I did not yet know that when you fire a gun, the bullet goes out through the front and flies away from you. After pulling twenty of us out of the crowd by the shirt, *Ngungunyana* dismissed the rest. He said we were the ones going to Mwenezi. Others were chosen for the work of repairing the fence broken by *twiza* [giraffe] trying to cross, because some were this side, some had been shut out on the other. Their task was to patrol the fence, killing or driving away any *mhuka* that they found endangering it. Meanwhile, repair teams with machines would follow these *magocha*, mending the fence where broken.

Figure 9.1
Mugocha Julius Mavasa.
Source: BPP.

In the morning, we were taken before *mudzviti* (magistrate). He said we should get fingerprints, which consumed our entire day till sunset. After that we were deployed to Chipinda Pools. We came back. Our drivers were Mazhau, Langton, Kingston, and *Peturo* [Petros] the [four] drivers at the local tsetse camp. *Peturo* was short, Langton had a big body, Kingston was heavily built, while *Mazhau* was a giant, so tall that he would start a [Nissan] UD truck while standing outside, with his foot on the accelerator and revving. He was tall.

We were driven back by these men. We were egging them on, oblivious of the fact that a *motokari* [motor car] can kill. People ran to Mazhau's lorry because he was a fast driver. We wanted to get to Chipinda early. At this time we had not yet been issued with rifles.

So we returned to Chipinda. The Runde was in flood; we got off and rested on the west bank. The next day Ngungunyana arrived to cross us [in a boat] to Chipinda tsetse camp. He said: "*Magocha mauya!*" ["Welcome Magocha!"]

We said: "E-eh."

He said: "Now that you are all here, let's go to the ground." So, we had left our vehicles this side of Rundé; on that side there were similar lorries. So we were taken into that ground at Chipinda Pools, the one close to Sevenjeke, between Guluje and the rubbish dump. Little did we know that the guns were here, many guns. 303s.

When we arrived, we are issued 303s. Gillett—*Ngungunyana*—was pulling the trigger numerous times, but the chamber was of course empty.

Then he called out: "One, come here!"

Figure 9.2
A typical tsetse field officer's headquarters before 1960 would, like this one, include his hut, office, store, and hospital, all built using 100 percent local materials and designs.
Source: The Rhodesian Annual 1932.

We remained standing there, our minds fearful about what the guns could do to us, what to do with the guns.

Ngungunyana took one *mbumburu* [round]. They had put up a shooting target, far away, almost at the mid-point of the ground, one of those drums used for mixing *mushonga* [insecticide] for spraying. He started hitting the target. Then he started calling us one by one to do the same. Some coming from Savé knew guns, they were smacking the target. There was one, of the Chudhu family, there was one short one ... he was very good. We were now under training. So the short one took the gun, fired, hit the target. Gillett told him to stand to one side.

"Next!" *Ngungunyana* bellowed.

It chimed 12 o'clock [noon] *tichingodzokera shure, kuti vatange vakuru, vatange vakuru* [and we kept going back to the end of the line, so that the older ones go first, the elders first] until we finally arrived.

"*Vuya!*" [Come!] Ngungunyana again called in his *xitshanganized fanakalo* access.

So we went up. Some were running away from even touching the gun. When they pulled the trigger and it fired, they threw it down and ran for dear life. I was selected for the task of patrolling the tsetse fence, when I had no knowledge of guns. I fired I think 4–5 times without even looking up to see where the rounds were going.

> So they left me alone, and took my companions, whom they said were being assigned to protect the game fence. Another group was selected to go *kuhugocha*. I was deployed for one month in the fence protection.
>
> Then I returned to Chipinda; on my second deployment I hit *mhene* [duiker]. So they said, ah, the boy who is assigned fence protection duties is a good shot, he must be recalled and taken to Chifukwa at Sevenjeke to hunt with other *magocha*. So I was removed from fence duties. Those that were doing badly were removed from *hugocha* and redeployed to fence protection.
>
> I started shooting.
>
> Now, we were allowed to kill fourteen *mhuka* per month. If you hit your fourteen *mhuka* before the month ended, you'd sit out the rest of the month. We were paid on the 24th each month. I hit the fourteen *mhuka*, I think, on the 5th or 6th, long before the month had gone anywhere. ...
>
> Again they said this guy is too good; he must be removed to Guvulweni where there were still many *mhuka*. Here the animals were no longer easy to find. So I was taken there to stay with *Chifukwa*; his name was Peter but every white man we gave our own name. Our own name based on his behavior or looks, or the places we first met him, where he stayed. Chifukwa is where Peter usually hunted. So we called him *Chifukwa*. I hunted with *Chifukwa* and a young man named Aaron who came from Mozambique, who was *Chifukwa*'s cook.
>
> At some point I realized that this was not working out well for me. I had been bitten by a snake on the inner lower shin of the right leg. I said I am done with hunting. I am going. I went to my homestead.[1]

They were called *magocha*—black men the Branch of Tsetse and Trypanosomiasis Control (BTTC) employed to kill *mhuka* to starve *mhesvi*. *Magocha* was not a name of insult—just an acknowledgment of a fact—for *men who were always barbecuing.*[2]

As discussed in *Transient Workspaces* (Mavhunga 2014), members of the older generation were not novices like Mavasa when they began *hugocha* (game destruction work); they were already professors of the hunt, especially the Korekore of the Zambezi valley and Hlengwe and *vedzimbahwe* of the Savé-Runde and Limpopo valleys. The government issued such men guns and deployed them in *mhesvi*-infested *masango* (forests). "They would test us to see if we could fire a gun," recalls Willias Chabata of Nembudziya, Gokwe. "Some could, so they were issued firearms, and moved around with them, shooting *mhuka*. The likes of Saira, this old man from Machichiri, were *magocha*. Miriyoni and Misheck were *magocha*. They hunted this side, and later crossed the Sanyati into Hurungwe."[3] Some hunted until all *mhuka* were cleared from designated areas. Others, like Mugocha Mavasa, retired prematurely after being bitten by *nyoka*. Some were city boys visiting *kumusha* (home), who joined to make a quick buck—men like Raymond

Muzanenhamo of Nembudziya.[4] Others, like Mavasa and Chabata, were based in *mumusha*.

But where did the idea come from—that by exterminating *mhuka*, hunters like Mavasa, Chabata, Muzanenhamo, and countless others could also help Rhodesia annihilate *mhesvi*? In this chapter, we shall return to the pivotal moment in the transition from a period before and into Rhodesia to follow the mobility of *ruzivo rwevatema* into so-called *ruzivo rwevachena*, thus shaping its very core, complete with the dependence on *vatema* to execute such indigenous stratagems. Then we will reflect on the conduct of hunting and its meanings and purposes in *vatema*'s lives. I argue that whatever the Rhodesian apparatus added to the hunt as a method of pest control, credit is due to *vachena* only for repurposing, retooling, and renovating within the context of what was already present among *vanhu vatema*. The glossary at the back of the book will allow the reader to reference *chidzimbahwe* and other regional keywords.

The Origins of *Magocha*

Hugocha emerged out of *ruzivo* of *mhesvi*'s association with big *mhuka*. Those *vachena* we saw in chapters 1 and 2 interacting with *vatema*, learning from and surviving on their ideas and practices and documenting such *ruzivo* in their journals, became the bridge that took *ruzivo* into *vachena*'s scientific communities and official Rhodesian state policies.

The perfect starting point for this discussion is *A Monograph of the Tsetse Flies* (Austen 1903), in which British entomologist Ernest Edward Austen agreed with the positions that the medical missionary David Livingstone, the big game hunter and explorer *Serowe*, and many other nineteenth-century travelers had taken—namely, that *vatema* were correct to say that *mhesvi* was found wherever big game was present. The rinderpest epizootic seemed to have put the matter to pasture when exterminating *mhesvi* in most parts of southcentral and eastern Africa by destroying its most reliable food source: *mhuka*.

By 1900, however, the *mhuka* population was picking up again, and *mhesvi* too was recovering as its "food" recuperated. This twin development triggered a fierce debate in the *Journal of the Society for the Preservation of the Wild Fauna of the Empire* and *Field* from 1907 to 1908 between Europe-based laboratory scientists, on one hand, and *vapambepfumi* who were ex-hunters, on the other, regarding the association between *mhesvi* and *mhuka*. Opinions crystalized around Austen, who now dismissed the premise from the perspective of laboratory *experiment*, and *Serowe*, with twenty years of hunting *experience* in *dzimbabwe* and virtually a mobile encyclopedia on anything related to "nature," who had already endorsed *ruzivo rwevatema* on the *mhesvi-mhuka* association (JSPWFE 1907).

Few who had experienced *mhesvi* in its natural haunts could doubt *Serowe*. Fewer still could disagree with him that the only way to ensure complete freedom from *mhesvi* was to slaughter *mhuka* and prevent *mhesvi*'s revival in the future (Selous 1908). Austen had not experienced but had experimented on *mhesvi*, and he regarded the slaughter of *mhuka* as a danger to "faunal preservation." *Serowe* had found that other kinds of game might take the place *nyati* (buffalo; see figure 9.3) had vacated and yet be unsuitable hosts for the savannah-loving types of *mhesvirutondo*, which *vachena* called *G. morsitans*. He believed that "exterminating game of all kinds in a country in order to get rid of tsetse fly would not only be an abominable crime, but an absolutely unnecessary one." *Serowe* concluded: "You cannot have buffaloes without having tsetse flies as well" (*Field* 1907).

Austen conceded that no type of *mhesvi* could survive without blood of some sort, but insisted—based on experiments—that it did not have to be that of *mhuka*.

Figure 9.3
Nyati the buffalo, lover of water, companion of *mhesvi*.
Source: Author 2011.

Experiments in Lake Victoria had revealed that *mhesvi* also fed on riverine *mhuka*, specifically *mvuu* (hippos), *makarwe* (crocodiles), *mbeva* (mice), and *shiri*. To single out big game for execution while leaving these blood sources was a waste of time.

Writing in *Field* in 1907, *Serowe* reinforced his thesis of a strong connection between *nyati* and *mhesvi*:

> It took many years before the fly had completely died out, but today there are neither buffaloes nor tsetse flies in a part of the country where less than five and thirty years ago both literally swarmed. If there is no connection between the buffalo and the tsetse, why is it that, not in one district alone, but everywhere in Africa south of the Zambesi, in countries as far apart as Delagoa Bay and the district of Victoria Falls, as soon as buffaloes have been completely extirpated, tsetse flies have at once diminished very rapidly in numbers, and sooner or later have become completely extinct? (Selous 1907)

Because they moved slowly, loved water and were found in habitats also favorable to *mhesvi*, *nyati* came to be seen after that debate as a sure sign of *mhesvi* presence.

The Austen-*Serowe* debate happened when *mhesvi* was still a remote risk—worth watching, but nothing to start fighting against in Southern Rhodesia. By 1909, the first serious signs of a return of *mhesvi* to its prerinderpest haunts appeared. The few *mhuka* that had survived the plague in the hinterland had been all but hunted out. Most that survived the rinderpest were to be found far from areas of concentrated human occupation—principally along and east of Southern Rhodesia's border with Portuguese East Africa (Mozambique) and along the Zambezi River. As their numbers grew due to natural increase, *mhuka* began ranging outward, with *mhesvi* in tow.

At the same time, to cope with new challenges to their livelihood and to escape from—and find money for—paying taxes, *vatema* were traveling across *mhesvi*-infested lands to seek work in the mines of central Rhodesia and South Africa. *Vachena*, native commissioners, and illicit labor recruiters were raiding *vatema* for taxes and labor.[5] All these mobilities attracted *mhesvi*, always alert to anything that moved; it wasted no time catching a ride and traveling as carried *mhesvi* to new places, where it alighted; then, while feeding on new hosts, it deposited *hutachiwana* it had carried into them. This is the mobile work of *mhuka*, *vanhu*, and *mhesvi*.

In those days, any spotting of the *chipukanana* itself or outbreaks of *n'gana* could only mean one thing: A *mhesvi* "invasion" or "advance" was under way. Those like *Serowe* who had seen it all before made it very clear that the *chipukanana*'s recovery had everything to do with the return of *mhuka*—in particular, *nyati*—to their former stomping grounds. *Serowe*'s words need to be weighed in the context of who exactly among these *vatema*—the scientists (from) overseas, or the "men on the spot" facing *mhesvi* who dismissed the scientists as "armchair faddists"—held influence over state policy (Mavhunga 2007). Unlike other colonies, in which the British Imperial Government had direct responsibility for policy, Southern Rhodesia was a self-governing white settler state that made and executed its own government policies (Jeater 2005, 1).

Government entomologists answered to British South Africa Company (BSAC) shareholders prior to 1923; after that, they answered only to the settler government, which believed not just in *expertise by experiment*, as experimenters like Austen did, but also, pragmatically, in *expertise by experience*.[6] White farmers, mine owners, and administrators became important conspirators in tsetse policy; the government was accountable to the white public.

From Intellectual Debate to Actual State Policy

All *ruzivo rwevatema* was considered *manyepo* (falsehoods), *zvenhando* (trivia), or *ngano* (fable) until proven as *zvokwadi* (truth) or *ruzivo* (knowledge) using *vachena*'s method. Thus, despite what *vatema* had told *Serowe* and others, and despite confirming the truth value of such knowledge through their own experiences, game elimination had to be put to the test. The shooting began on an "experimental" basis in 1919 in the Gwaai and Shangani Native Reserves of Sebungwe District as "a practicable method of fighting this terrible scourge" of *mpukane*. One experiment comprised the systematic shooting of *nyamazana* in a selected area, the other destroying the evergreen trees and other refuge vegetation around the winter habitats of *mpukane*.[7] By 1920, the shooting experiment was showing "a considerable reduction of fly."[8] On December 15, 1921, the Gwaai-Shangani experiment was concluded, with the chief entomologist declaring that it had resulted in "a very marked reduction of tsetse." That same year, plans were drawn up to carry out similar experiments in the Munyati River area around Chegutu, stretching north into Gokwe.[9]

In his annual report of 1922, Chief Entomologist Rupert Jack announced that *hurumende*'s (government's) game elimination had "proved" *mhesvi*'s dependence on large *mhuka*. His department thus authorized, effective 1922, an "experimental policy" of granting concessions to individuals to hunt in twelve areas delineated for the purposes of controlling *mhesvi*. Of these, two were set aside for hunting by *vanhu vatema*, one for the Entomology Division, six for private white hunters, and three left unassigned.[10]

In 1923, following deaths of numerous cattle belonging to *vatema*, shooting operations were extended to the Kandeya area of Rushinga (Darwin) under charge of the Native Department. In November, the acting Native Commissioner opened to free shooting a wider area than the outbreak radius. As the rains intensified and forests became impenetrably thick, the operations were suspended until the dry season. That very same year, in November, the Nemakonde area was subjected to anti-*mhesvi* shooting.[11]

From 1923 on, elimination of *mhuka* was extended to the rest of the country as the first line of defense against *mhesvi* invasion. From the perspective of white cattle ranchers, whose herds *mhesvi* killed, and *hurumende*, the revenue of which it ruined, shooting operations were important not for their experimental value but as the best solution available for eliminating *mhesvi*. From a third standpoint, that of *magocha* (see figure 9.4a, b), the shooting operation was an unlimited source of *nyama* (meat) now that *vatema* were banned from owning guns or hunting.

By the 1940s, the Chegutu operations had reached Gokwe. "Meat was eaten here in Nembudziya," observes a *mugocha* who participated in game elimination in the area. "That's when the name *magocha* was bestowed

a) b)

Figure 9.4a, b

A *mukorekore* hunter, master of the bow and arrow, and other hunting arts (left). The young apprenticed with older, more seasoned hunters, including their brothers (right).

Source: The Rhodesian Annual 1932 and 1934.

upon us by the people."[12] Initially, *hurumende* did not restrict *magocha*; they encouraged them to kill as many *mhuka* as they could. At the camp, the TFO reserved the liver and other special cuts for himself, and designated everything else for *magocha*. Later, to cater to the dietary needs of teams of spraymen, who carried only their *mishini yekupureya* (hereafter simply *mishini* [knapsack sprayers or spraying machines]), *hupfu* (meal), *nyemba* (beans), *nzungu* (peanuts), and *nyimo* (Bambara nuts), the tsetse authorities required all *nyama* to be dried for distribution as rations.[13]

It was not just the craving for *nyama* but also grinding poverty that led many *vatema* to enlist as *magocha*. Many just "went there to see if it would work out."[14] By the late 1960s, the earlier generations of hunters that had apprenticed in using guns and bows and arrows, making *chepfu* (poison), and digging and setting pits with their fathers and grandfathers had thinned out. The tsetse people abandoned their custom of preferring practicing hunters and started recruiting any man willing to hold a gun, run through the bush, and eat *nyama*. "They would just say come," explained one *mugocha* in Gokwe. "You want work, e-e, come. You, you, you, go this side—you are hired."[15] As he recounted, this is how Mugocha Mavasa was hired. *Mugocha* was a nickname colleagues and locals gave him for his hunting prowess.

To give a sense of how much a typical game elimination operation might have cost in the 1950s and 1960s, the amounts shown in figure 9.5 are the official cost estimates for the Gonakudzingwa shooting operation in 1959. The budget items included materials and equipment costs for 213 miles' worth of fences; two hundred rifles; 87,500 rounds of ammunition; salaries and housing for seven TFOs and eight thousand miles of fuel per month for two lorries and four Land Rover vehicles, and African Assistants (*baasboys*), flyboys, gate guards, fence guards, and any ancillary personnel.[16]

Tsetse Field Officers and *Magocha* in the Bush

On paper, the powers of the TFO over game elimination were nothing short of impressive: He directed all hunting operations, prevented unauthorized hunting, and disposed of hides, meat, bones, tusks, and so on.[17] Under him were other TFOs (all of them *vachena*, and many ex-police), each in charge of twenty to thirty *magocha* armed with Martini-Henry and, later in the late 1920s, .303 rifles.[18] Once these hunters had cleared their areas, the TFO ensured they automatically reverted to the role of "native police" and assumed patrol to prevent the reentry of *mhuka*. The TFO inspected the hunting grounds to check on the hunters and keep them to heel.[19]

Estimated Costs of Shooting out Gonakudzingwa N.P.A.

(1)	Fences. 213 miles at £100 p.m.	= £21,300
(2)	@ 1 hunter p. 5 sq. miles, 1350 sq. m. will cost, at £5 p. man p. month for 4 years	= £64,800
(3)	Ammunition, 25,000 head of game @ 3.5 rounds each @ 7d. per round	= £ 2,500
	200 Rifles at £5	= £ 1,000
(4)	Seven European Houses @ £400	= £ 2,800
(5)	Transport @ 7,000 landrover miles (1/2) and 1,000 lorry miles (2/4) per month	= £19,500
(6)	Fence maintenance, Fly boys, Gate guards gates, Native Housing, etc., etc., Say	= £10,000
		£121,900
	Plus seven European Salaries @ £1,000 for 4 years	= £28,000
	Total	=£149,900
	Say	£150,000

Figure 9.5

Estimated costs of shooting out Gonakudzingwa Native Purchase Area 1959.
Source: "Sabi Programme." SACEMA/TA.

In practice, things were not that simple. *Magocha* were expected to report to the TFO, to submit the tails of all *mhuka* killed and have their permits stamped to indicate that they had submitted their returns. These were the tails the TFO collated for statistics on game reduction. Raymond Muzanenhamo's father, Lot, was one of the pioneer *magocha*. His team started destroying all *mhuka* from Hurungwe in the 1930s, crossed the Sanyati, and reached Nembudziya. Soon after, the senior Muzanenhamo retired on the land the government had resettled him on. His generation of *magocha* operated from their homes, engaged initially in destroying *makudo* (baboons), which were quite plentiful in Nembudziya and were hosts of *mhesvi* and destroyers of crops. "The white men wanted the tails of the baboons you have hit," explained Raymond. "If you hit *gudo* [baboon], bring the *muswe* [tail], if you hit *dzoma,* bring the tail, if you hit *ngwarati* [sable], bring the tail, if you hit even any animal, bring the tail. Those were our fathers. ... All they were supposed to take to *murungu* [white man] were the tails; the rest they left behind."[20]

Having already distinguished himself on Lone Star Ranch (in southeastern Rhodesia) as a hunter destroying "vermin" (predators and baboons), John Piet was recruited as *mugocha* and armed with a .303 rifle to cull *nyati, nhoro,* and *ngongoni* (wildebeest) in the 1940s and 1950s. He would shoot, gather the tails of all kills, and take them to Mwenezi, where *Chibwechitedza* (Allan Wright, the district commissioner) was based. Before leaving, Piet usually sent word to his village for people to come and skin the meat.[21]

However, as one entomologist found in 1953, it was "possible that Native hunters keep what may be termed a 'tail bank,' into which they deposit spare tails when they have a particularly successful month, withdrawing them when not so successful."[22] *Magocha* picked ammunition rations and returned to the "battlefield," or went straight home to do their own personal work in *musha,* returning only at the imperative of *nyama* or to pick up tails from their "banks" to submit to *murungu.*

By the time that Lot Muzanenhamo's son, Raymond, became *mugocha, mutemo* (law) had changed. They were now restricted to just four types of *mhuka*: *njiri, nguruve, nhoro,* and *dzoma.* They were also no longer required to submit tails.[23] They were issued twelve rounds, and they hunted with the TFO under supervision; they accounted for every round expended.[24] Each kill was recorded against the *mugocha* who had shot the animal, which "would enable *murungu* to determine how many animals were killed in that specific area."[25] When each hunter's kills were tallied, they showed the number of rounds per kill, how many *mhuka* still remained in an area, and how much work still needed to be done. If the kills matched the amount of ammunition issued, then such hunters were reissued firearms and ammo; if two months in a row passed without any kills, the operation was concluded and the hunters redeployed further ahead. Usually, *magocha* were based at one camp for four to six months before being redeployed to another camp, with another team replacing them from somewhere else. By the time these strangers established local connections, they were being moved again to a new location far away.

Varungu prohibited *magocha* from venturing into *misha* when off duty to limit the deliberate shooting of *mhuka* for the purpose of selling *nyama*. This measure was also designed to stamp out the previous problems of tail and ammunition banks and underdeclaring of kills:

> If you fired too many bullets, you would not kill any animal. That was also questionable to *vachena*; they knew that we were conning them. *Muchena* knew we were killing *mhuka*, but instead taking the meat to our *misha* and also selling. And it was true: we did those things. We got good money. The wage we got was not enough for our needs. There were people who would come to us and say, "Hey, can't you just spare me a piece of *nyama*. I will give you whatever price you ask." Instead of a little animal you went there and hauled a whole eland to the village. *Munhu mutema* was wiser than *vachena* gave him credit for.[26]

As was standard Rhodesian practice, every government activity required the supervision of *muchena* (singular of *vachena*). Initially, the TFOs who were supposed to supervise *magocha* were located at camps by the roadside far from where they were deployed. They had no vehicles—let alone lorries or 4 × 4s; even if they did, thick thorn trees dominated the roadless terrain where the actual shooting was supposed to take place. Engine wear and tear was the order of the day. The government allowed just 150 miles mileage (or sixteen gallons of fuel) monthly, but TFOs always used double that figure expensed from personal coffers to conduct government work. TFOs would visit the nearest store "once a month to obtain monthly supplies of food" and so on, but their life of isolation would mean that once they went to town, they never came back. Fewer *vachena* enlisted.[27] And even with the best of roads (a rarity in the 1940s), inspecting tsetse control corridors was always footwork, and it was "impossible to get men to work under the present conditions of service."[28]

The situation was simply dire, as the senior entomologist explained in 1953: "Half of these [TFOs] have proved to be unreliable, most have stayed for only two or three months and either resigned or been dismissed for mischief or incompetence, and in the intervals the area has been entirely without a resident TFO to supervise it."[29] The departure of TFOs who had gained some experience and the introduction of virtual novices further exacerbated the government's dependence on local hunters, who became the only chain of consistency tying *mhesvi* control operations together.[30]

In June 1971, Glossinologist Ted Davison put the problem down to the TFOs' unwillingness to exert physical bodily effort and stamp out poaching by *magocha* under their command.[31] In an effort to seize the initiative and end the bottom-up agency of *magocha* over tsetse control operations, Chief

Glossinologist Desmond Lovemore sent Davison in July "to investigate and check on the hunting operations and to report on his observations accordingly." "Our hunting operations were fast settling into a rut," Lovemore said. "In fact in some areas this had already occurred, and there was therefore the virtually desperate need to have someone constantly prodding to maintain the enthusiasm" among TFOs.[32]

Lovemore was concerned that game destruction might easily degenerate into "literally nothing more than reporting each month, in our various monthly reports, that so many animals were destroyed in this or that hunting section or operations area, with no further attention being paid to the work. ... Under these circumstances the technique will never work satisfactorily."[33] To avoid that, the TFOs had to stay constantly in touch with *magocha*, directing them "in order to ensure that the tsetse is starved with the minimum of delay." The work of starving *mhesvi*, Lovemore stressed, could only succeed if every person in the branch, from top to bottom, was "constantly stimulated to approach their task intelligently and with enthusiasm."[34]

For Davison, this was not just a question of labor management; hunters needed training on the shooting range, they needed checks against poaching, and that would take a hands-on level of man management, "the will and physical ability to get out into the veld and control the movements of our hunters." He particularly discouraged hiring as TFOs "candidates who express an early desire to shoot animals," who were "infirm," and those "of an age or disposition" unsuited to physical exertion.[35]

In interviews, former *magocha* concur with this damning assessment: "We had *kabhunu* (a small white man) that came from Gweru. I worked with him as his assistant. Maybe because he was always drunk so maybe that's why the tsetse bit him so much. He was always drunk. He would put a crate of beer in the Jeep, under the seat, I turn around and look at his arms and hands, I find so many engorged fly, *slap!* I'd kill them and tell him: 'You have been bitten by tsetse,' and he would be busy gulping his beer, not even noticing the flies."[36]

Some of these TFOs had not even a single feather of knowledge about *mhesvi* when arriving in the field. The man referred to in the previous quotation is said to have come to Nembudziya to resume his post, thinking that *mhesvi* had a tail. These *vachena* were here only for the meat and the money, only too grateful to get a job and the reward of killing *nzou* and keeping the ivory—until that rule changed.[37]

Some TFOs were distinguished marksmen. Former *magocha* in Nembudziya remember a man they called *Chiwoko* (Short Hand), who told them he had lost his arm during World War II. "But he was such a beast of a marksman in the bush," recalls one *mugocha*. "*Waidhuura murungu iyeye* [he was a good shot, that white man]. When he saw *mhuka* coming from there stampeding, he would just balance the rifle on what was left of his hand, and pull the trigger with his full hand."[38] There was another white man *magocha* called *Peturu Bhomba* (Petrol Bomber), who shot only *nzou*. "That one was a Satan. If he saw forty, fifty elephants, all of them would not live long. All by one person. ... All he wanted was the ivory."[39] *Peturu Bhomba* was of several white huntsmen the government contracted to shoot for ivory as reward for their services. "He would tell us: 'I am the government. It is not me who is killing and taking the horns; it is the government. When you see me, you see government.'"[40] Then there was *Bhatani* (Button), who came from Mutoko. "He would fire only once, and the elephant would not rise where it had fallen."[41]

Most of them were very cruel to *vatema*, just like the majority of *vachena* in Southern Rhodesia, and the isolation of the bush gave them complete impunity to exercise the worst of it. "They called us *bobojani* [*bobjaan*; Afrikaans for baboon]," one *mugocha* said. "'*You fucken bobojaan, baboon wena*' ['You fucking *bobjaan*, you baboon'], they would call us." In those situations, *magocha* just took instructions and followed without question: "'*Yevo nkosi, yevo mambo*' ('Yes my king, yes my lord'). You've done nothing wrong, but you find yourself being given a thorough hiding."[42] A story is told in Nembudziya of one scrawny white man local *magocha* called "*Kamadendere*, a man with not an iota of dignity, so thin that if you nudged him, he would fall easily." He was one of a group of soldiers that President Kenneth Kaunda had expelled from Zambia at independence in 1964. He was a sadistic racist. One day, he ordered a *mugocha* named Machazura to get into the Land Rover (see figure 9.6), drove out into the bush, ordered him into a prone position, and severely thrashed him. Then he ordered him again into the truck and drove him back to join others.[43] Another time, Kamadendere "yoked us all to pull a wagon. Sixteen of us. And it had no tubes like today's carts—iron wheels. He ordered '*mhuka*' (us) to drive 'other *mhuka*' (each other, i.e., other blacks), taking turns on the yoke, swinging the whip on each other's bodies as one would upon oxen, but without hides, the lashes lacerating our flesh."[44]

Some of these *vachena* stole the pay intended for *magocha* under their command. Former *magocha* in Gokwe talk of TFOs like Razmas (Erasmus), who received far more money from headquarters than they paid out as wages to the hunters. "Because the crookery was in the pen," recalls one hunter, a veteran of the Copper Queen Mine game-elimination drive.

Figure 9.6
The Land Rover was a symbol of white racist oppression to the black majority in Rhodesia.
Source: National Archives of Zimbabwe.

"When Razmas says that he gave you so much money and here is where you signed, the superiors will never argue with that." Such complaints led the government to send the paymaster in person to hand each *mugocha* his pay, instead of just giving it all to the TFO, and to go through the process of payment with him.[45]

The white man would be clad in government-issued long-sleeved camouflage or khakis, a cap, a warm jersey, a trench coat, and military-style boots, and he was housed in a roomy canvas tent or mud and thatch hut. He was equipped in the best manner possible for the ragged mountains, thorny bushes, and cold nights.[46]

By contrast, *magocha* had no uniform; the government issued them none. In the bush, they wore their own clothes, the ones they wore at home when going to the fields, which were *marengeny'a* (tattered clothes). They were going into thorn-filled bush, to engage in rough-and-tumble work;

their thin shirts and trousers would soon be tattered. Then, when returning home, *magocha* put on more decent clothing so that they reappeared from the bush as respectable fathers and husbands returning from fending for their families, not victims of *vachena*.[47] Most were barefoot and later wore *many'atera* (sandals with soles from car tires and straps from cowhide processed in the village); others used boots they had brought from migrant work in South Africa.[48]

Finally, it was easy for the TFO or private hunter to shoot *nzou* and other "whites-only" *mhuka* and outside permitted areas and blame it all on *magocha*. *Munhu mutema* was only allowed to skin the carcass and chop out the ivory, carry it, and do whatever *vachena* commanded. In Nembudziya, one such white hunter was *Makaingidze* (the Spoiler of Happiness), who in 1967 accused Raymond Muzanenhamo of killing *ngwarati* (sable). In fact, Makangaidze's nephew, John, had killed the *mhuka* in Gandavaroyi. The young boy then lied, saying that he was with Raymond, who he said had shot the animal. Muzanenhamo was promptly arrested, only to be saved by the testimonies of his *magocha* colleagues. He had had enough; he immediately quit and went to look for work in the city.[49] The TFOs at the Chipinda Pools camp in the 1960s were not just supervising; they were also poaching *nhéma* (rhino).[50]

Interviews with former *magocha* and *vanhu* who lived through the 1960s in Gokwe and Chibwedziva show that *nyama* was so abundant that it was free. It was made into *chimukuyu* (dried meat), bags and bags of it.[51] The wives, mothers, and sisters of *magocha* would pick up the dried meat and carry it on their heads back to *misha*, while even more was stashed, sack upon sack, in the hollow bellies of the baobab trees that punctuate the Gonarezhou and Savé-Runde river valleys. *Magocha* became the vehicle for *vatema* to access the protein of the forest, entry to which was forbidden to them under Rhodesian law. In Nembudziya, *magocha* camps were sited at boreholes at Vashe, Madzivanyika, Mawere, Magumise, Tengdam, Misi, and Magocha, the last one named after the hunters who frequented the camp. Many of these hunters were *vechishangwe*, the most well-known being their supervisor Kaingidza Mudimu, who hailed from Goredema. Locals befriended these men to obtain *nyama*.[52]

From Indiscriminate to Selective Game Elimination

Fema Ngonda was a *mufrayi* and *mugocha* at different times in Nembudziya during the 1960s. He remembers clearly the *mhuka* he was required to kill—four types: *nhoro* (kudu), *dzoma* (bushbuck), *nguruve* (bushpig), and *njiri* (warthog). "*Idzodzo ndidzo dzakanga dziine ropa raitandiwa nemhesvi*" ("those ones had blood most coveted by *mhesvi*"), Ngonda recalls. The hunters were not allowed to kill anything else.[53] Ngonda's account illustrates the implementation of a recommendation made after the 1954 Commission of Inquiry to move from indiscriminate toward "discriminative game elimination."

The supervision tightened. From 1958 to 1963, all *magocha* were required to base at fixed camps known to the TFO and accessible by road. In Sebungwe, *magocha* were now deployed in rotation in groups of twos and threes to "fixed camps located so as to achieve optimum coverage of the area, taking into account places that were particularly attractive to animals."[54] At night, the hunters slept in semipermanent camps, hunting from dawn to noon, then breaking for lunch before hunting again from 2:00 p.m. to dusk.[55] These camps were called *vhuka lala* (*vhuka* = wake up; *lala* = sleep; see figure 9.7), located in a radius of 7–10 km from the next water point. They were rendezvous for the night only, the hunters making camp toward sunset and starting early the next morning. They had only *moto* (fire) and

Figure 9.7
A *vhuka lala* camp, with a TFO's vehicle parked close by and *magocha* busy with cleaning and cooking chores. The TFO is in the background, apparently stripping and cleaning his firearm.
Source: The Rhodesian Annual 1932.

no blankets, and they roasted *nyama* from their kills.[56] Some worked from home;[57] others from camps inside the bush.[58] The primary determinant for the *magocha*'s base was the availability of drinking water, usually from boreholes or perennial pools nearby.[59] Later, in the late 1960s, most *magocha* were from the hunting area; they were supplied with firearms, ammunition, pots, meal, and salt. They were also issued two-man tents, and deployed to specific areas and to camps around each area.[60]

From 1963 on, another move was made away from "hunters ... dispersed in groups of 3" toward a new system involving "hunters operating from one camp, thus enabling stricter control." Twenty *magocha* operated from one camp under a TFO, returning there every night. A senior tsetse field officer (STFO) administered a cluster of these twenty-man teams. Each *mugocha* was supposed to "account for every shot fired" with his TFO; in turn, the TFO submitted a monthly return to the STFO detailing the number of *mhuka* killed and shots expended on each. This enabled the field officers to determine the average number of shots fired per animal killed "as a measure of the efficiency of the hunting teams."[61]

In 1970, the principle that "a high density of hunters is necessary for the removal of the last few hosts" was adopted but not uniformly applied.[62] The unfavorable results led Davison to advise in 1973 against deploying big teams of hunters, preferring smaller ones instead. He decried the increase in hunter strength per team in Guruve to over twenty as a bad idea, proposing instead "to widen the distribution of effort each month with more teams" to provide "constant effort over the entire front."[63]

After centralizing control of the hunters, the next step was to improve their shooting skills so that they shot as well as they tracked—the latter a skill gained from childhood through *ruzivo rwevatema*. In 1970–1971, the Tsetse Branch instituted a comprehensive "hunter training programme." The glossinologist's report from 1971–1972 describes the program involving, inter alia, marksmanship with "specially designed 3-inch by 3-inch targets ... obtained on an exchange basis with a city packaging firm." He notes that hunters and "some" officers had "enthusiastically received" the frequent target practice and periodic competitions, which early on exposed "a chronic weakness in our system in the form of appalling shooting ability amongst the majority of hunters." Davison was disappointed but hopeful:

> Even some of those employed for several years and who produce kills regularly were found to be very poor shots. That this deficiency can be remedied has been demonstrated by some officers who by a concerted effort have raised the standard of shooting in their teams to a very high level. The post of Learner Hunter recently instituted

will, it is hoped, create an incentive for new hunters to work hard at the basic skills of shooting in the first few months of employment. If as previously suggested the Bonus incentive can be increased to $2 per kill, I envisage a three tier hunter force. Learners at $7 per month and little chance of earning bonus money, Hunters at $9 per month getting some bonus money, and Good Hunters who by killing 4–8 animals a month earn most of their cash from the bonus scheme.[64]

The branch was enmeshed in a catch-22: By reducing incentives, it would also inadvertently be promoting poaching. In the 1971–1972 season, numerous cases "both out of the area and [of] shooting non-selected species by our hunters" had been reported. Davison did not need to search far for an answer: "There is little doubt that with more money in the tribal areas than previously, and more widespread settlement, the incentive to poach has increased sharply in recent years." Davison had established during hunter training programs that *magocha* with particularly bad marksmanship had turned in higher rates of kills than their (lack of) shooting skills deserved. To him, these were the poaching culprits. "This method can in my opinion be used to almost stamp out poaching amongst hunters," he declared triumphantly. "If we constantly monitor their shooting ability any undue misses should arouse suspicion." There was only one small glitch: "Unfortunately the next step, to back track and check spoor at the sight where the miss was fired, is not readily accepted by the field officers-in-charge. It of course involves physical effort which is beyond the ability of some of the older members of the staff, and beneath the dignity of some of the younger officers."[65]

In 1973, the branch installed a policy stating that a TFO must meet certain standards "within a few months or he is replaced." These standards included conducting range (target practice) to improve the hunters' shooting skills. Few TFOs enthusiastically greeted this extra work of conducting range for their *magocha*. Davison was a bit heartened that shooting had improved overall, but the shot per kill ratio was "still alarmingly high in some areas and at certain times of the year." Although BTTC policy required TFOs to check on shots *magocha* declared as "misses," "very few" such investigations ever took place, leaving "some doubt as to whether all these shots were fired at selected species or at others."[66]

Training counted for nothing so long as staff turnover remained high. From 1971 to 1972, Davison observed that many *magocha* were increasingly "of extremely poor quality," with new hunters "heavily" outnumbering the old and not staying long at all, unlike the old-timers. Only 20 percent of *magocha* hired for tsetse control operations in Sebungwe, Gokwe, Hurungwe, and Guruve from 1972 to 1973 stayed the whole season. Half the hunters hired worked for five months or less, although Sebungwe and Hurungwe fared better. The vacancy rate for Sebungwe was 57 percent and Hurungwe 44 percent, whereas Gokwe and Guruve were at 68 percent and 66 percent, respectively. Data on previous seasons was not available, but these figures convinced Davison that field staff "grossly exaggerated" staff shortages as an alibi for poor results.[67] Almost to a person, former *magocha* interviewed in Chibwedziva and Nembudziya confirm these shortages, stating the cruelty of TFOs, extremely poor wages and conditions of service, agricultural chores at home, and greener pastures in cities and mines as reasons for quitting.

Davison attributed the reasons for departure to the rise of the rural bus transport system bringing the city closer to previously remote areas. "The lure of higher wages and bright lights in towns—recently rendered 'nearby' by bus services, are attracting many tribesmen to leave the Tribal Trust Lands in search of jobs," he said. "The greatest incentive we can offer is cash earned near home. Whilst it is beyond our duties to strive for all round pay increases, the Bonus Scheme is one system we can manipulate relatively easily to fulfill our objectives."[68]

A study of figures from the northern regions shows that many *magocha* returned home from October to November to plow the fields so that their wives and children could plant crops, then would leave them to handle the weeding while they returned to hunting soon after. They returned home for Christmas (when they had their "off") with fresh meat, then (some) returned to their bush workplaces in January, and a few more in February. Davison's conclusion that "the payment of a bonus on the basis of a full season's attendance may go some way in preventing good hunters drifting in and out of employment as they wish" illustrates the agency of *magocha* in influencing conditions of service and the entire tsetse control operation—just by going home to perform their duties as fathers, husbands, sons, and siblings.[69]

The first attempt to control hunters through money-denominated incentives or restrictions arose in 1955 with the introduction of a salary. Chorley had reasoned that this "would enable the TFO i/c [in charge] to exercise greater control over the hunters and ensure a better distribution of the rifles."[70] However, as the 1960s wore on, it became clear to *magocha* that they were being poorly paid deliberately to induce them to hunt more and sell.[71] The bonus scheme introduced in 1970 had the double objective of "increas[ing] the welfare of the long serving hunters ... in the form of a bonus on kills" and creating an incentive "for all hunters to put more effort into the task" of hunting. To fund the initiative, the number of hunters per team was reduced by four to create a savings of eleven dollars per head, or Rhodesian dollars (R$)11,760 for the whole year for all *magocha*. By the end of the 1970–1971 hunting season, only 23.7 percent of the bonus (R$3,027) had been paid out, which to Glossinologist Davison was "a dismal failure" that had "not provided the desirable increase in the hunters' welfare." In some areas, "so little money [was being] earned and by so few that the scheme [had become] worthless." The whole plan boomeranged after the initial success, and in 1971, Davison noted that "latterly the effect has worn off and even some dissention has arisen due to the unequal chances of earning bonus amongst the teams."[72]

Chief Glossinologist Desmond Lovemore did not agree with Davison's conclusion that the bonus scheme had failed. The kills from June 1970 to January 1971 showed that despite reductions in the number of hunters per team from twenty-four to twenty in three areas, the total number of kills made in each case was "well in excess" of that for the previous season's corresponding period from June 1969 to January 1970.[73] Davison, meanwhile, conceded that during the 1971–1972 hunting season, the total bonuses paid to hunters under the bonus scheme was $300 less than the previous year's figure of $2,720. Instead, the amounts paid out were "split evenly amongst the [460-strong] hunter force [and] amount[ed] to a benefit of $5,9 each over the year." However, he maintained that the distribution was not even, with many hunters getting no more than one or two dollars from the scheme for the whole year. The glossinologist found the scheme problematic because it did "not provide any incentive to the majority of our hunters." His solution was simple enough: Increase the bonus per kill. In the likely event of kills exceeding the bonus budget available, the expenditure would then be limited "by raising the qualifying number [of kills] for teams in these sections."[74]

From 1972 to 1973, the bonus scheme for *dzoma* was discontinued. The result was a steep drop from 612 to 295 *dzoma* killed, a rise in the numbers of *nhoro* from 91 to 150, and constant figures for *njiri* and *nguruve*. Davison concluded: "It is also to some extent a demonstration of what a cash incentive can do amongst hunters."[75] That same season, the bonus on each kill was raised to two dollars "to increase the incentive and to ensure better distribution of the available cash."[76]

The changes in bonuses and the mechanism and criteria for payouts, as Davison explains, unleashed interesting kinds of subversion from *magocha*, felt by the government as follows:

> The total amount paid out this year ([R]$4,963) is almost double that of last year, ([R]$2,720). In this manner 80,1% of the funds were used as intended, as opposed

to 35,9% in the preceding year. The adjustment to the amount to be paid per kill has therefore achieved its objective, and greatly increased the welfare of successful hunters. The over-expenditure in the [Hurungwe] area was the result of the high number of kills obtained in the [Matsikiti] Section when the area in fact had the lowest monthly allocation—[R]$120 which allows for only 60 kills. In the Gokwe Area some resistance and disobedience was encountered when the qualifying number was set high to curb expenditure. Under these circumstances hunters were more likely to sell products to the local [villagers] than declare kills in the hope of qualifying for bonus payments. This practice, and that of forming syndicates to qualify more rapidly, has exposed a weakness in the present system when applied to areas yielding a lot of kills. Since no such further areas remain ... to be cleared however, the problem is resolved. However, the low earnings in the [Guruve] area indicate a problem which lies ahead, namely a decrease in the availability of hosts which will of course, decrease benefits paid out. One anticipates sharp reductions in the Gokwe and [Hurungwe] sectors in the coming year. If this prediction comes true it is suggested that in future years, part of the available money be spent on an "attendance bonus" as practiced with spraying labourers. This may go some way in solving the problem [of] the turnover of hunters.[77]

Conclusion

Because *magocha* operated virtually on their own (singly or in groups), they acquired such latitude as to determine every outcome of game elimination: the statistics the department produced, the conclusions they made of them, and the policy and "scientific" decisions arising from such conclusions. TFOs who were supposed to supervise them were located at camps far from the hunters, and most did not stay on the job long enough. This meant that the only chain of consistency tying the *mhesvi* operations together was *magocha*.

The hunters were not here for *murungu*'s "scientific research"—far from it. They were here for *nyama* and *mari* (money). Yet in killing for meat, they also killed for science and policy. *Hurumende* cared little for the nutritional needs—let alone welfare—of *magocha*. Yet in deferring to and availing them of space, guns, and ammunition, it provided them the instruments to become *magocha* and a rich vein of *nyama* supply to *musha*. The name *Mugocha* or *Magocha* stuck to some of these men and became their first or even last names—for example, Mugocha Julius Mavasa of Chibwedziva. Every animal shot for *nyama* was, in any case, one less potential carrier of *mhesvi*. That, too, was good policy.

10 The Coming of the Organochlorine Pesticide

The methods of the Gaza king Mzila—from whence Rhodesian authorities derived "Umzila's principle"—had been organic. *Hurumende* maintained his methods of clearing and keeping the land free of *mhesvi,* but the materials they used were toxic to the environment. The principles had shifted: *vatema* managed *mhesvi* through strategic deployment within the environment, rather than exterminating *mhesvi,* its habitat, and *mhuka,* whereas the Rhodesian state sought to control and eradicate it. We have already discussed the trail of forest and animal destruction left behind in pursuit of this goal.

We now add another dimension: the extensive poisoning of the environment to exterminate *mhesvi.* The deployment of *chepfu* to eliminate or control *mhuka,* both big and microscopic, that were dangerous to *vanhu, zvipfuyo,* and *zvirimwa* (crops) from the beginning of the Rhodesia project in 1890 is the subject of this chapter. It argues that the use of synthetic chemical poisons, while marking a significant turning point, was not entirely a new phenomenon in the history of chemical production and usage in local societies. Such innovations in *chepfu* (poisons), as will be shown, had existed before, but—tellingly—in organic, biodegradable, hydrodegradable, and photodegradable forms.

Given the extensive destruction of trees via both mechanical and chemical means, the massive amounts of organochlorine pesticides (OCPs) dumped into the environment to kill *mhesvi,* and the poisoning of *mombe* in massive chemotherapeutic interventions, OCPs immediately present an opportunity to explore the question of pollution and its health effects. Substantial research has been done on occupational health issues related to mining in Southern Rhodesia and South Africa (Van Onselen 1982a, b; Packard 1989; Phimister 1994), but mostly concerned with the political economy of migrant labor located at the intersection of race and class. Some studies have begun to explore citizen mobilization and activism relating to asbestosis in the Northern Cape and Swaziland, with a sustained focus on the scientific-medical aspects (Waldman 2007; McCulloch 2002, 2005). They echo Sheila Jasanoff's (1995) "science at the bar" approach (Meeran 2003), with its emphasis on locating occupational health and wellness at the intersection of law and science. In recent years, STS scholars have focused on the materiality of contamination and how it happens, with close attention paid to uranium, asbestos, and the chemicals used to extract them (Hecht 2012).

The purpose of this very brief chapter is a modest one: to introduce and account for the specific circumstances by which OCPs arrived in Southern Rhodesia. In fact, by the time organochlorines like DDT, BHC, and dieldrin and organophosphates like Thallium were deployed in combat against *mhesvi, hutunga, hwiza* (locusts), and *zvimokoto* (quelea birds) after World War II, Southern Rhodesia's farmers had been dispatching *mhuka, shiri, zvipukanana,* and *hutachiwana* with *chepfu* through ingestion, inhalation, and skin contact for over fifty years. The chapter therefore starts from this earlier history, well before DDT and its peers, in search of antecedents that profoundly shaped and offered a broader context for the use of OCPs.

Chapters 10, 11, and 12 build on and contribute to the global histories of pesticides, with the present one setting up the discussion of DDT and its organochlorine associates BHC, dieldrin, and, to a lesser extent, endosulfan. This chapter is concerned primarily with pre- and non-OCP history that places the coming of DDT and other OCPs in context. The story of DDT itself has received distinguished attention (Russell 2001; Stapleton 2005), with emphasis placed on experts and expertise, regulation, environmentalism, chemistry, and entomology. Most of the discussion centers on agriculture, armies, and indoor use, however (see also Russell 1999).

The discussion that follows does something different. First, it will introduce OCPs as material things, before discussing their coming to Southern Rhodesia, with emphasis on the circumstances of their arrival. The argument advanced is that this is not merely a case of knowledge transfer—that is, of a knowledge already developed and proven workable in Europe and North America. To start, like all other ingredients local and inbound, OCPs were coming in as raw materials to assemble stratagems that local actors were designing against *mhesvi*. OCPs were subjected to local experiments precisely because of the nature of the pest that Southern Rhodesia was fighting: a mobile *hutachiwana*-carrying *chipukanana* that inhabited climatic, geophysical, vegetational, and human environments different from those in the United States and Europe, where the chemicals had been designed.

Unlike preceding ones, this chapter has no conclusion, because it is the opening dialogue for a story continued in the next two chapters.

DDT, Dieldren, and BHC

Three organochlorine pesticides (OCPs) arrived in Southern Rhodesia just after 1945 and dominated *mhesvi* control operations for much of the late Rhodesian period. They were DDT (dichlorodiphenyltrichloroethane), lindane (gamma-hexachlorocyclohexane), and dieldrin. Later, they were largely replaced with two OCPs: thiodan (also called endosulfan) and deltamethrin. This section first explains what these chemicals were and how and for what they were originally designed in the United States and Europe, as well as the circumstances of their travel and deployment in *mhesvi*-occupied Africa, focusing on Southern Rhodesia (later Zimbabwe).

In the global OCP discussion, Africa does not exist; an impression might be created that the chemical was never applied. The most extensively used of the three OCPS in Zimbabwe, DDT is a colorless, crystalline, tasteless, and almost odorless organochloride first synthesized in 1874 through mixing chloral (CCl_3CHO) and chlorobenzene (C_6H_5Cl) in a sulfuric acid catalyst. When it decomposes and loses some of its constituent parts, DDT becomes DDD (dichlorodiphenyldichloroethane) or DDE (dichlorodiphenyldichloroethylene). DDT is a hydrophobic chemical that is highly soluble in fats and oils. It acts through skin absorption and ingestion; once inside the body, it attacks the nervous system by interfering with normal nerve responses. Its immediate effects in *vanhu* and *mhuka* are less toxic because it is poorly absorbed through thick skin, but laboratory animals express hyperexcitability, tremors, incoordination, and convulsions, with fatal doses producing liver lesions. In *vanhu*, exposure leads to a prickling sensation of the mouth, nausea, dizziness, confusion, headache, lethargy, incoordination, vomiting, fatigue, and tremors. Its effects in *zvipukanana* are far more rapid and lethal because the chemical is easily absorbed through its outer covering (the exoskeleton).

DDT was first used in a pesticidal role in 1942 by the US Army to kill *hutunga*, *zvikwekwe* (ticks), *inda* (lice), and *mbeva* (mice), and from 1946 onwards it became the signature pesticide for malaria and *mhesvi* control throughout Africa. After World War II ended, the chemical became a civilian pesticide in agricultural pest control and in campaigns to kill *hutunga* and stop malaria in Europe and North America (WHO 1979; de Zulueta 1990). The antimalarial role of DDT went global in 1955 under the World Health Organization's program covering North Africa, the Balkans, the Caribbean,

Northern Australia, Southeast Asia, and the South Pacific (Gladwell 2001; Chapin and Wasserstrom 1981; Sadasivaiah, Tozan, and Breman 2007). Today, DDT remains in use for malaria control in South America, Africa, and Asia; in 2016, calls intensified for its use against Zika virus–carrying *hutunga* in Brazil. In addition to its use in the control of *mhesvi,* the chemical has been extensively deployed in Africa to protect crops and fruit trees against worms and other pests and indoors to kill *masvosve* (ants), *mapete* (cockroaches), *nhunzi* (houseflies), and *makonzo* (rats). DDT has remained in use for as long as it has because it is inexpensive to make, is effective, and has a long residual effect in the environment.

First believed to kill only *zvipukanana* and liberally deployed for such, DDT has since proven to be both a pollutant and a health hazard for plants, *mhuka,* and *vanhu* alike (Allen et al. 1979a, 513). It is very "lipid soluble" (can easily dissolve in fatty substances), it has a very long "half-life" of six to ten years in *vanhu,* and it is very slowly released from fat once absorbed. That is why the Stockholm Convention on Persistent Organic Pollutants (2001) moved to limit its use to the control of public health disease vectors like *nyong'o* and *n'gana* through killing and repelling *hutunga* and *mhesvi,* respectively. DDT did not become a pesticide overnight. Even as its postwar use shifted to civilian pest control, scientists in the United States—and newspapers such as the *New York Times*—were already saying that the chemical was a health hazard (EPA 1975). They were largely ignored; however, the awareness campaign attracted the attention of the naturalist Rachel Carson. From an op-ed in the *New Yorker,* Carson produced the seminal 1962 book *Silent Spring,* which argued that OCPs were poisoning humans and the biotic environment (Lear 2007). The popularity of the book spawned the rise of the postwar environmental movement in the United States, forcing President John F. Kennedy to institute a commission of inquiry into Carson's findings. The subsequent Scientific Advisory Committee advised the use of DDT and other OCPs to be discontinued (Greenberg 1963).

In the antipesticide and antichemical environment of the 1960s, DDT became a focal point of attack, with Carson the ammunition and rallying point. The Environmental Defense Fund (EDF), an organization composed of scientists and lawyers formed in 1967, produced increasing evidence of the lethal effects of OCPs on bird populations and successfully petitioned the government to ban the chemicals altogether. DDT use was duly suspended (*Time* 1971). Accusations of bias created such controversy that the Environmental Protection Agency (EPA) was forced to convene seven months of hearings from 1971 to 1972. In 1972, the EPA finally banned all uses of DDT barring a few public health ones, and even those only under stringent conditions. The lawsuits from the DDT firms began; countersuits from the EDF followed, one seeking a reversal, and another a total prohibition, until the Supreme Court came down on the side of the EPA in 1973 (EPA 1975). After the decision, DDT was used only in exceptional public health cases, such as outbreaks of potentially epidemic-causing pests, like an outbreak of fleas in 1979.

One of the major paradoxes of US biosafety and industrial regulations is that the US government bans certain dangerous chemicals and pharmaceutical drugs and institutes recalls of automobiles and other products at home, but allows US companies to continue making and exporting such dangerous products and exposing countries of the Global South to them. Also striking is the fact that the United States has, for pragmatic reasons in the national interest, refused to sign the Stockholm Convention on Persistent Organic Pollutants, and, after the domestic ban, still allow its companies to make these chemicals and export them for use outside the United States despite their known health and environmental effects ("Report of the Expert Group" 2010).

In the aftermath of the DDT ban, production for the Global South market continued at an average of three hundred tons of poison a year, in response to the pestiferous mobilities of *mhesvi*, *hutunga*, and other insects (USDHHS 2002). Twenty-six countries—among them Cuba, Singapore, Chile, and Korea—had banned DDT by 1986. Over 170 countries later ratified the Stockholm Convention, which became international law in 2004 and limited DDT only to WHO-approved *hutunga* and *mhesvi* control (UNEP 2001).

The second OCP is lindane, also known as Gamatox or benzene hexachloride (BHC) in Southern Rhodesia, but more appropriately as hexachlorocyclohexane (γ-HCH). *Lindane* is a neurotoxic organochlorine that interferes with the central nervous system, the liver, and the kidneys in *mhuka*. Lindane was registered in the late 1940s. In *zvipukanana*, it kills through skin and egg contact and absorption and through ingestion. Lindane first acquired global "fame" as a pesticide used against *mhesvi* and *hutunga* in Africa in the 1950s. However, the OCP was also lethal against sucking and biting pests, grain sores, soil pests like fleas, beetles, and mushroom fleas, and *zvipukanana* that attacked crops in the soil (Kumar and Kumar 2007, 2). It was considered deadlier to *zvipukanana* than vertebrates, although severe exposure in *vanhu* resulted in nausea, vomiting, diarrhea, muscle fibrillation, tremor, and convulsions. BHC also damaged tissue in testicles, kidneys, skin, and liver in *mhuka* (Videla, Barros, and Junqueira 1990).

Lindane was first synthesized in 1825 by the English scientist Michael Faraday (1791–1867) and is named after Teunis van der Linden (1884–1965), the Dutch chemist who first isolated and described γ-HCH in 1912. Only in 1942 were its pesticidal properties first discovered; thereafter, Imperial Chemical Industries Ltd (ICI), the largest manufacturer of chemical products in twentieth-century Britain, started manufacturing it for wider-scale use. Based out of London, from ICI's formation in 1926 the company specialized in chemicals, explosives, fertilizers, pesticides, dyestuffs, paints (specifically the brand Dulux, dominant in southern Africa), the cloth-making fibers Terylene (also called *tererini* among *vatema*) and nylon, and nonferrous metals. In World War II, ICI participated in Tube Alloys, Britain's nuclear weapons program (Smith et al. 2008). Without doubt, the company's "gift" to Africa remained lindane, used to treat grain harvests against weevils and for mass killing of *mhesvi* and *hutunga*. In its long history, lindane was made in Britain, Europe, the United States, China, Brazil, India, and Russia (Commission for Environmental Cooperation 2006). Only in 2009 was its production and use in farming banned under the Stockholm Convention. In the United States, lindane continued to be manufactured by Morton Grove Pharmaceuticals, which as late as 2007 received a warning from the Food and Drug Agency (FDA) to correct information on its website that omitted and minimized the health risks of its products (CEC 2006; "Report of the Conference" 2009).

The last OCP, which was also used as an anti-*mhesvi* pesticide in Rhodesia from the late 1950s until independence, was *dieldrin*, a much more toxic organochlorine compound than DDT. It was first industrially produced as an alternative pesticide to DDT in Denver, Colorado, by the US company J. Hyman & Co. in 1948. In 1987, it was discontinued worldwide when the manufacturer canceled its registration. Alongside chlordane, aldrin, heptachlor, and endrin, dieldrin belongs to a family of chemical *chepfu* called *cyclodienes* that are "less acutely toxic but of greater potential for chronic toxicity than the organophosphate and carbamate insecticides" (Allen et al. 1979a, 518).

Like DDT and lindane, dieldrin killed *zvipukanana* through skin contact and ingestion of contaminated matter. This OCP was widely used in agriculture and insect pest control from the mid-1940s to mid-1960s, but, like DDT, its persistence in the environment would lead to its termination

(Boryslawsky et al. 1985). Dieldrin was named after the Diels-Alder reaction process, through which norbornadiene and hexachlorocyclopentadiene are synthesized into a chemical pesticide. It is extremely persistent in the environment and does not easily break down—again like DDT and lindane, a characteristic that made it effective as a pesticide but also dangerous. Dieldrin was used against parasitic *zvipukanana* like termites, blowflies, ticks, and lice, and widely in cattle and sheep dips, to protect fabric from moths, beetles, and carrot and cabbage root flies, and as seed dressing against wheat and bulb fly (Kumar and Kumar 2007, 2). Although dieldrin is now banned in most of the Global North, aldrin continues to be used as an anti-termite in most of Africa.

The Coming of OCPs to Southern Rhodesia

OCPs arrived in eastern and southern Africa right after the end of World War II. The East African Tsetse Research Organization conducted experiments to find out how best to use OCPs to kill *mhesvi*. The first involved finding out if the *chipukanana* could die when exposed to OCPs through skin contact. In this investigation, the pesticide was applied directly on the skin initially, and then on traps. The answer, contained in a report in 1947, was affirmative: OCPs could kill *mhesvi* through skin contact (Vanderplank 1947). These experiments have already been discussed in chapter 4 alongside other traps and need no further mention here. The second experiment involved ground and aerial spraying of the insect's hides, breeding places, and shelter. This is our concern in this and in the next two chapters.

The general concept underpinning OCP use remained the same, however; the insecticide worked through a residual killing effect built up through a lethal deposit of sprays or dust, or aerosol smokes, fogs, or fine sprays. At least until 1951, residual and aerosol spraying were not combined. Spraying was divided into two types depending on the killing effect. The *residual killing effect* was achieved through selective application to vegetation from the ground using sprays, non-selective application to vegetation from the air with sprays, impregnated screens, dusts, and bait animals sprayed with insecticide. The *immediate killing effect* was achieved using smokes from generators, aerosol fogs from the Todd Insecticidal Fog-Producing Apparatus (TIFA), and aerosols and finely atomized sprays from *ndege* (aircraft).[1] Residual killing using Four Oak machines had been tried on islands in Lake Victoria against *mhesvirupani*. DDT and BHC oil solutions were applied initially, but the residual effect was lost quickly because of absorption by leaves. Later, DDT and BHC emulsions were used with a 98 percent kill rate during the peak *zvikukwa* period. However, this was a method ill-suited to *mhesvirutondo,* a savannah type ranging all over that could only be sprayed in its resting and breeding sites, along the game tracks, and at waterholes. Nonselective application on four Lake Victoria islands had shown that canopy and unpredictable meteorological conditions "made the difficulties of achieving anything approximating a complete kill as difficult as they could be."[2]

In a bid to find a less laborious method than spraying by hand, "impregnated screens" were stationed at likely places on Lake Victoria and sprayed with a weekly dose of DDT and BHC. The 50 percent success rate, already unimpressive in *mhesvirupani,* was declared virtually useless against *mhesvirutondo,* for which stationary screens and traps were unattractive. Nonetheless, it could still be useful against the thicket-favoring *mhesvirupani.* DDT and BHC dusts had been used in the mountains and broken areas of KwaZulu unsuited to fixed-wing *ndege*; hence, they were sprayed by hand and with mechanically operated dusters. However, only DDT ended up being used because BHC "proved irritating to the operators."[3]

By 1950, DDT had already begun to be sprayed indoors in both "European areas" and "African areas" against malaria in the Mazowe area. DDT was used in the former, whereas BHC was deployed for spraying huts (cottages and compounds)—66,712 huts of *vashandi vatema* (black workers) in total—and barring the "objectionable odour," it had been very effective. Interestingly, the archives show *vatema* pleasantly surprised by the mass destruction of *hutunga, zvikwekwe* (ticks), and *mapete* (cockroaches; Alexander and Ranger 1998, 209). On the face of it, a Daniel Headrick–style "tools of empire" scenario, in which *ruzivo rwevatema* were enabling *vachena* to colonize the territory, seemed to be under way.

Not so with *mhesvi,* at least initially. The most intriguing aspect of residual *mushonga* involved bait oxen dipped or sprayed with *mushonga.* Both at Shinyanga in Tanzania and in KwaZulu, such *mombe* were driven everyday into a small block of bush infested with *mhesvirupani.* The Shinyanga experiment failed because *mombe* were driven in after and taken out before *mhesvi*'s feeding times (early morning and evening, respectively) to save them from marauding *shumba.* It is possible that the experiment would have succeeded against *mhesvirutondo,* which fed by day. The South Africa Division of Veterinary Services' KwaZulu experiment succeeded against *mhesvirupani,* but the residual films (DDT) were only effective for five to six days, then became harmless. A Uganda trial with BHC had shown that "the dose needed to produce toxicity in the ... tsetse is too near that point harmful to the animal."[4] The South Africans switched to BHC because it was more economic (du Toit 1954). Rhodesia then invited the South Africans in 1953–1954 to help stop the *mhesvirutondo* advance in Hurungwe with BHC.

Dieldrin, from dieldrex 15 percent, product of Shell Chemicals of Central Africa (Vale 1968), inaugurated a new era in *mhesvi* control operations, "in which the basis of control was dependent on the residual properties of the insecticide."[5] The first operation of this kind in Southern Rhodesia was conducted at Kapondo in the northwestern Hurungwe District in 1958. On February 6, an experiment was completed that tested the reaction of blowflies in the laboratory to dieldrin applied as a residual film and compared it to the effects of spraying residual films in the field.[6] The conclusion was that "a 4% concentrated dieldrin applied as a residual film is the most efficient in causing death provided the flies are exposed to it for more than 0.5 minutes."[7]

The dieldrin experiments of 1958 were followed with another hugely successful one in the Maseme River drainage area of Binga in 1960. The method used was taken from west and east Africa, and involved ground teams using pneumatic or motorized knapsack sprayers to apply *mushonga* to dry-season resting places and refuge sites of *mhesvi*. To succeed, it was critical to spray *mushonga* before the extreme dry season began—that is, before the end of September. Thus, all ground operations were conducted from the beginning of June to the end of September so that "the insecticide used ... remain[ed] strongly lethal for a minimum of six months after being placed in position."[8] Successful ground application of dieldrin on a selective basis relied on the theory that during the hot, dry season, *mhesvi* concentrated in areas of evergreen vegetation to survive. These habitats were riverine fringes, vlei edges, and hills. *Mhesvi* retreated from open woodland to the shade and humidity of the forest's edges and riverine thickets, making them the perfect foci for selective spraying.

The success of the Maseme and Hurungwe experiments paved the way for the extension of 3.1 percent dieldrin spraying in all tsetse control areas of Southern Rhodesia until 1967, when the move to the much cheaper 5 percent DDT was made. The switch "was only made after the longevity of the residual properties" of DDT had been demonstrated in the baking-hot conditions of the Zambezi valley. The *mushonga* remained effective "well in excess of eight months and, in fact, DDT proved to be superior to dieldrin, which was being tested simultaneously." Comparatively so much cheaper was DDT that "it became possible to operate over very much more extensive areas of country than had hitherto been possible, which went a long way to overcoming the problem of subsequent reinvasion."[9]

In 1970, tests ongoing at Rekomichi Research Station since 1969 sought to establish whether poles treated with white DDT attracted more or fewer flies compared to those treated with gray DDT or to untreated trellis poles. The trial was inconclusive. Experiments also switched to finding the most effective spraying techniques. Further trials were conducted on whether a sticker might be used to prevent deposits of DDT WP from being washed away by rain. In October 1972, baobab and mopane trees were sprayed with several formulations of 5 percent DDT WP suspension and the sticker, but it made little to no difference. Trials were continued in 1973 using stronger DDT concentrations.[10] In 1974, the experiment was discontinued altogether because the sticker proved to be unsatisfactory.[11]

From 1971 to 1975, a total of 48,168 km^2 were sprayed with a total of 1.2 million kilograms of DDT. At the same time, on farms, farmers were using 1,470 kg/ha of the 75 percent wettable powder or 1,102 kg/ha of active ingredient, four times every season, year after year. The powder was also sprayed in foliage and fruit trees, for cutworm control in maize and tobacco, for stock-borers in maize, in vegetable gardens, and to control bollworm in cotton production. Most of the pesticide fell to the ground or simply drifted. The defenders of DDT use in *mhesvi*-control work pointed out that farmers were using a far higher DDT concentration compared to the "only" 5 percent DDT BTTC was using. Desmond Lovemore, assistant director of Tsetse and Trypanosomiasis Control, mounted a spirited defense of DDT in 1976–1978, citing the latest studies performed since 1970:

> In tsetse control spraying, ... the minimal quantities applied are selectively placed, very often in situations where little wash can occur. It is also usual for there to be only one application ever to an area, although certain areas had to be re-treated several times over successive years before elimination was eventually achieved. ... No obvious effects on animal, bird, reptile, fish and other insect life have been observed during the very large scale spraying operations which have been conducted with DDT in Rhodesia. Field staff have been instructed prior to each operation to pay particular attention to this important aspect over the years, but nothing of interest has been recorded. Similarly, as regards the more insidious effects of the chemical the work done by Phelps and others at the University of Rhodesia has shown that no serious problem has developed as yet from tsetse control operations in Rhodesia. ... It is also noteworthy that in Nigeria where DDT has been used very extensively in tsetse control operations, in fact, probably very much more so than in Rhodesia, no serious "side-effects" have been noted.[12]

For these reasons, Lovemore remained adamant that DDT would continue to be the preferred choice for *mhesvi* control operations. However, concerns about its use could no longer be ignored.

In the late 1970s, therefore, the Tsetse Control Branch sought alternatives to 5 percent DDT, partly for economic reasons (it was too expensive) but also in response to public (white) outcry against DDT. The branch focused on two chemical alternatives. One was to reduce the strength of DDT from 5 percent to 2.5 percent, based on what other African countries were doing and on research at Rekomichi. In general, the research found "very little difference" in overall effect between 5 percent and 2.5 percent mixes. Flies lived marginally longer when sprayed with the latter, but still died; invaders were indeed caught, but "did not apparently survive long enough in the 2.5% block to re-colonise it." Although investigators were optimistic, the experiment did not conclusively resolve the question of whether 2.5 percent was adequate or equivalent to the usual 5 percent, precisely because the two blocks sprayed with each dosage had different features and "required different treatment techniques."[13]

Sounding tentative and guarded, the director of Veterinary Services concluded in his annual report:

> It seems there are grounds to reduce the concentration of the suspension used to some degree, provided planners ensure all likely habitats are well treated—the objective of good planning anyway. If by reducing the concentration it is necessary to raise the application rate per unit of area, there may be no benefit in terms of active ingredient dispersed. In this trial it did seem that the lower amount of active ingredient was adequate, so routine planning principles as applied by locally trained glossinologists may not have to be modified to compensate for a decrease in concentration. It is important to ensure this point, for, if indeed a weaker suspension does require a wider distribution to be effective, then other significant cost factors such as transport and labour, currently running at approximately 34% and 15% of total cost respectively, will be elevated![14]

The second move was toward replacing DDT altogether with a rapidly degradable substitute. In 1973, five years after DDT replaced dieldrin as the first-line *mushonga* against *mhesvi*, the BTTC commenced experiments to compare the effectiveness of DDT and endosulfan. The latter *mushonga* was aerially sprayed using an ultralow volume of 20 percent endosulfan, applied to a 261 km^2 woodland infested with *mhesvi* in Chirisa Game Reserve. Six cycles at intervals of lengths between seventeen and twenty-two days and approximating to seven liters per km^2 were applied between June and November. The results were impressive; 99 percent control was obtained. The tests continued into 1975, but the final lines of Assistant Director

Desmond Lovemore's 1974 report are instructive: "Effects of the insecticide on a variety of insects, fish, mice, frogs, and geckoes were also studied. All test creatures, except the fish, which were rapidly killed (but their situation was highly artificial and exposed), appeared unharmed."[15]

Even in the chemical-intensive agricultural sector, the importation of three hundred tons of DDT for the 1977–1978 growing season belied an already unfolding shift to the more readily degradable OCP, endosulfan. Until then, trials to determine the viability of endosulfan as an alternative had proved it to be not persistent enough and more expensive.[16] The ultralow volume (ULV) application of *mushonga* using fixed-wing *ndege* operating at night was still at the experimental stage, "some years" away from being perfected enough to replace DDT (Chapman 1976). Even then, it would only complement ground spraying with DDT in areas where selective application was unavoidable.[17]

11 Bombing Flies

The first aerial spraying operations in Africa were conducted in South Africa right after World War II. In 1945, DDT became the first synthetic hydrocarbon *mushonga* to be produced in the Union of South Africa and available for the control of *mhesvi*. The war in Europe had ended. Pilots and *ndege* (fixed-wing aircraft) serving abroad returned home and became surplus to air force requirements, just when the *mpukane* situation in KwaZulu was getting out of hand. No tried and tested method yet existed for applying pesticides using *ndege* (du Toit 1954; du Toit and Kluge 1949).

The first experiment was undertaken in Mkuze Game Reserve, a high-density *mhesvirupani* area, from December 1945 to January 1946. Finely atomized droplets of DDT were sprayed at two- to three-week intervals, resulting in drops in the weekly totals of *mhesvi* caught in the Harris traps from seven thousand to between six and seven hundred per week. The numbers remained constant for three months, and then escalated sharply as the summer progressed. It was clear that six weeks of spraying was adequate to destroy adult *mhesvi*, but too short to destroy those emerging from *zvikukwa* after spraying. The method of application needed improvement—for example, adding course markings to aid navigation, when more than one pilot was involved and *ndege* shared reciprocal parallel boundaries. Finally, the spraying could be more discriminate and based on concrete intelligence about *mhesvi* presence (du Toit 1954, 1959).

An extended campaign in the Umfolozi Game Reserve began in April 1947, covering all permanent breeding areas. The fixed-wing *ndege* failed to access the mountainous and bushy northern parts of Hluhluwe Game Reserve and areas along the western reaches of the Mkuze and Pongola Rivers, however. In 1951, heavier military *ndege* (fixed-wing) gave way to light commercial Piper Aztecs (Cruiser and Super Cub) capable of maneuvering ragged mountain terrain hitherto negotiable only with *zvikopokopo* (helicopters). The operation ended in 1953 with the total eradication of the *mhesvirupani* (du Toit 1959, 237–238).

The Southern Rhodesia aerial campaign against *mpukane* was born in KwaZulu. *Ndege* were adopted for pesticide spraying because of their ease of use, capacity for large-scale coverage, very few personnel required, and capability to reach *mpukane* habitats otherwise inaccessible by ground spraying. State officials in Rhodesia considered that aerial applications of *umuthi* (pesticides) were likely to result in greater environmental contamination than ground spraying. The solution was to reduce the dosage rates and include ambient pesticide-monitoring techniques. Aerial spraying required higher initial financial outlay than ground spraying, but the spraying itself was cheaper than ground spraying per unit area covered, depending on the type of *umuthi* used and the ground sprayed. These savings came from the use of fewer personnel, reduced logistic requirements, and simplified operational planning. Finally, *flymachina* (flying machine, airplane) could take off from and return to airbases in the city for servicing and repairs, whereas it was difficult to do the same with dust-corrupted, bumped and bruised ground-spraying machines in the middle of remote areas. Apparently, this was also the expectation and experience when airplanes were first used in pest-control work in the United States in the 1930s (FAO 1977).

In Southern Rhodesia, aerial spraying developed along two trajectories—namely, nonresidual (beginning in 1948) and residual spraying (1969 onwards). The first method involved the sequential application of tiny droplets of concentrated *mushonga* into savannah woodlands to kill *mhesvirutondo* while they rested in the tree branches or flew around in a panic (Hadaway and Barlow 1965). By contrast, residual spraying—the second method—was simply ground spraying adapted to aerial methods and involved spraying *mhesvi* in their habitat, leaving a residue that killed them through skin contact long afterward. This method was favored particularly for riverine vegetation, drainage lines, and *ecotones* (areas where two vegetational communities converged; FAO 1977).

Southern Rhodesia's aerial spraying began in 1950 when Dr. Rene du Toit, subdirector of the Union's Division of Veterinary Services at Onderstepoort, visited Southern Rhodesia to advise staff on the application of *mishonga* from *ndege*. Chorley secured a *ndege* from the Southern Rhodesia Air Force for a reconnaissance flight over Hurungwe District.[1]

In 1951, a master plan was drawn up for "a large field-scale experiment using *ndege* for the application of insecticide in aerosol form."[2] The operations only began on November 30 the following year, when the rainy season was under way, and they continued through the end of March 1953. The

South African contractor and the pilots that flew the three planes involved brought to Rhodesia vast experience from KwaZulu. The techniques were also based on this prior experience; their opinions were "accepted without any preliminary experimental work."[3]

The spray area was chosen because it was where *mhesvi* congregated during the dry season. It lay between the lower Msukwe and Badze Rivers, just west of Hurungwe—"a natural line of dispersal into the reserve"—as well as into the white-owned Karoi Block. Detailed tests were conducted to determine aerosol behavior under different temperatures, wind, and other conditions. Du Toit had originally planned for six applications at twenty-eight-day intervals; in practice, it was discovered that three planes could cover the originally targeted area in just fourteen days. Therefore, the area was widened to include the Rongatutu River system to the north, and the spraying cycle was reduced from twenty-eight to fourteen days.[4]

The heavy rains made communications a nightmare. Wireless signals were initiated, but contact between the pilots and the aerodrome "could not be made before the aircraft took off at dawn." The coordination between the pilots flying and spraying above and the flyround teams measuring the effectiveness of the spraying below proved tenuous, the aerosol coverage—and kills—uneven. The "kill" on the Msukwe-Badze system was "reasonably satisfactory," that on the Rongatutu "poor." This unevenness in results was attributed to the difference in vegetation and poor pilot ground observations due to the heavy rains. The director, Du Toit, and the general manager of the South African spraying company visited the area and decided to suspend operations in late March until the rains had subsided.[5]

In the 1953–1954 operational year, in cooperation with the Departments of Civil Aviation and Irrigation, an airstrip (see figure 11.1) was constructed at Zvipani with *vatema* commandeered by the Native Commissioner.[6] From this advanced airstrip, two to three sorties could be conducted every day, taking off and landing, refueling and replenishing *mishonga* supplies, thus eliminating the dead time that existed previously when fixed-wing *ndege* had to return to Salisbury to perform such tasks.[7]

The spraying operations started in early July 1954 with the objective of covering all the river systems and dry-season concentration areas of *mhesvi* between the Badze and Kanyati Rivers. During the first cycle, the planes, taking the Badze as their starting point, failed to reach their target, the Kanyati; subsequent sorties had to be abandoned.[8]

By the end of September, only the original Badze-Msukwe area had been covered. Meteorological conditions were worsening with every subsequent

Figure 11.1
Airstrips like these were used for speeding up operations by creating a field base for fuel and ammunition supplies and for landing and taking off without having to return to urban-based airports. The picture shows insecticide drums (foreground) and fuel.
Source: Allsopp 1990.

cycle. In the hot, dry season, the high winds in the morning and the heat-induced atmospheric turbulence made flying and the control of aerosol discharge impossible.[9] The operations doddered for two more months, but by the end of November conditions had deteriorated beyond flight safety. The operation was abandoned.[10]

As this chapter will now show, these operations were only the beginning. For the rest of the century, BTTC airborne operations would become more sophisticated and dynamic. This chapter will discuss the technical aspects of aerial spraying, treating them as an example of the extension of means and ways designed in the United States for agricultural or military purposes to deal with a *chipukanana* and conditions for which they were not originally designed. It will focus on fixed-wing *ndege* first and then turn to *zvikopokopo*. Figure 6.1 shows all the areas in Rhodesia mentioned in this chapter. The glossary at the back of the book will aid in the understanding of *chidzimbahwe* and other local keywords.

The Tiger Moth and Avro Anson XIX: Kariba, 1955–1956

In the 1955–1956 operational season, plans were drawn up to conduct "block application" and "linear application" to areas adjacent to the township of Kariba and outlying rivers, respectively. The aerial spraying target was an eight-square-mile area including the Kasese River, the main access road, and the township. Eight applications were scheduled, involving a 4 percent BHC solution mixed in diesoline (to make it stick to and penetrate target) sprayed at an average rate of 0.1 gallons per acre. The operations were scheduled to begin on May 1, 1955. A local contractor, Messrs. Skyworks (Pvt.) Ltd., was hired to do the job, with four Tiger Moth *ndege* powered by Gipsy Major Series I engines and each bird carrying a fifty-gallon tank of *mushonga* (see figure 11.2).[11]

The de Havilland Tiger Moth was a 1930s biplane named after its designer, Geoffrey de Haviland, and a product of de Havilland Aircraft Company (UK), which saw service in the UK Royal Air Force until 1952. Following the adoption of the de Havilland Chipmunk as the preferred primary trainer, the Tiger Moth became excess ordnance and was decommissioned for civilian

Figure 11.2
Two fixed-wing *ndege* on an aerial-spraying pass over northern districts.
Source: Proceedings and Transactions of the Rhodesia Science Association 1960.

use. The Tiger Moth itself had retired the de Havilland Gipsy Moth which had made its maiden flight in 1931 in response to the British Air Ministry's request for a *ndege* with a more accessible cockpit. It was standard training procedure that the front seat occupant of this bird must have ease of escape with a parachute strapped on in case it was going down. The Gipsy Moth's fuel tank was directly above and severely limited access to the front cockpit (Hotson 1983; Bain 1992). By contrast, the Tiger Moth was powered by the Gipsy Major, a four-cylinder, air-cooled, inline engine, standard for 1930s light *ndege*. Its cylinder pointed downward under the crankcase, thus keeping the propeller shaft in high position, so that the cylinders steered clear of the pilot's view past the bird's nose. Early on, the *ndege* consumed too much oil, and the tank (located outside) needed constant refueling, forcing frequent landing and takeoff. To remedy the problem, the piston rings were simply modified (Bransom 2005).

In the 1955–1956 operations, the Tiger Moth was mounted with fifty-gallon tanks of 4 percent BHC solution in diesoline, delivered by electric pumps to the exhaust stacks. The droplet sizes of aerosol emitted from the exhaust stacks were tested and adjustments made to spray nozzle sizes until the required size was achieved. The spray unit was then standardized so that the nozzle size was set for this type of *ndege*. In several sorties, tests were repeated on the droplet size and aerosol delivery rate, using magnesium oxide plates as indicators of droplet size.[12] Unlike in 1953–1954, the 1955–1956 operations went smoothly throughout the entire eight-cycle spray routine thanks to a prolonged spell of suitable weather conditions. The wet season had gone on longer than normal, and leaf fall had been delayed, allowing the completion of the eighth spraying cycle before *mhesvi* had concentrated in the riverine vegetation.

Also in action over Kariba in the 1955–1956 operation was another British exmilitary *ndege*, the twin-engined Avro Anson XIX, again operated by Skyworks. This *ndege*, named after British admiral George Anson, made its maiden flight in 1935. Avro was a British *ndege* maker established in 1910 in Manchester but based in Lancashire; its birds saw action in both world wars and in the Cold War—the trainer Avro 594 in World War I, the Avro Lancaster in World War II, and the Avro Vulcan in the Cold War. *Avro* is an acronym formed from the name of the company founder, Alliott Verdon Roe. The company initially was called A. V. Roe and Company. The Avro Anson was designed for maritime reconnaissance, only to prove virtually worthless in that role, so it was redeployed as a multiengine aircrew trainer instead. The earlier version of the Anson, the Mark I, had a wooden wing made of spruce and plywood and a fuselage made of steel tubing wrapped in fabric, and the nose of the *ndege* was coated with magnesium alloy. Two Armstrong Siddeley Cheetah IX seven-cylinder, air-cooled radial engines, each with 350 horsepower (260 kW), powered the *ndege*. In its maritime recce role, the Anson had a three-man (later four-man) crew composed of the pilot, navigator or bomb-aimer, and radio-operator or gunner. Its wings could carry 360 lbs. worth of payload, whereas the front fuselage had fixed Vickers machine guns dashboard-operated by the pilot. In addition, the gunner operated another machine gun mounted on a turret (Holmes 2004; Jackson 1965). By the time the Anson was discontinued in 1952, 11,020 birds had been manufactured.

What changed? Spray nozzles replaced gun barrels. The Avro Anson XIX *ndege* was powered by two Cheetah XIX engines loaded with two one-hundred-gallon tanks of 4 percent gamma BHC in diesoline. The *mushonga* was introduced into the *ndege*'s exhaust stacks using electrically driven impeller-type pumps to produce the aerosol.[13] By 1955, the *ndege*'s spray system had already experienced several innovations.

In the first East Africa trials of the 1940s, Avro Anson XIX *ndege* were fitted with four fifty-gallon spray tanks. They were fitted in such a way that the spray came out through gravitation force from two wide pipes extending some thirty-five centimeters below the *ndege*'s fuselage. At the end of each pipe was an Iris diaphragm to adjust the emission rate of the spray. The pilot could make these adjustments in flight. To atomize the spray, the pilot worked with the slipstream to break up the liquid coming out of the pipe into spray droplets. However, too many droplet size options often presented problems of evenness and effectiveness, and this spraying system was subsequently terminated (FAO 1977).

Another early spraying method of the 1940s, first tried in South Africa and Kenya, was to turn the Avro Anson's thermal exhaust into a spraying machine. This method simply involved letting *mushonga* move down into the exhaust system through a narrow pipe, so that it was then emitted in an upright position down into the slipstream thirty centimeters below the rear edge of the *ndege*'s wing (du Toit 1954). Meanwhile in Kenya and Tanganyika, major advances in air-to-ground and air-to-air insect spraying were achieved against locust swarms (Gunn et al. 1948a, b).

Entomologist R. J. Phelps oversaw the Southern Rhodesia operation. From May to September 1957, Phelps's job was to standardize the dosage rate and droplet size, decide when and where to spray, and record the *mushonga*'s effects on the *mhesvi* population. The workday started at four in the morning and ended at about ten o'clock in the night, the planes taking off whenever weather conditions allowed.[14]

Terrain preordained the complementary deployment of the Tiger Moth and the Avro Anson in Kariba. The two were intended to spray the flat areas, flying at 120 miles per hour over a swath seventy-yards wide, each bird dumping one gallon of *mushonga* in 27.6 seconds. The Tiger Moth, meanwhile, was assigned to the more rugged country and along the riverine fringes, flying at about 80 mph and with a twenty-five-yard swath. Each *shiri* discharged one gallon of *mushonga* in 62.5 seconds. Aside from the terrain, there was a problem of far-from-ideal weather conditions, characterized by very strong northeasterly winds that restricted the amount of flying time available.[15]

The spraying operation was divided into a twenty-one-day cycle, which was the standard for treatment, corresponding to the breeding cycle of *mhesvi*. The weather had other ideas, and the cycle was accomplished only once, with the effect that "a female larva deposited immediately before an application of *mushonga* could mature, become adult and mate, but would encounter an application of *mushonga* before dropping its first larva."[16] The aerial sprays were sustained for 135 days, enough time to cover the phase of *zvikukwa* deposited prior to spraying.

In all, spraying constituted just 8.5 percent of the Tiger Moths' flying time and 35 percent of the Avros'. The low efficiency was not blamed "on pilots or ground crews, who exhibited great skill and patience at all times, but [was] an indication of the difficulty of flying along the narrower river courses," of time lost while maneuvering the *ndege* after making a spray run, and of "obsolete aircraft."[17]

Overall, the treatment was declared a success. The valley floor application had been effective, even though the Chikomba vlei traverse had shown that the linear treatment was unreliable for achieving satisfactory kills from May to September. At 14s. 22d. per acre for six applications, including *mushonga* and hiring the *ndege*, it was less than half the cost of the Kariba aerial spraying, in which ten applications were made.[18] Even after the Tsetse Branch felt that its shift was done, the Federal Power Board, concerned about the impact of *mhesvi* and *n'gana* on the construction crews building the Kariba dam and power station, continued the operations for a further three cycles, focusing on the riverine vegetation.[19]

To do this, more *mushonga* supplies were required instantaneously. The South African company Klipfontein Organic Products undertook to send the concentrated BHC solution by railroad in three days to Kariba. Further delays in transport meant that breaks occurred between the application cycles, and when the ninth, tenth, and eleventh cycles were finally deposited on the riverine vegetation, the trees were already in leaf and *mhesvi* made good its escape. Still, the objective of the spray was achieved within the budget, and by 1956 the acting director concluded that the concentrations of *mhesvi* between the Kasese River and the Kariba Gorge construction site had been vanquished—barring the upper headwaters, where the vegetation was too dense and flying conditions too severe for effective aerosol application. Still, along the riverine thickets of the Chavaru and Nyanyanya Rivers seven and nine miles from the construction site, *mhesvi* had been reduced only, far from eliminated.[20]

The 1955–1956 aerial-spraying operations at Kariba ended in late October 1956 on a high note. Subsequently, flyrounds were maintained throughout the year to keep *mhesvi* under surveillance, track its postspraying behavior, and respond according to the intelligence gathered. As expected, right after the treatment was concluded, a rapid buildup of *mhesvi* threatened to overrun the sprayed areas from the unsprayed ones nearby. The recovery in the *mhesvi* population of the sprayed area from the Kasese River eastward was swift; between that river and Kariba township, however, there were no *mhesvi*. At the dam construction site and Nyamhunga township, *gopé* (sleeping sickness) was recorded in *imbwa*, but none in *vanhu*.[21]

The aerial spraying succeeded in significantly reducing the *mhesvi* population based on pre- and post-treatment catches, only for the numbers to build up rapidly again, exceeding the pre-spray figures. The entomologists concluded this was due to immigrants from untreated areas following *mhuka* now roaming freely after the removal of *vatema* who had hunted them to make way for the Kariba dam, especially in the Kasese River area. Overall, the aerial spraying had reduced the *mhesvi* population in the dam site area and averted an impending sleeping sickness hazard to workers. As the waters began to fill the entire area, *mhuka* would either drown en masse and die or flee to small islands, severely limiting *mhesvi*'s food source and transport. Such habitat was being systematically destroyed, the bush sprayed; the *chipukanana* would have nowhere to hide from the air and ground assault.[22]

On one point, entomologist Rawdon Goodier was clear. The aerial spray over Kariba left more questions than definitive answers:

> There has been a rapid reinvasion of tsetse from the surrounding infested country, at a rate that was foreseen. It is now evident that to have achieved elimination of tsetse between the Nyanyanya River and the dam site, and to have maintained it free for a period of 18 months, it would have been necessary to treat a far larger area. How much larger the area would have had to have been to achieve the desired result one can only guess but it may well be something in the region of at least five times the block spray area and probably considerably greater. Riverine spraying alone must be

considered unsatisfactory as the concentration of tsetse on the rivers cannot be relied upon for more than a brief period and this period occurs at a time when weather conditions are far from ideal for spraying.[23]

The Piper Aztec (PA-23 Pawnee): The 1974–1975 and 1982 Operations

In 1974, a twin-engine Piper Aztec *ndege* fitted with a single Micronair spray system was used to spray the Chirisa Game Reserve in Gokwe, having been successfully used in clearing *mhesvi*-infested areas of the Okavango Delta of Botswana the previous year (Kendrick and Alsop 1974; Lee et al. 1975; Chapman 1976). Also called the Piper PA-23 or simply the Apache or the Aztec, this *ndege* was a four- to six-seater twin engine initially designed by Stinson Aircraft Company of Dayton, Ohio. The company was established in 1920 by Edward Stinson and later moved the bulk of its operations to Detroit, Michigan, under the name Stinson Aircraft Syndicate. The site of its factory was what is now Detroit Metropolitan Wayne County Airport (established 1920). After World War II, Stinson Aircraft Corporation entered into several buyouts by bigger corporations, eventually being sold to the Piper Aircraft Corporation in 1950. It was at this point that first the Piper Apache and then the more formidable Piper Aztec entered the scene. These four- to six-seater twin-engine light *ndege* were designed for the US Navy and for air forces of friendly countries as late as the 1980s. When Piper acquired Stinson's Consolidated Vultee Aircraft Corporation, it also took over the latter's Twin Stinson design and developed the Piper Apache (later Piper Aztec) 23 (PA-23). On its test flight in 1952, the *ndege* was a four-seater, low-wing, all-metal monoplane equipped with two Lycoming O-290-D piston (125 horsepower) engines. It failed the test, prompting a new design with a single vertical stabilizer, all-metal rear fuselage, and 150 horsepower engine in 1953 (Peperell and Smith 1987).

Designed for agricultural purposes, the PA-25 Pawnee became (along with the Cessna) the signature aerial insecticide spraying *ndege* throughout the world, including Africa, from 1959 to the 1980s. Before 1949, the bulk of *ndege* deployed for agricultural purposes in the United States were retrofitted military birds, but in 1949, Fred Weick of Texas A&M University designed the AG-1, dedicated specifically to agricultural spraying purposes. The following year, the bird successfully completed trials. In 1953, Piper made Weick its consultant on a project to create an agricultural version of the PA-1 capable of distributing pesticide dust and seed; that is how the PA-18A was born. Another Piper grant later, Texas A&M developed the AG-3, a fusion of compatible AG-1, PA-18A, and PA-22 elements.

Smaller than the AG-1, with steel-tube fuselage, fabric covered, this single-seater, low-wing monoplane was equipped with conventional landing gear, a tailwheel, a 135-horsepower engine, and an 800 lb. capacity hopper in front of the cockpit. The pilot's high seat in the fuselage allowed for clear visibility. The bird was tested successfully in 1957 and was renamed the PA-25 Pawnee, now outfitted with a 150-horsepower Lycoming O-320-A1A engine. Other generations of the Pawnee followed (Peperell and Smith 1987). Today, its design rights and technical support are (since 1988) owned by Argentina's Latino Americana de Aviación, again showing the Global Southernization of companies or their artifacts (Peperell and Smith 1987).

The 1974 operations over Chirisa Game Reverse were conducted using an ex-military PA-23 and intended to spray endosulfan (thiodan) at a strength of 20 percent active ingredient. These were small-scale spraying trials, not synchronized or mutually complementary to adjacent ground-spraying operations.[24] The *mushonga* was delivered at 5.62 liters per minute to a wind-operated Micronair AU 3000 rotary atomizer fitted to the wing of a twin-engine Aztec *ndege* operated at 8,500 revolutions per minute. The *ndege* flew at 150 miles per hour and twenty-five to thirty meters above the ground (treetop level), depositing swaths of *mushonga* at two-hundred-meter intervals over 130 square miles. Judging the prevailing wind direction was important to accurately predict wind drift relative to the positioning of the spray. The start and end points were clearly marked with pencil flares and twelve-volt spotlights operated from roads or other features visible from the air. The best time to use aerosols was when the *mhesvi* was at the *chikukwa* and *chiguraura* stages of its life cycle.[25]

The spraying campaign may have knocked back the *mhesvi* population in the experimental area, but it failed to attain the objective of eradicating the *chipukanana* in the spray area. Just why this was the case was not certain, but one explanation may have been that the spray area covered was too small relative to possibilities of reinvasion. Another reason may have been that the concentration of *mushonga* was insufficient to achieve a total—or at least effective—kill of all the adult female flies, especially *mhesvirupani*. Overall, the method was found to be impractical and inefficient, and further trials were ordered in 1975 just to be sure.[26]

The Tsetse Branch now cleared a 732 km^2 area inside Chirisa Game Reserve for the new trials. The purpose of the spray was to "reduce the probability of rapid invasion"; to achieve this, the two-hundred-meter swath was doubled. To cater to the rough terrain, the maneuverable single-engine Pawnee was used in place of the Aztec. The spraying was conducted at night, except in the more broken terrain, which was sprayed in the morning and early evening to capitalize on daylight. Small flares and lights were replaced with "a very bright light (adapted from a photoflash unit) flashing every 3-5 sec and carried up to 200m by a hydrogen-filled meteorological balloon," itself "raised or lowered rapidly with a rod and reel and ... transported between marker stations in a protective cage fixed behind a Land Rover." The speed of the Micronair atomizer was increased to 10,000 revolutions per minute (rpm) to reduce spray droplet sizes. It had been found that the smaller the droplets were, the more effective the spray was because "the small appendages of the flies collected such droplets."[27]

The spray area was divided into a central region and a perimeter region. The central region was subjected to intensive spraying of 70 percent solution applied in swaths 200 meters apart to monitor the effects of *mushonga*. The perimeter region was treated at half the rate (35 percent applied in swaths 400 meters apart) as a perimeter "fence" to protect the central region against reinvasion from the surrounding bush. The core area was divided in two again, with one section given a 20 percent thiodan treatment, the other 25 percent thiodan. The operation started on July 10 and terminated on September 19. Five applications of *mushonga* at intervals of nineteen, nineteen, sixteen, and thirteen days were deposited, making the most effective use of the first deposition of *zviguraura* and late phases of *zvikukwa*. Very good tsetse control was accomplished overall.[28]

In 1982, BTTC started a program of postwar aerial-spraying operations to arrest the advance of *mhesvi* in the Zambezi valley area of Gokwe and Sebungwe adjacent to Lake Kariba. Later, the operations moved northeast along the shoreline, the idea being that the lake was a hydro-defensive shield against a *mhesvi* invasion (Allsop 1991, 7). The 1982 operation in Sengwa and Sesami was a combination, for the very first time, of ultralow volume (ULV), nonresidual aerial spraying and normal 5 percent DDT suspension ground spraying to address a rapidly deteriorating *mhesvi* situation in Gokwe. The Department of Veterinary Services engaged the services of a contractor, Messrs. Agricair (Pvt) Ltd. of Harare, and used thiodan (endosulfan) made by Hoechst, Zimbabwe (Pvt) Ltd. Two Piper Aztecs and a Turbo Thrush were used. The latter was a low-wing, single-seater monoplane specifically designed for agricultural purposes by Leland Snow and flown first in 1956. It was manufactured by Ayres Corporation of Georgia in the United States. The turboprop engine was a 1980s development, prior to which the Thrush had been powered by a radial piston engine (Green 1964; Simpson 2005).

The differences and similarities between the two planes are important. The Thrush was a conventional crop sprayer; the Aztecs were originally designed as light twin-engine military personnel carriers and were thus modified to carry *mushonga* and navigation equipment. The Thrush was virtually brand-new—the latest means there was on the market. The Aztecs, by contrast, were old birds, "and this together with the fact that they were being used outside their designer's intentions generated a lot of technical problems." The crop sprayer was designed for navigation at night, when meteorological conditions "favoured the sinking of near weightless micro-droplets down into the woodland."[29] The terrain too was difficult: escarpments and plateaus like Domwe Hill required that pilots climb steeply from treetop height to around two-hundred to one thousand meters before reaching the clear sky above them. The turbo-charged Thrush could do this without any problem; not so the lumbering Aztecs, which had to commence the climb in good time. To aid night vision, the obstructions were marked clearly with flashing strobe lights inserted by a Bell 47 helicopter (*chikopokopo*) a day earlier. Premarked baselines also guided the *ndege*'s direction of flight and spray. Each *shiri* was also fitted with a track-guidance system to allow parallel runs.[30]

The mission of the Aztec was clear: to penetrate and spray areas of the drainages inaccessible on foot, extend the area targeted for treatment, and "give the operation greater depth."[31] The combined strategy required commencing the ground spraying well before the Aztecs took off to give the DDT enough time to take effect. This ensured that the adult flies were killed in the areas surrounding the vleis or river lines targeted for aerial spraying; if any were present, they would move in and deposit *zviguraura* after the planes sprayed the first cycle. Subsequent cycles were intended to cover just one *chikukwa* period, and females invading between cycles would deposit their *zviguraura* in time for them to hatch after the fifth and final aerial treatment, thus rendering the entire effort null.

The operations began on the night of July 27, 1982, and were completed without incident by the end of September. Between 14.7 and 25 grams of active ingredient were deposited per hectare from the air. The droplet pattern each *ndege* emitted was established by collections of droplets on rotating magnesium oxide–coated glass slides before and after the spraying, with each night's work monitored by droplet collections and by three mobile ground teams. Total eradication was achieved: Not a single female adult *mhesvi* was caught; all flies caught were recently hatched. The young females dissected showed a severely disrupted mating or insemination due to residual effect, thus achieving delayed reproduction and buying *mhesvi* operations more time.[32] The only drawback was the loss of twenty-nine atomizers that broke, burned out, or simply dropped off in flight, severely hampering the free flow of operations (Allsopp and Hursey 1986, 34).

The 1983 operation was a continuation of the 1982 combined air and ground attack, the objective being to drive *mhesvi* toward Lake Kariba, this time targeting the area between the Chizarira escarpment and the shoreline. The 1982 operations had concentrated on the Simchembu *mhesvi* refuge; those of 1983 focused on a 2,100 km^2 area of Binga, with no prominent features, yet still rugged, undulating terrain that worsened as the planes headed east. The gorges of the Rwizirukuru River valley marking the eastern edge of the spray area were quite steep, thus making the topographical conditions in 1983 much like those of the previous year. Unlike in 1982, however, the first cycle spray used deltamethrin in diesoline solvent at 0.25 g/ha, as per an agreement between the Tsetse Branch and Wellcome, which was testing the insecticide as a possible alternative to DDT. The *mushonga* made "a respectable reduction" in *mhesvi* but did not achieve 100 percent success. The failure to eliminate *mhesvi* from the Rwizirukuru valley was attributed not just to deltamethrin, or subsequently endosulfan, but was an indictment of fixed-wing aerial spraying as a method in general (Hursey and Allsopp 1984).

The Cessna 401 in the Chizarira and Matusadona Operations: 1984–1988

The third fixed-wing *ndege* deployed in Rhodesia was the Cessna, in its several varieties. Elsewhere in Africa, the Cessna 180 and 310 were used (Lee 1969; Lee and Miller 1966; Baldry 1971; Lee et al. 1975; Lee, Pope, and Bowles 1977; Park et al. 1972; Hocking et al. 1966). Our focus here is on the Cessna 401 used in spraying the Chizarira escarpment near Kariba in 1984, four years after independence. The *ndege* was manufactured by Cessna Aircraft Company, a US general aviation aircraft-manufacturing corporation based in Wichita, Kansas (Phillips 1986). The 1984 aerial spraying operation in Chizarira was conducted using two such turboprop-powered Cessna 401s.

With its distinctive four small oval windows, the six- to ten-seater, light-twin, piston engine Cessna 401 was one of the business jets named Businessliner or Utiliner that Cessna had been making since 1966 with affordability as a key selling point. The seats were detachable, and the *ndege* could be used for other utility purposes—hence the name Utiliner. Cessna 401s and 402s were nonpressurized and rather slow in speed, being powered by three-hundred-horsepower turbo-charged Continental engines with three-bladed, constant speed, fully feathering propellers. Models built from 1966 onwards were limited to 75 percent cruise power, and some were fitted with propeller synchrophasers to reduce cabin noise. The turboprop-powered conversion of the *ndege* began in 1969, with the objective of increasing fuel tank capacity, gross weight, and speed control; it was completed in 1974 (Plane and Pilot 1978; Montgomery and Foster 1992). Two such turbo-propelled engines were involved in the 1984 campaign.

The two fixed-wing *ndege* covered 1,700 km^2 north of the Chizarira escarpment, stretching west toward Mcheni River and the Ume to the east. Accurate aircraft navigation relied on Decca Doppler equipment connected to a tactical air navigation system (TACAN) computer, complemented by three ground-based "marker parties" using ground-to-air radios, 15 mm signaling flares, and elevated flashing beacons. This method was used in the previous operation; it worked "satisfactorily." The pesticide was sprayed through wind-driven Micronair AU 5000 atomizers attached to the fuselage, behind and below the wing, with the pesticide drawn from tanks suspended beneath the fuselage.[33] Thiodan was to be applied in five-cycle sequences at dosage rates of twenty-five, eighteen, fourteen, fourteen, and fourteen grams per hectare. The results were inconclusive "because of the inability to determine whether the old flies captured ... survived treatments or had immigrated from the surrounding ground sprayed area."[34]

The blame for the failure of the 1984 operation was placed partly on the almost total absence of localized night winds, which made droplets fall directly down instead of sweeping sideways to penetrate *mhesvi* hideouts between cracks in the bark or underneath leaves and logs. This is where ground spraying excelled, so the teams swept in (Allsopp and Hursey, 1986; Allsop 1991, 8). Even after the dosage strength was increased from 14 g/ha to 18 g/ha for cycles 4 and 5, a low-density residual population of *mhesvi* still remained. In fact, the combined operation was even less successful than those in 1982 and 1983, and a Bell 206 *chikopokopo* had to be brought in to re-treat the Umi valley (Allsopp and Hursey 1986, 16–17).

Several theories were put forth to explain the failure; they reemphasize what happens at the site of encounter between incoming things and local conditions. They are significant because the 1985 aerial campaign was designed to test those hypotheses. One was that the population of *zvikukwa* was too dense and so too were the emerging flies after application cycles 1 and 2; thereafter, there was faster contact and thus mating between male and female. The post-ovulation speed of *zviguraura* development was contingent upon prevailing temperatures: faster if warmer, slower if cooler

(Allsopp and Hursey 1986, 15). The campaign had been delayed, and the closing cycles of spraying had coincided with rising August–September temperatures (23°C–25.4°C) as the southern African summer beckoned. Thus, the cocktail of rapid ovulation and faster development of *zviguraura* unleashed large numbers of heavily pregnant females even before cycle 3 began. They easily resisted the 14.8 g/ha dosage (16).

The second theory was that whereas in the 1982–1983 spraying campaigns two concentrations of 30 percent emulsifiable concentrate (EC) and 20 percent EC had been applied, in 1984 30 percent was used throughout, bar a few drums of old 20 percent stock sprayed in cycle 3. The effect of the dosage reduction was that the number of droplets that might drift through *mhesvi* habitats also decreased. At the time, the role of the spray cloud's physical structure in spray effectiveness was not yet understood, but authorities speculated that smaller droplets in sufficient numbers were "the most lethal component of the spray cloud" (Allsopp and Hursey 1986, 16).

The third theory was that failure was not to be assessed just at cycle 3, but for the entire combined ground and air operation. The ground sprayers had failed to rid their assigned tactical area of responsibility of *mhesvi* and consequently to protect the aerially sprayed area from reinvasion. It was thus impossible to tell whether the old females in sprayed areas were survivors or invaders. Precedent had shown the river-hugging *mhesvirupani* to travel much further and more rapidly than previously believed. In other words, the problem did not lie with the aerial spraying itself; ground spraying seemed "too slow for it to be entirely effective in this role" (Allsopp and Hursey 1986, 16).

The fourth hypothesis was the meteorological effect—that is, the presence or absence of specific wind conditions determined the effectiveness of the spray. At a mean wind speed of 2 m/s, 20–30 micrometer (μm) droplets usually traveled between three and nine kilometers downwind, so the wind carried the aerosol sideways into the hidden sides and undersides of trees, logs, rocks, and leaves. With weaker or zero wind speed, the droplets fell vertically, thus leaving *mhesvi* resting under leaves and logs or in bark and rock crevices untouched (Allsopp and Hursey 1986, 17).

By 1985, a low-density *mhesvi* population was building in the northern section of the area sprayed the previous year, growing much heavier to the east, between Sebungwe and Omay. Rather than simply targeting these residual populations, the combined operation sought to eradicate dense *mhesvi* buildup in a much larger area of the Matusadonha Game Reserve extending into the strip between the Sengwa River (west) and Siakobvu.

The purpose of the aerial spraying was to establish why the technique had failed in areas like Siakobvu—albeit without jeopardizing the campaign, which was generally a success.

The operations started in late July with two Cessna 401s and were conducted exactly as in 1984—with two twin-engine Cessna 401s flying in formation, starting at six in the evening and ending at six in the morning during moonlit nights, restricted to three early evening and one early morning sorties. Then, for selective treatment in difficult terrain—especially the Sengwa and Umi escarpment—the Jet Ranger was deployed. The Jet Ranger and Bell 47 were also deployed to position and service the warning beacons on dangerous obstructions like hills (see figure 11.3). However, they were fitted with the latest Micronair AU 4000 rotary atomizers, which still had the cage diameter of the original AU 3000 (i.e., six inches) (see figure 11.4a, b) but were shorter and faster. The *mushonga* used was endosulfan at 30 percent for cycles 1 and 2, reduced to 20 percent thereafter.[35] In other words, the modification of the spraying technique was an experiment to assess the capability of fixed-wing *ndege* for spraying to treat "stubborn" *mhesvi* presence. The high-density *mhesvi* population was concentrated in mopane

Figure 11.3
Chikopokopo: A Bell 47 positioning a beacon prior to spraying.
Source: Allsopp 1990.

a) b)

Figure 11.4a, b
Unmounted Micronair AU 4000 (left), and mounted Micronair AU 4000 (right), with a metal or fiberglass shroud to protect the fuselage in case the blades break during flight.
Source: Allsopp 1990.

woodland in flat to undulating terrain—a perfect testing ground for the fixed-wing technique (Allsopp and Hursey 1986, 17).

To test this hypothesis, two adjacent blocks totaling 1,700 km^2 were chosen. One block overlapped the eastern half of the area sprayed the previous year, including the Sengwa River (reinvaded after treatment) and the Siakobvu area (never completely eradicated). To be certain that no distortions to results occurred due to carried *mhesvi*, *vatema* were stationed as "deflying pickets" on roads leading into Siakobvu.

For all the care put into the operation, it still failed to eliminate the *chipukanana* from Siakobvu. The old females apprehended all over the aerially sprayed area had clearly survived the misty bombardment. The theories put forward in 1984 to explain the failure to completely eradicate the pest were now confirmed under experiment to be false—bar one: namely, that the absence of wind or breeze had reduced the droplet efficiency, a conclusion that triggered the start of "wind tunnel studies" in the United Kingdom, which later confirmed this theory to be a fact (Johnstone 1985; Johnstone, Cooper, and Dobson 1987; Johnstone et al. 1988). To make aerial spraying more effective, the meteorological parameters needed to be well understood and the spraying technique adapted accordingly. Night spraying was now to be limited to continuously flat terrain and selective spraying to daytime and to deep river valleys and high escarpments only (Allsopp and Hursey 1986, 33).

There were positives, however. The Cessna 401 was a far superior bird than the aging Aztecs deployed in 1982 and 1983. The spray equipment had been vastly upgraded, with external custom-built tanks and well-secured and efficient Micronair AU 4000 atomizers. Not even one was lost during operations, compared to 1982. The navigation equipment and metering systems had also improved, as had loading, refueling, and servicing procedures (Allsopp and Hursey 1986, 34).

East of Matusadonha, the BTTC, in partnership with the Wellcome Research Laboratory (UK), was conducting a small trial with deltamethrin in the farming areas of Rushinga—specifically, the Chesa small-scale commercial-farming area. *Mhesvirutondo* literally had come to pasture in these farms. The planes used the Rushinga airstrip as an operational base. The sorties began on July 19 and terminated on October 1, using thiodan (endosulfan) applied in five cycles at dosages of 22, 18, 14, 14, and 14 g/ha. Except for the winds that delayed cycle 5 by four and half days, the weather could not have been better. In the end, the results were good, and barring a few holdouts thereafter (which were swiftly cleared) the operation was declared a success.[36]

Wellcome Research Laboratories also conducted another trial in a *mhesvirutondo*-infested bush straddling the Mudzi River near Nyamapanda, funded by the European Economic Community (EEC). Its objective was to determine whether deltamethrin might be a substitute for endosulfan.[37]

The 1987 operation was designed to identify a large, continuous, flat to gently undulating area for aerial spraying using fixed-wing *ndege*—the Cessna 401. Because hills could not be moved aside, they were marked with flares throughout the operation. The target of the spray was a 4,700 km^2 block between the Mozambique border and the Muzarabani-Mukumbura-Chiswiti-eastern Dande area. The Cessnas took off from Rushinga airstrip; the ground control center was at the foot of the escarpment by Musengezi River. Thiodan was the preferred choice of pesticide in five cycles in successive dosages of 22, 20, 16, 14, and 14 g/ha. This was a night-only operation, spraying five cycles from July 13 to September 19, and the weather conditions were generally conducive, allowing a westerly drift of *mushonga* droplets beyond the Manyame River. The *mhesvirutondo* population was wiped out, but the *mhesvirupani* remained even after a sixth cycle was applied. The verdict was that these *zvipukanana* had survived the aerial spray.[38] BTTC then undertook a follow-up operation to remove this residual population—without instant success, even though the *zvipukanana* died out a few months later. No *mhesvi* was captured until August 1988, and then only as the result of reinvasion from Mozambique.

The 1988 operation, also with Cessna 401s, had two aims. The first was to clear *mhesvirutondo* and *mhesvirupani* from a 2,000 km^2 area north of the Zambezi escarpment between the Angwa and Manyame drainages, composed of western Dande and parts of the Dande and Chewore safari areas. Second, the operation was also an experiment to ascertain the effectiveness and environmental effects of deltamethrin as a possible alternative to endosulfan in aerial-spraying operations. The contractor was supposed to use *ndege* capable of taking off from and landing at Mashumbi Pools. Such planes were not available, so the airstrip had to be upgraded to suit the *ndege* available.

The night operations began on June 30 and finished on September 4, having deposited five cycles of pesticide layers at 0.25 g/ha throughout. The weather was good, but still air made for suboptimal conditions in the north and center of the block; consequently, the drift was poor, reducing the effectiveness of the spray. Meanwhile, deltamethrin proved a "highly effective" pesticide against the heavy *mhesvirutondo* concentration; total eradication was achieved. However, as with endosulfan the previous year, the proposed substitute was far less effective against *mhesvirupani*—99 percent at best. As impressive as such a kill rate was, that 1 percent remaining necessitated future retreatment—a negative mark from an economic point of view. *The verdict:* Deltamethrin was neither better nor worse than endosulfan. It could be used in future operations.[39]

The Bell 206 Jet Ranger II

By 1980, *chikopokopo* the helicopter had become a popular instrument for discriminate treatment of *mhesvi* in continuous thickets, riverine forest, and tough-to-reach places. This was not by design; as one researcher noted in 1977, "the choice of a particular technique has been determined to a large extent by the nature of the habitat and the topography of the land" (Lee 1977, 6).

The Bell 206 Jet Ranger (see figure 11.5) was first deployed in *mhesvi* operations in the 1984 campaign in Chizarira, but only in a complementary role to *ndege*. From 1989 to 1990, it was put on trial as the principal sprayer. The *chikopokopo* was a two-bladed, single- or twin-engine craft, made at Bell's Mirabel plant in Quebec, Canada, but it started its life as the Bell YOH-4, intended as a light observation *chikopokopo* for the US Army, which did not adopt it. The company redesigned it as the Bell 206A Jet Ranger, which the US Army then accepted and turned into the OH-58 Kiowa.

Figure 11.5
A Bell Jet Ranger spraying in hilly terrain.
Source: Allsopp 1990.

Several models and generations of the Bell *chikopokopo* were used in a residual spraying role throughout Africa's *mhesvi* flashpoints. For example, in Lambwe Valley, Kenya, in 1968, the Bell 47G was used effectively to apply invert emulsions (oil-based mud) (Le Roux and Platt 1968). A Bell G4A was deployed to spray DDT, dieldrin, and HCH in Niger in 1969 (Spielberger and Abdurrahim 1971). In Zimbabwe's case, the Bell 206 was available, not the Bell 47G.

The site of the spray in 1989 was a block of 126 km^2 at Shamrock Mine in Hurungwe. The Bell Ranger was equipped with two Micronair AU 4000 atomizers spraying 30 percent thiodan surplus from previous operations. This concentration was maintained throughout the spraying to enable the *chikopokopo* to lift the required volume while minimizing the number of sorties. To distinguish the droplets for experimental records, hostasol yellow 3G was added to the pesticide. The maximum safe amount of 280 liters (twenty shy of the absolute maximum) for the pesticide payload was preferred to ensure safe climbing up and away from the Shamrock Mine loading bay. This translated to 260 liters usable load and twenty remaining in the spray gear system; at the rate of 24 g/ha, that amounted to 31.5 km^2 per sortie. The Bell Ranger had no sophisticated navigation equipment like the Doppler or the SGP 500 attached, so the pilot and co-pilot navigated from maps and first reconnoitered and then followed recognizable ground features, like hills, rivers, and roads. In addition, a line marker was placed at 250-meter intervals along the line separating the block into two; from there, a marker party used flares to direct the *zvikopokopo*. The percentage of total flying hours dedicated specifically to spraying was 43 percent, compared to 44 to 94 percent for fixed-wing *ndege*; this "efficiency" would have been 50 percent if the time needed to deploy and maintain the *chikopokopo* in operation was excluded from flying time. At 11 km^2 per hour, the number of square kilometers treated per hour was far inferior to the 28.72 km^2 per hour achieved during the 1988 operation (Allsopp 1991, 23).

Effectiveness of aerial spraying remained elusive. The physicochemical monitoring showed good droplet size and distribution. The litmus test of the *chikopokopo*'s utility as a technology of spraying depended on whether there was good aerosol penetration and distribution in different terrains compared to fixed-wing *ndege*. The valley floor droplets were significantly fewer than those on hillsides and ridges; still, the ridges and hillsides were exposed and undulating, and more maneuverable with a *chikopokopo* than a *ndege*. The results in practice were overwhelmingly in favor of the *chikopokopo*: 1,500 to 1,800 droplets/cm^2 versus 276 droplets/cm^2. The explanation was not difficult to determine: *Ndege* had to maintain a safe height of between 150 and 1,000 feet over such delicate areas, whereas *chikopokopo* simply followed the terrain and rarely climbed above 300 feet. The use of *zvikopokopo* (plural of *chikopokopo*) in rugged terrain had been confirmed to be "a viable technique"; to maintain an accurate flight path without sophisticated navigational equipment was all the more impressive. With satellite navigation, ground support would no longer be necessary (Allsopp 1991, 24).

Conclusion

This chapter illustrated the interesting link between *ndege* and OCPs above with *mhesvi* below. It has shown that this was no straightforward transfer of ready-made means and ways—that is, of *ndege* and *zvikopokopo* from Europe or the United States—straight into combat against *mhesvi* in the sense of *kupa* (giving) or *kupihwa* (receiving) proven means and ways. On the contrary, aerial spraying was a site of experimentation contingent upon the very specific vegetation preferences and habitat of *mhesvi*, the geophysical nature of such habitat, and the climatic mobilities (temperature, humidity, wind speed) and seasons amenable to it. The procedures and techniques of spraying vegetation inhabited by *mhesvi* were developed through experiment in the field; that alone constitutes the spray area as a site of knowledge production. We are here, as its analysts, because of *mhesvi*, which by its presence and inevitably pestiferous mobilities forced *hurumende* to deploy this untried, expensive machinery on an experimental basis, the work of which we now write about.

The production of knowledge and standardization of the spraying modus operandi was an incremental process of informed trial and error, contributing to a more sophisticated and dynamic one. This is the re-Africanizing power of this chapter: to say that *vachena*'s knowledge and means and ways were not Houdini acts or well thought out and stable from the beginning. Instead, like *vanhu vatema*, *vachena*'s methods started from shaky premises, often on a trial-and-error basis, *until they justified themselves in practice*. This mobility of knowledge from a shaky to a stable place is the story of Rhodesia: When we set aside the morally repugnant racism and oppression, the one important lesson to be learned from the Rhodesia project (1890–1980) in Zimbabwean history is that of its creative resilience. Planes that were otherwise conventionally designed for military or passenger-carrier purposes were retrofitted into *weapon vehicles*—part transport, part weapon, transporting and bombing *mhesvi* with deadly chemicals in one move. The airspace became a test site for trying out new things and perfecting existing ones—be they *ndege*, *zvikopokopo*, spraying nozzles, chemical solutions and their strengths, or manipulations of terrain and wind conditions to achieve optimal results.

Still, the resilience of *mhesvi* shines through the treetops, as if pointing a mocking finger at the *ndege* up above. The *chipukanana* invades from adjacent areas and hatches from its shell after the spray is complete; timing and strict surveillance of the *chipukanana* and its organic vehicles becomes key. Success only has one measure: when *mafrayi* go out with a black ox and no *mhesvi* mobilities are detected.

Here, over the remote borderland forests, *ndege* deposits clouds of aerosol that land on anything below: *zvipukanana*, vegetation (including fruits growing thereon), rivers, and even people. Far from their sites of manufacture, these planes have become the face of *vachena*'s superiority over *dirt* for some, mass polluters of the environment for others.

12 The Work of Ground Spraying: Incoming Machines in *Vatema*'s Hands

Ground spraying is different from aerial spraying in that it is not a one-person job; it is teamwork. Unlike a *ndege* (fixed-wing aircraft) or *chikopo-kopo* (helicopter) in flight, as a spraying machine—a mobile workshop for spraying a mobile *chipukanana*/insect—the sprayman and knapsack sprayer are subject to the control and supervision of *vanhu*, whereas the pilot, once airborne, is alone in the cockpit. Moreover, whereas the pilot is a *murungu*, the ground-spraying team is composed entirely of *vatema*, with the exception of the supervisor.

Ground spraying is thus better analyzable as a transient workspace, the meeting point between *mushonga* (spraying machine), translated and known among *mafrayi* as simply *mushini wekupureya* (hereafter *mushini*), and the sprayman (*mufrayi*; literally, "fly man"). The discussion is no longer simply one about inbound means in local hands, but about the transient work of spraying and the workspace populated by *vatema* who see themselves as *vanhu vari kubasa* or *pabasa* (people at work). By contrast, the *murungu* sees the same *vatema* as instruments for the dispensing of OCPs into the environment, just like the *ndege* above. Barring a few cases in which *murungu* was good to them, these *vatema* were subjected to the vilest *rusaruraganda* (racism), *hutsiny'e* (cruelty), and *hudzvanyiriri* (oppression—but better translated by the late reggae great Peter Tosh as *downpression*).

The chapter thus first takes the reader inside the workspace of *kupureya* (spraying) to appreciate *kupureya* as a *mufrayi*'s intellectual engagement, at once hazardous and experimental, with *mushonga* and *mushini* on the move. Understanding what *mushini* wanted, what it could and could not do, was an act of reading. When this engagement has been adequately explained, including from the perspectives of *vatema* who were once *mafrayi*, the discussion then narrows to one giant campaign in three neighboring countries: Southern Rhodesia, Portuguese East Africa, and the Union of South Africa. The purpose of the campaign was to stop the advance of *ndedzi* from the Rio Savé region of Mozambique into the Savé-Runde junction area of Rhodesia, potentially threatening northeastern South Africa. Four aspects are considered: the racial politics and relations of power, the location of the workspace in *sango*, the transience of this workspace, and the materials involved in dispensing toxic chemicals to kill *ndedzi en masse*. Having outlined that campaign, the chapter then ends with spraying as experienced by *mafrayi* themselves. The glossary can help the reader navigate *chidzimbahwe* and other local keywords.

Inbound Things and Their Strategic (Local) Deployment

Whatever else they may have been designed for, the government imported the sprayers that arrived in Southern Rhodesia for spraying a specific *chipukanana*: *mhesvi*. This was a local and new role for which the equipment had not been originally designed. Indeed, the sprayers arrived as *potential* or *possible* rather than *actual* and *proven* means and ways of dispensing *mushonga* effectively against *mhesvi*. *Potential*, because no inbound thing arrives as means and ways a priori; only through deployment and performance—moreover, those that accomplish the purposes set for them to locals' satisfaction—does potential become actual means and ways. Pieces of *mushini* discussed here are the Motoblo, two versions of the Colibri, and the Schefenacker.

The earliest *mushini* deployed in Southern Rhodesia's *mhesvi* operations were the Four Oaks, the product of a British firm called the Four Oaks Spraying Machine Co (FOSMC; see their brochure in figure 12.1a and the Four Oaks in use in figure 12.1b). After filling the Four Oaks with *mushonga* to two-thirds capacity, its tank was pressurized using a hand pump, *mufrayi* adding more pressure as *mushonga* reduced in volume. The effect was to create uneven spray: more at the start, less as the tank receded toward empty. In April 1959, a small-scale spraying operation of 5 percent DDT wettable powder was undertaken in the Chiredzi riverine woodland using four Four Oaks pressure knapsacks spraying *mishini*. However, because of the receding pressure problem, the machines "had no great effect and one *mhesvirutondo* and one *mhesvirupani* were taken on the Chiredzi shortly afterwards."[1] Weight was also a concern for *mafrayi* maneuvering through thick and rugged forest.

A switch was made quickly to the Motoblo, a motorized, shoulder-mounted mist blower, the use of which was especially pronounced during the spraying operations in Chiredzi, Humani Ranch, and Devuli Ranch from 1959 to 1960. The mist blower was a product of Kent Engineering

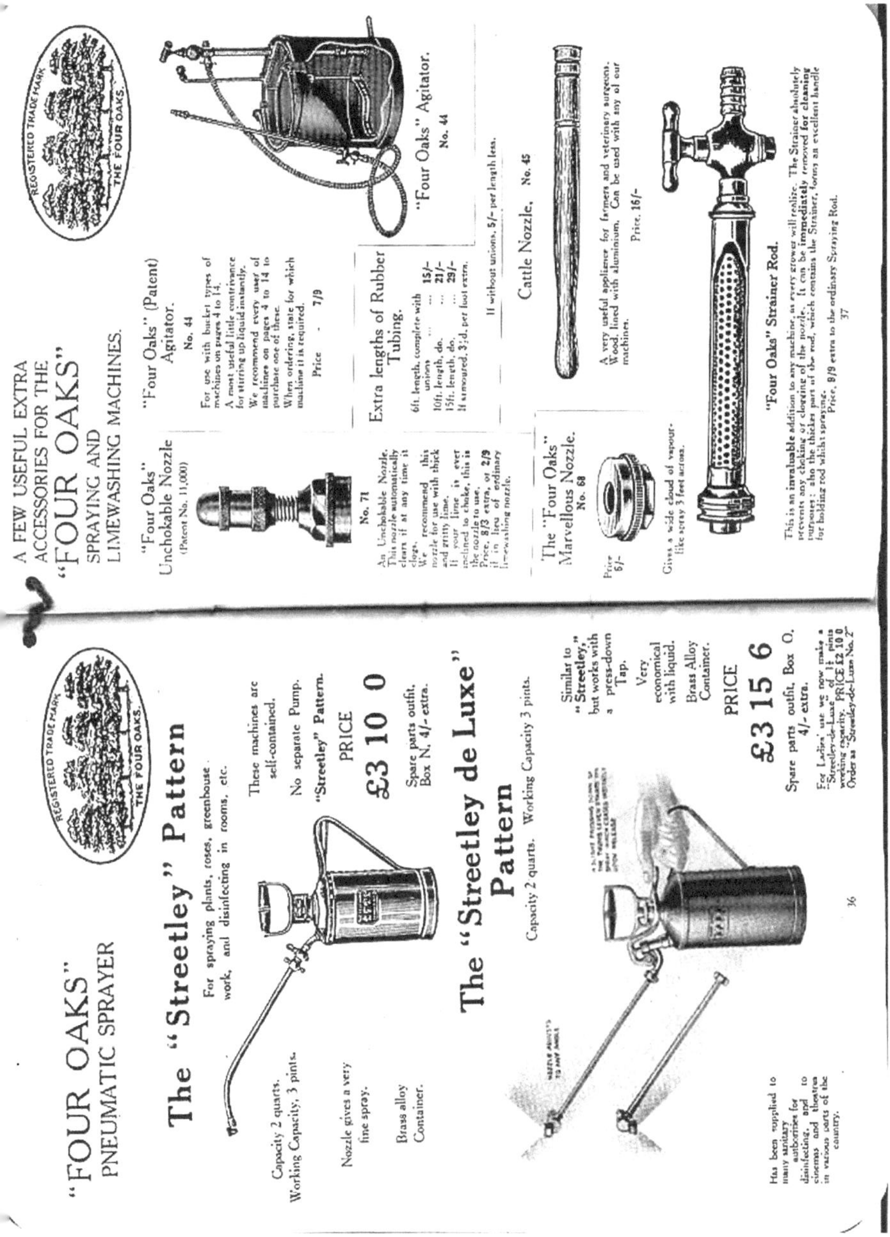

a)

Figure 12.1a, b (continued)

b) 

Figure 12.1a, b
The Four Oaks Spraying Machine Co. brochure for the Four Oaks pneumatic sprayer (page 249); ground spraying with a pressurized knapsack sprayer (page 250).
Sources: www.sutton-coldfield.net (12.1a); Allsopp 1990 (12.1b).

and Foundry Limited (Kent, United Kingdom), equipped with a 70 cc two-stroke engine and ten-liter tanks of *mushonga*. It came in two models of different sizes, the Motoblo 60 and the Motoblo 90, the former preferred for its lighter weight and ease of carriage (Chadwick et al. 1964; Tarry 1967). Chiredzi, Humani, and Devuli were dense riverine vegetation areas, and the Motoblo 60 was considered more portable, allowing for flexibility of spraying. The 70 cc engine took a 25:1 petrol and oil mixture, carried in four-gallon jerry cans. The nozzle sprayed very fine *mushonga* droplets, which the two-stroke motor-powered fan blew into the dense vegetation and up steep banks to a height of twenty feet (FAO 1977).

The Motoblo's performance was "generally satisfactory." However, there were four main operational faults that *mafrayi* noticed during *kupureya*. First, the Motoblo's pressure pipe, located inside the spray tank, constantly came loose from its connections. The team of *mafrayi* improvised by securing it with wire, but if the wire was too long, it damaged the agitator; if it was too short, the tank cap could not be raised sufficiently to spray normally. Second, there was a problem with the clamping bolts that secured the blower elbow to the fan casing: if the elbow became loose on its seating, it could not be tightened up again, and the entire blower tube waggled loosely or even came off. Third, after just forty hours of work, the agitator gland housing underneath the spray tank began to leak. To solve this problem, the large hexagonal nut was tightened—but if the gland itself was the problem, then there was nothing *mafrayi* could do. Finally, the clips securing the plastic nozzle to the metal blower tube were too weak and often broke when used.[2]

In addition, the Motoblo displayed problems very specific to the intense, humid, dusty, and ragged conditions of the southern African lowveld. The spark plugs required cleaning and resetting every two days. Fan belts needed adjusting every three to four days. The adjusting lever on one *mushini* became unserviceable. Petrol filters had to be cleaned every three to four days, and "considerable attention [had to] be given to cleanliness of fuel" in the future to avoid the "considerable, though minor trouble ... caused by dirty fuel, in spite of the use of filter funnels in mixing." Finally, the 25:1 petrol and oil mixture caused "the oiling up of the plugs," and a 32:1 mixture was quickly adopted in situ with positive results.[3]

Another model, the Schefenacker, was a power mist sprayer manufactured by H. D. Hudson Manufacturing Company of Chicago, Illinois. The Schefenacker was advertised as shown in figure 12.2. In 1960, a trial was conducted in the Zambezi Valley using ten motor-powered, low-volume Schefenacker mist sprayers that proved "extremely satisfactory" barring "a few minor difficulties."[4] Interestingly, the dusting valve and control and the dust air delivery hoses were removed from all *mishini* "because there was a tendency for [black] operators to fiddle with and to open the dusting valve."[5] In their place, rubber bungs were used to close the holes in the delivery bends.

It was not long before the *mushini* succumbed to the demands of a highly humid, dusty climate. First, the polythene pesticide tank tended to swell in "the extreme Zambesi valley temperatures," such that by the time the spraying operation was complete, it was "extremely difficult to remove the tank from the dust hopper."[6] Instead of polythene, "a metal tank might [have been] a better proposition" to meet the hot conditions the Schefenacker was being imported to work in.[7]

The tank was not the only problem, however; pesticide tank lid gaskets "perished very rapidly," their broken pieces causing blockages in pesticide delivery systems. The problem was partly blamed on the *mushonga* used—in this case, Dieldrex 15, which was deemed incompatible with the Schefenacker. The air filter elements were found to be too expensive; as *mafrayi* maneuvered the *mushini* through the dusty terrain, the poor thing clogged up after only thirty-five minutes and needed replacement. The manual

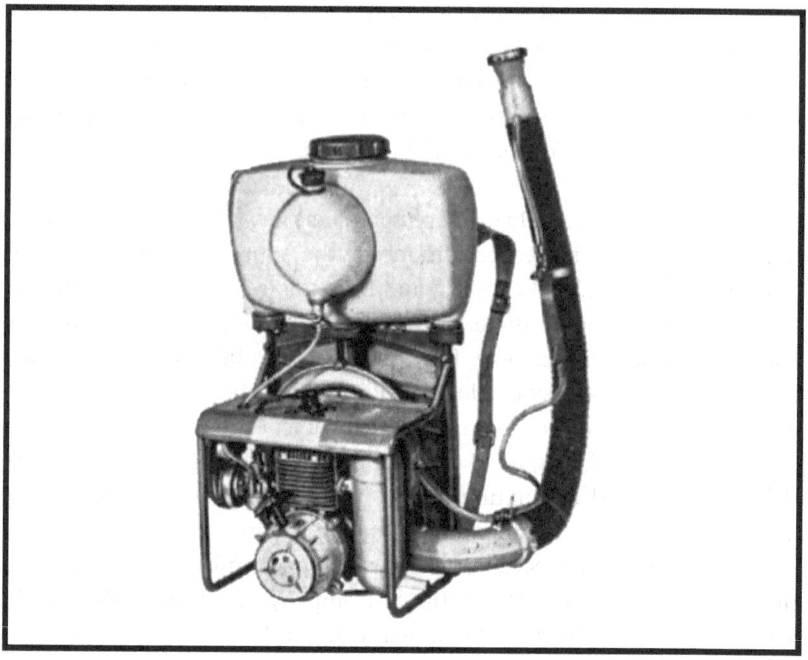

Figure 12.2
The Schefenacker power mist sprayer as advertised by Hudson Manufacturing Co. of Chicago, Illinois, in 1963.
Source: Weeds and Turf 1963.

that came with the machine said that the filter element could be cleaned with a simple tapping on a flat surface; in practice, that method did not work. But when a *mushini* was started and it did not run properly, *mafrayi* would immediately know that it had developed a blockage; modifications were necessary if the filter was to work in local conditions. *Mafrayi* were, in one sense, "demonstrating the interpretive flexibility of an artifact by deconstruct[ing]" it (Bijker 1997, 76).

However, let's not lose sight of their intellection, their reading of the *mushini*'s materialities in motion, their thinking-doing of *kupureya*, and their fixing-experimenting with the *mushini*'s problems. There was a problem with the ignition mechanism itself: "The hole in the cover plate ... for the spark plug [was] too small," the *murungu* wrote in his report. As a result, when pull starting, "the spark plug bakelite cap ... hit against the cover with resulting damage to the former." A remedy could have been found, but for the difficulty of getting a wrench through the hole to remove the plug, which was generally housed off-center.[8]

The Schefenacker did have something that no other *mushini* offered at the time, however: the capacity to spray far and wide, covering large areas at once and penetrating into thick vegetation.[9] *Mafrayi* loved that. They could finish faster, get their money, and go back home to their families. This capability was particularly welcomed for drainage-line spraying. The *mushini* was thus extended in 1961 to spray the 1,000 sq. mi. Nagupande-Manyande-Mzola drainage systems of Sebungwe. Forty Schefenackers were deployed in the operation.[10]

The Achilles' heel of the Schenefacker motorized sprayer was its fuel consumption and, consequently, its sheer expense to the government. Hence, it later was replaced with the Colibri, the product of a family-run French company called Vermorel, which was in the agricultural and automotive manufacturing business between 1850 and 1965 (Georgino 1975, 2001). In the post-1945 era, as white governments lurched onto DDT's power as a weapon of mass pest destruction, proven in the United States, Vermorel found a ready market for its *mishini*, the Leo-Colibri and Favori-Colibri. The machines had numerous selling points: ability to maintain pressure after discharging spray, less effort to operate, rapid recharge with a pump, and a floating-ball valve to prevent loss of pressure.

Two types of Colibris were used in Southern Rhodesia. One was the Leo-Colibri, with a detachable hand pump. The other was the Favori-Colibri, which was charged using a manual or motorized pump. The motorized pump could charge several *mishini* simultaneously, saving time, particularly when *mafrayi* were working near tracks or close to each other. The detachable pump reduced the weight of *mishini* and thus operator fatigue—a critical factor, given the rugged, steep, thorny and inhospitable terrain of the Zambezi and the Savé-Runde valleys (FAO 1977; Davies and Blasdale 1960; Kirkby and Blasdale 1960).

The Colibri sprayers were therefore deemed cheap and effective for the 1963 operation covering the Mzola-Dongamusi drainages and the vegetation surrounding Cewali Pan. These pneumatic knapsack sprayers consumed approximately eleven gallons of 3.7 percent dieldrin per machine mile, much less than the Schefenackers had expended.[11] In 1964, they were also used to spray the Nyamusanzara, Ruenya, and Kaerezi drainages, as well as road verges and *mombe* paths in Nyanga North.[12] From 1965 to 1969, the machine was also used in the Chipinge, Savé-Runde, and cross-border operations into Mozambique between Rio Savé and the Malvernia-Lourenço Marques railway line (Thomson 2001, 49).[13]

The Southeastern Spraying Operations into Mozambique

To understand how this massive, multiyear operation unfolded, we need to first explore the political and institutional context within which it was undertaken. Put another way, this is an example of environmental engineering in context; the focus is trans-border spraying of OCPs.

The first steps were taken in 1929, just after the conclusion of the pan-African Agricultural and Veterinary Conference held in Pretoria, when Chief Entomologist Rupert Jack visited KwaZulu to learn about the Harris fly trap.[14] In June 1943, Rene du Toit and E. B. Kluge of the South African Division of Veterinary Services visited Southern Rhodesia and "inspected [the] clearing operation on the eastern border." They reported back to the director of the Division of Veterinary Services at Onderstepoort, who sent their reports as feedback to Chorley.[15]

The following year, Jack's successor, Chorley, consolidated these relations when accepting an invitation from South Africa's director of Veterinary Services, P. J. du Toit, and visiting KwaZulu in May to "inspect the work in progress" in the Umfolosi and Mkhuze forests and the Hluhluwe Game Reserve. Sub-Director Rene du Toit and Kluge accompanied him to the research site to observe and learn.[16] In 1947, Chorley again returned to observe the large-scale experimental DDT-spraying operations at the invitation of the Union Government.[17] In 1953, Rene du Toit in turn visited Southern Rhodesia "to advise on the application of *mushonga* from *ndege* and to fly over parts of Hurungwe."[18] The success of the KwaZulu spraying campaigns opened up possibilities for an interterritorial spraying campaign involving Southern Rhodesia, Portuguese East Africa, and the Union of South Africa in the 1960s.

Meanwhile, Southern Rhodesian exchanges with the Portuguese had begun well before the outbreak of World War II. As early as 1925, the Rhodesians had asked the Portuguese to ascertain the position of *ndedzi* in the Rio Savé border area and urged them to act. However, the Portuguese were busy with the even more serious *ndedzi* and *n'gana* situation in the Tete and Cabo Delgado provinces. In any case, *ndedzi* was to all three neighbors of less importance than foot-and-mouth disease (FMD) and African Coast fever (ACF), with the border areas of Masenjeni, Mahenye, and Chitsa the worst affected. So serious was the problem that plans in 1926 to create a game reserve joining together Gonarezhou forest, the adjacent Mozambican areas, and Kruger had to be shelved.

In 1937, the Rhodesians alerted their Portuguese neighbors about a disturbing southwesterly spread of *ndedzi* from Manica Province in Portuguese East Africa (PEA) into the Honde-Rupembi River borderlands. Discussions about possible responses continued with little Portuguese urgency. The situation continued to deteriorate. In 1942, during the Portuguese minister for the colonies' visit to Salisbury, a memorandum was handed to him concerning the *ndedzi* threat from Portuguese territory. In response, Lisbon opened a large area along the border east of Chipinge to indiscriminate game elimination. However, Chief Entomologist J. K. Chorley was skeptical about whether the measure was a permanent solution.[19]

As the situation on the Rio Savé unraveled, Dr. de Sousa, Mozambique's *Chefe de Missao a Combate as Tripanossomiases* (Chief of the Sleeping Sickness Mission), informed his Rhodesian counterpart of another *mhesvi* presence in Catandica, very close to the adjacent border area of Nyanga. Chorley now feared that both the Ruya and Kaerezi rivers north of Chipinge might already be infested. The Southern Rhodesia Trypanosomiasis Committee (SRTC) tasked the Native Commissioner of the Nyanga District to investigate. The Portuguese government's point men for this interterritorial cooperation—de Souza and Veterinary Surgeon de Souza Santos, the latter based at Chimoio—did likewise.[20]

Out of the growing rapport emerged a bilateral approach to the *ndedzi* problem in Rio Savé. In July 1945, Southern Rhodesian and Portuguese authorities conducted a joint investigation to ascertain *ndedzi* positions in Mozambique south of Rio Savé and assess the possibility of it crossing over into Gonarezhou Game Reserve and Kruger National Park. Chorley, Conservator of Forests E. J. Kelly Edwards, Director of Veterinary Research D. A. Lawrence, and government ecologist R. R. Staples represented Southern Rhodesia. That same year, Chorley also participated in a meeting with the East African Standing Committee on Tsetse and Trypanosomiasis.[21] In 1948, two officials of the *Missao,* Dr. da Andrade Silva and Dr. Jose Marques da Silva, visited Salisbury to review the fly problem in Catandika and Rio Savé.[22]

I did not find any archival record for the period from 1946 to 1957 concerning these tripartite consultations. The next entry is from 1958, when Andrade Silva and Rhodesia's new director of Tsetse and Trypanosomiasis

Control, John Ford, held two meetings. The second of these was a tour of the Savé-Runde under the guidance of Dr. R. J. Phelps, the entomologist in charge of the area. Both sides made a sizeable professional staff available for technical evaluations of possible strategies on the ground.[23] Two years later, at Andrade Silva's invitation, two entomologists from Phelps's Savé-Runde area of jurisdiction "made a most useful visit" to the Sitatonga Hills inside Mozambique, earlier the scene of Swynnerton's groundbreaking study. We now know from interviews with Timothy Sumbani and Mugocha Mavasa, two veterans of the spraying and game-destruction campaigns, respectively, that several *mafrayi* were recruited locally and armed with nets to accompany Phelps.[24] In one hour of flyrounds, they collected specimens of both *mhesvirutondo* and *mhesvirupani*—not just on the Rio Savé, but also on the Guvulweni.[25]

Based on this research, South Africa, Mozambique, and Rhodesia convened a meeting in Lourenço Marques in November 1960 that established the Interterritorial Standing Committee for Tsetse and Trypanosomiasis Control in South-East Africa (ISCTTCSEA). This was the body that would coordinate cross-border pesticide spraying operations going forward among the three countries. South Africa, Mozambique, Northern Rhodesia, and Southern Rhodesia were members, whereas Angola, Botswana, and South West Africa (now Namibia) would send "observers" (Robertson and Kluge 1968, 19). The hosting triangulated between Lourenço Marques, Pretoria, and Salisbury. Subsequently, ISCTTCSEA convened before every spraying operation to receive updated reports on the previous year's results and to plan for the next. Among other things, these meetings focused on logistics: the quantity of *mushonga* required, numbers of spraying team leaders, spraymen, numbers and types of vehicles for each of the three parties, and agreement on precise limits of spray areas (Robertson and Kluge 1968).[26]

In 1961, ISCTTCSEA convened in Lourenço Marques to, inter alia, review the southward movement of *ndedzi* on either side of the Southern Rhodesia-Mozambique border and its threat to Kruger. South Africa immediately offered to dispatch a team to search *zvikukwa* (chrysalises) as a way of determining the altered presence and extent of *ndedzi* in the Savé-Runde area since Phelps's Guvulweni-Rio Savé study the previous year. The team started searching for *zvikukwa* on either side of the border in November 1961 and completed its mission in January 1962, recommending that a joint spraying campaign with dieldrin be commenced in the late dry season (September–October). By then trees would have shed their leaves, bar the evergreens of the Guvulweni drainage system, where *ndedzi* were forced to concentrate to survive the heat and predators.[27]

The subsequent ISCTTCSEA meeting approved this recommendation. Three preliminary steps were also drawn up as a blueprint for all subsequent annual sprays. First, the maintenance of old and cutting of new access tracks would be assured so that spraying teams could get as near as possible to the spraying area using motor vehicles. The mapping was conducted with aerial photographs and then translated to the ground. Second, the siting and construction of temporary camps were planned for spraying personnel for the duration of the operation. And third, a detailed plan was prepared for the next year's spraying, entailing the stereoscopic examination of aerial photos of the entire spray area, marking all "spraying lines" the teams would follow. Such lines included drainages, contact zones between vegetation types, motorable access tracks and roads, and paths of *ndlopfu* and *vanhu* in areas of heavy *ndedzi* density. Large-scale spraying maps were also prepared; on these, the spraying team supervisors marked the ground their team had covered for the day (Robertson and Kluge 1968, 20).

Throughout the 1960s operations, 3.1 percent dieldrin emulsion, prepared from specially formulated 18.6 percent emulsifiable Dieldrex 15 T concentrate was used. The deposits remained lethal to *ndedzi* for four months or longer after spraying. Until 1964, motorized Schefenackers were used, "but these proved to be highly susceptible to stoppages and breakdowns and, owing to their extreme noisiness, made difficult the control of movement of the operators" (Robertson and Kluge 1968, 21). The South African team supervised this work, and the Rhodesians provided fuel, engine oil, Schefenacker motorized knapsack sprayers, and *mafrayi* recruited locally.[28]

The spraying in the Savé-Runde basin began in 1962 as a joint operation involving South Africa and Rhodesia. This and similar operations in Mozambique were designed to reduce the threat of invasions of *ndedzi* across the Limpopo into the Kruger National Park and to deflate *ndedzi* pressure on the Savé-Runde basin corridor. G. K. Gillett was in charge; locals in Masivamele area remember him well by his nickname, *Ngungunyana*, after the king of Gaza.[29]

By October 1962, it was hoped that *kupureya* would reduce fly pressure building up on the forty-mile long, ten-mile wide protective barrier of discriminative clearing along the eastern side of the Chikombedzi-Malvernia stretch of the Lourenço Marques railway line. To facilitate the cross-border spraying operations against *ndedzi* inside Mozambique, the Tsetse Branch had bulldozed parallel lines five miles apart through the very sandy forest country between the Rio Savé (north) and the Malvernia-Lourenço Marques railway line (south). In between these strips, tracks had been cut at five-mile intervals to allow vehicles to deliver *mushonga* and *mafrayi*, who were then unleashed onto the blocks of forest with their DDT-filled *mishini* (Thomson 2001, 49).[30] The Nyamasikana, Guvulweni, and Chepfu drainages were sprayed, with the treatment confined almost entirely to vegetation along and around rivers, pans, road verges, and contact zones.[31] The following year, as some teams re-sprayed the Nyamasikana, others descended on the Benji drainage line.

In 1964, the switch was made from motorized sprayers to handheld Favori-Colibri, prepressured, knapsack sprayers: "Once the minor teething troubles with these latter had been overcome, they were found to be much more satisfactory and free from breakdown" (Robertson and Kluge 1968, 21). In 1964–1965, the international spraying operations south of the Runde raged on for the third successive year. For eleven months, no *ndedzi* was caught on the flyrounds, barring that part of the sprayed area where only two annual treatments had been meted out. In March 1965, the South Africa Division of Veterinary Services officers "generously offered" pesticide, increased staff, and transport to cushion the Rhodesians' "heavy commitments" on the Zambezi fronts. In preparation for spraying, two hundred miles of roads were opened with *mabhurudhoza* in the areas lying between the international border, the Guvulweni River, the northeast to southwest game fence, and the Runde River. All the drainages, roads, and vegetation contacts targeted for spraying were demarcated on aerial photographs, with the information used to produce working maps. The total area to be sprayed was 800 square miles (sq. mi.).[32]

The operation began on July 27, 1965, and ended on October 12. It involved five teams of local Hlengwe and Ndau *mafrayi*. Three of them worked under *Madhebheni* (*Durbanites*, as the South Africans from the KwaZulu operation were called locally) supervision and covered the Savé Runde area. *Madhebheni* would "take the medicines in the tins, line up teams recruited from our villages here *pawaya yemakurundundu, vachipureya* (at the crude fence, spraying), heading in the direction of Mupfichani."[33] The other two worked under Rhodesian supervision and sprayed the north Runde area from Chipinda Pools to the Savé and from the Chivonja Hills to Hippo Valley Estates. They all used the Favori-Colibri pneumatic knapsack sprayer.[34] On July 12, 1967, six teams of *mafrayi* operating under *Ngungunyana* set off again. Two of them operated for some time inside Portuguese territory. The operation pioneered the use of "a parallel line spraying technique in the rather ill-defined tsetse habitat" of the Centre Road, Kapiteni, and part of the Chivonja Hills. These parallel lines were spaced four hundred yards apart to allow for command and control.[35] No further spraying was considered necessary, and the operation was ended.[36]

In March 1968, ISCTTCSEA resolved to confine all spraying operations to the south of the Runde, with the Kapiteni-Masenjeni border area as the major focus. Southern Rhodesia would do the planning and cutting of the necessary trace lines and then upgrade them into tracks, the *Missao* would provide two *mabhurudhoza*, and the South Africans would contribute transport and fuel. In April, the serious work began. *Ngungunyana* and *mafrayi* took to the bush first, cutting the new access-track system. Seven teams of *mafrayi* (two under Rhodesian and five under South African supervision) were deployed with 5 percent DDT.[37] In 1969, Kapiteni was re-treated, with a focus on the Mutezwa, Bepi, and Hyfananga drainages.[38]

Meanwhile, on the Mozambican side, flyrounds and survey catches showed the need for more intensive treatment of the Upper Mahungwe drainage and the featureless country within eight miles of the border. The 5 percent DDT suspension applied in grids four hundred yards apart had failed to eliminate *mhesvirutondo* from the area. The branch became anxious about DDT's efficacy compared to the 3.1 percent dieldrin emulsion used prior to 1968 and about what rate of application was best suited to parallel line spraying. Thus, at its annual meeting in Salisbury in 1969, ISCTTCSEA endorsed the modification of the parallel line spraying technique and a large-scale field comparison of 5 percent DDT and 3.1 percent dieldrin. The experiment showed that both treatments were highly effective, but DDT was the much cheaper option and just as efficient as dieldrin. The parallel line spraying operation applied DDT suspension at the rate of 381 gallons per square mile on the Rhodesian side and 263 gallons per square mile in Mozambique, compared to 284 gallons of dieldrin. These rates were all very much higher than those achieved in conventional spraying, and they proved far more effective than the older method used in 1968.[39]

After the eighth annual joint spraying operations of the Rhodesia/Mozambique border area in 1969, 898 square miles had been treated: 141 in Rhodesia, 757 in Mozambique. Test herds were deployed at strategic points in the sprayed Gonarezhou area. The Upper Pombadzi, Lower Pombadzi, and Masanya test herds stationed on the north bank of the Runde (an area not sprayed in 1968 and 1969) did not contract *n'gana*. South of the Runde, six test herds had long been established at Chepfu, Guvulweni, Chivonja, Lisodo, Songwene, and Kapiteni, and they had also tested negative for disease by April 1970. Cases were recorded in two new southern Runde test herds established in November 1969, confirming results from catches in parallel bait ox surveys and cycle flyrounds.

Re-spraying was ordered. The ninth joint interterritorial spraying operation from June to September of 1970 thus focused on re-spraying the

Mutezwa-Kapiteni region and the Chitove-Tembwehata Pool area of the Runde. All eleven test herds in the Savé-Runde area tested negative in August and September.[40] The operation's success owed much to the sinking of six new boreholes in the spraying area by the Rhodesian Department of Water Development and the joint interterritorial team's extension of the geometric system of new parallel access tracks. D7 bulldozers were involved in the latter task, cutting "avenues" north from Main Road to the Kapiteni-Masenjeni Road and extending some of them south to the Mozambique game fence.[41]

The 1971 joint operation began in June and concluded in October and involved fifty-one Rhodesian and three Portuguese teams of *mafrayi*, nine more than the previous year. The Rhodesian spraying in the Chipinge, Bikita, Nuanetsi (Mwenezi), and Gaza West districts did not commence until August, because all teams were assisting in Mozambique, where the required spray area proved to be more extensive than previously thought. When it finally began, the operation in Rhodesia continued into October and fell afoul of the rains.[42] In total, 757 square kilometers had been sprayed.

Subsequent spraying with far smaller teams was aimed at maintaining the status quo in the Masenjeni-Chigamane consolidation line, the eastern limit of the campaign inside Mozambique.[43]

Kupureya Ndedzi: Doing Ground Spraying

Piet Barnard of Mapikule Village was one of the local *mafrayi* involved in several of the annual spraying campaigns. He recalls: "We were working under whites who were coming from South Africa—there was *Kambombo*, a white man whose name I don't recall, a tall very well built man, who was well-liked by all of us. We carried *zvigubhu* (tanks) for spraying. I did not know the name, but the *mushonga* was whitish in color. The spraying camps were in Mabote or further into Mozambique at Chinyezani. The headquarters was at Border B camp inside Zimbabwe."[44]

Staff Masungwini, a local villager, now a pastor, remembers seeing these spraying teams: "The tsetse people were at Chipinda near the river Guluje. They moved around spraying *ndedzi*, following *zvikoronga* (drainages). They sprayed here in Mambile going down all the way to Mpfichani beyond the Runde."[45]

Mafrayi say they were told to target places where *ndedzi* most preferred to refuge or rest, especially in large numbers. Such areas included woodlands and riverine fringe; along roads and paths of *vanhu*, *tihomu*, and *sviharhi* through *xanyatsi* or *musasa* woodland; heterogeneous vegetation covering a large *tshuka* (termite mound); tree vegetation surrounding cattle pens; the tree and bush cover surrounding waterholes and dams frequented by game; the vegetation growing among scattered boulders at the feet of kopjes and small hills. Inside these areas, *mafrayi* targeted tree bark, overhanging branches, rot holes in trees, holes dug by *mhandzela* (anteaters) or *nhonci-nhova* (warthog), and holes in river banks and under fallen logs. In the trees and branches, however, application was usually restricted to bark of no less than six inches in diameter and larger overhanging branches, and from ground level up to a height of twelve to fourteen feet.[46]

With their Favori-Colibris packed with *chefu* (*xitsonga* for poison; not to be confused with *chepfu*, *chidzimbahwe*—i.e., DDT and later dieldrin), *mafrayi* were lined up ten yards apart along the edge of a five-mile block; on a given command, they marched through the block, spraying all the tree trunks and thicket stems and every fallen log and every *mhandzela* hole with *chefu*. At the other side, their canisters were recharged with DDT. They were realigned farther along the cut line, and then they marched into the next section of the strip to repeat the process. On they marched, from the Rhodesian border east into Mozambique and toward the Indian Ocean.[47]

The basic unit of the operation was the spraying team, composed of one tsetse field officer (TFO) in charge, one senior "African assistant" commanding *mafrayi*, one driver, eight *mafrayi*, eight carriers of *mushonga*, and a few other odd-job men. The unit was equipped with a five-ton truck, a Land Rover, four *mishini*, two charging pumps, and a bicycle front fork and wheel fitted with a cyclometer to measure the distances covered. Also included in reserve was an extra *mishini*, a running repair kit, and spares for the less durable parts.

The procedure of spraying was as follows:

> With the officer-in-charge leading the way with his marked aerial photograph, the team moves along, spraying the drainage line or contact, normally four abreast, but along narrow spraying lines only two abreast. As soon as the first of the four machines is empty, the whole team is stopped and all machines are recharged with insecticide. The mileage covered and all other relevant data are carefully recorded on a special field record form before the team moves on to discharge its new fill of insecticide. As the team proceeds, each sprayer operator covers a 15-yard swathe, spraying the boles and undersides of overhanging horizontal branches of all the larger trees within the spraying zone, up to a height of about 12 feet. Other important sites of application are the undersides of fallen logs or large rocks, rot holes and those dug by antbears. Special attention is also paid to all vegetation in the vicinity of water-holes. (Robertson and Kluge 1968, 21)

To measure spray effectiveness, changes in *ndedzi* catch were measured using cycle and ox flyrounds and at traffic control points like tsetse gates, and tests were conducted for *n'gana* incidence as well.

Vachena called *vatema* "spans" (*zvipani*), the term given to oxen yoked up to pull a plow or wagon; this led to the view of *kupureya* as *kubopwa pajoki* (being yoked on a yoke). The teams "moved forward in two spans of five, each span stopping simultaneously for refueling, and refilling with insecticide."[48] The spraying operations along the Chiredzi River, for example, were conducted in that manner, in a militaristic way:

> The RM [right marker] of Span 1 walks down the initial path, and acts as right marker for the relief operators of each machine, who work in front, keep the 40 yards dressing, and chop a path where needed.
>
> A *panga* man follows in the wake of No. 5 machine, and blazes a path to be followed by the RM of the second span. In turn, a path is blazed behind No. 10 machine to guide the RM of Span 1 on the next leg.
>
> Thus 44 men are required for the ten machines, plus another 4 at the vehicles. When the machines are moving parallel to the river, No. 1 machine works at the bottom of the bank, No. 2 at the top, and No. 3 dresses off from the top of the bank, there being 4 machines in each span. Sometimes a third machine would work in the river bed if the vegetation warranted it.[49]

While one *mufrayi* sprayed, the other would be recording the distance traversed during the insecticidal discharge "by means of a bicycle wheel fitted with a cyclometer."[50] River courses or narrow vleis required teams to operate simultaneously in two pairs to ensure that both sides were sprayed.

The Humani-Devure operations were conducted using the Chiredzi spray as a blueprint. For twenty-six days (between August 15 and September 15, 1960), ten teams of *mafrayi* started early in the morning. Each team was equipped with one Motoblo shoulder-mounted motor mistblower, covering 400–500 yd. blocks. The woodland zones to be treated were marked clearly on the operational map. All ten teams were supposed to cover a total surface area a mile wide and four to five miles long per day. Just as in Chiredzi, two men operated one *mushini*, two carried *mishonga*, one cut the path ahead, and two carried fuel and refueled the *mishini*. Five *mafrayi* designated as "survey hands" supervised the teams.[51] Before the operation began, fifty men cut spray paths in the thicket vegetation along the Tugwi River. In nine days, they had cut over one hundred miles of path. Some areas were so thickly forested that they had to cut tunnels through them, with tapestries of canopy above and on either side.[52]

For transport between the base camp and spray site, the ten teams were supplied with one five-ton long wheelbase lorry with a driver and one Land Rover pick-up truck.[53] Each machine took up ten five-gallon drums of *mushonga* per day; only the five-tonner could carry enough for ten machines. Sixty-five to eighty empty five-gallon drums remained on the lorry. Dieldrex 15 was pumped from a raised point on the left, and one gallon was placed in each drum. From the bowser at the raised water point on the right, water flowed down an extension pipe to fill the charged drums up to the top.[54]

This proved to be a very rapid system; four men could load *mushonga* in two hours. All *mushonga* required for a full day's spraying sortie was mixed at the base camp in the afternoon the day before it was to be used and transported to the spray area on the five-tonner on the day of spraying. Gasoline for the *mishini* was carried in four-gallon drums, containing a 32:1 (gas to oil) mixture.[55] A gallon of 18.6 percent Dieldrex was pumped with the aid of a semirotary petrol pump out of the drum into another containing four gallons of water. A five-gallon drum of Dieldrex 15 T produced twenty-five gallons of 3.6 percent weight/volume (w/v) emulsion (the first coat of spray) and fifty gallons of 1.8 percent dieldrin emulsion (the second coat after two months). This figure varied; in the 1966 operation in the northeastern districts of Mudzi and northern Nyanga, the Tsetse Branch used a 3.1 percent emulsion formula.[56]

Thirty miles of track were cut to enable the five-ton lorries to carry *mafrayi, michini,* and *mushonga* along the Savé and Tugwi Rivers. On the latter river, a low-level bridge was constructed to shorten the distance between the operational camp and the sites of spraying. The lorry carrying *mushonga* followed the spraying team closely, on tracks when they were available or across country. Ready access to *mushonga* was essential; when the distance between the team and lorry widened to half a mile, supplying *mushonga* became a serious problem, because each *mushini* consumed two drums per hour. After loading their *mishini* and *mushonga,* the spraying teams clambered onto the back of the lorry. Sometimes, a couple of Jeep pick-up trucks were available.[57]

The workday started at 6:30 a.m., when all the men would board the trucks that were already loaded with *mushonga* drums mixed the day before. On disembarking on site, the teams—clad in protective overalls but no shoes, gloves, goggles, or headgear—began the technical and muscular work of *kupureya.* The *murungu* instructed the teams to line up at the start line forty to fifty yards apart. With all tanks loaded with *mushonga,* the teams started forward, each along a defined axis until they reached the end of the designated block, four to five miles away.[58] The spray tanks were usually empty within five to fifteen minutes depending on the type of vegetation; tree boles were sprayed on one side only.[59] The *murungu* strictly instructed the men to fill spray tanks without any splashing to avoid both self-poisoning and waste. Therefore, *mafrayi* were supposed to first pour *mushonga* into a bucket and then pour more carefully into their *mishini* using a funnel.[60] It was the duty of the TFO to closely supervise the application of *mushonga*, and he constantly moved among *zvipani* (spans), explaining and demonstrating spraying techniques whenever there was a problem. He was the on-site mechanical engineer, walking behind and driving all *zvipani*, repairing any machines that broke down, and seeing to it that distances between *mishini* were maintained, that all *zvipani* were kept in line, and *mushonga* properly applied.[61]

The *mushini bhoyi* (machine boy, as the sprayman carrying the pneumatic machine was called) was a very important player in this technical process. He was supposed to keep the largest aperture on the dosage slide, to obtain and dispense a coarse droplet capable of issuing a more persistent deposit compared to a fine droplet. A larger aperture had the advantage of releasing *mushonga* more rapidly, thereby increasing the spray coverage rate and making the rate of progress faster.[62] *Mafrayi* were supposed to achieve an even *mushonga* distribution, and each spray target demanded a different approach. For instance, a single tree bole or small bush required a brief discharge of mist. A large hollow thicket demanded a prolonged discharge, the bombardment creating a cloud of mist inside the thicket. The *mufrayi* deposited *mushonga* on all vegetation within three to four yards of the spray line, either side of it, and from the ground up to a height of eight to ten meters into the tree branches and thickets.[63]

Conclusion

In *Transient Workspaces*, I showed that the presence on the borderlands of poachers and illicit labor recruiters like Cecil Bvekenya Barnard forced neighbors Southern Rhodesia, Portuguese East Africa, and South Africa to cooperate in the 1910s and 1920s. In the 1930s, the mobilities of African Coast Fever, FMD, and then *ndedzi* from Mozambique—as passengers in/on *mhuka* and *mombe*—led to the abandonment of plans to create a vast transboundary game reserve (Mavhunga and Spierenburg 2009). The escalating movement of *ndedzi* from the Rio Savé area of Mozambique to the south and west in the late 1950s shocked the three territories into establishing ISCTTCSEA and commencing one of the biggest, most sustained spraying campaigns in the history of anti-*mhesvi* operations. This interstate cooperation was forced upon southern African territories by *mhesvi*.

As a site, *kupureya* thus became a transient laboratory in which *mishonga* and *mishini* became experimental material and in which new forms of *ruzivo* were produced. At that site, *vatema* were not just tools of empire but key intellectual agents and experimenters. The *mishini* and *mishonga* were placed in their hands to destroy *ndedzi* and end its pestiferous mobilities. The transient work of *kupureya* has been examined foremost as a workplace populated by *vanhu vatema* under the yoke (hence a span, *chipani*) of *murungu/muchena* who supervised them but did not personally handle *mushonga* and *mushini* (barring giving brief instructions). *Kupureya* has been analytically repositioned not just as the execution of *murungu*'s orders, but as work demanding new and preexisting modes of *ruzivo* on the part of *mafrayi*. We have seen *mishini* put to work in roles they were not originally planned for, coming in as *potential* rather than *actual, proven* means and ways of spraying *mhesvi*. The emphasis on *potential* is important because, as shown, no inbound thing comes into Africa as technology a priori; it proves itself as such in the hands of the people, as they deploy it to perform specific tasks. The rapid turnover of *mishini* shows that things that were technological (fit for purpose) in one part of the world (Europe or North America) were utterly useless in another (Africa).

It bears reemphasizing that the central actor in *kupureya* was not *murungu*; he did not do the spraying. Rather, it was *mufrayi*, the applicator of *mushini* and *mushonga*, who could tell whether *mushini* was working or not. Here too, the *ndedzi* presence in punishing terrain ensured that only *vatema* could do such exacting work, based on their longstanding presence and experience in the countryside. Tellingly, the fate of all these artifacts of North America's and Europe's industries—the Four Oaks, Motoblos, Schefenackers, Colobris, and so on—and their becoming or failing to become technology in Southern Rhodesia depended on the experience and opinions of *vatema*. Their *hunyanzvi* (expertise) was based partly on the experience acquired from continued *kupureya* and the experience with each *mushini*, the problems it caused, and how a new one solved them or made them worse. It also derived, more fundamentally, from *mweya* (spirit) of *husiki* (creativity) and *kushingirira* (resilience), or creative resilience drilled into *vana* (children) in the professoriate of *chivanhu* (custom—defined not simply as frequent repetition of the same act, but as means and ways and as methods of doing, living, and deploying).

The chapter has not only shown how these men were organized and conducted *kupureya*, but also engaged their views of what they were doing, as revealed in their memories. The combined written, oral, and field-observed evidence not only shows how dangerous a site of work *kupureya* was; this arduous work has been shown to be as much about exterminating a *mhesvi* as it is a site of applied knowledge production on the move.

13 DDT, Pollution, and *Gomarara*: A Muted Debate

Frantz Fanon once said: "I do not know; but I say that he who looks into my eyes for anything but a perpetual question will have to lose his sight" (Fanon [1952] 1967, 29). This is the intention of this chapter: to ask questions. Could the massive aerial and ground spraying of the Zimbabwean countryside be catching up with us? Is there a connection between *gomarara* (cancer) and OCPs?

So serious is the scourge of *gomarara* in Zimbabwe (formerly Rhodesia) that on June 23, 2012, the Ministry of Health announced that it was close to finalizing a national cancer prevention and control strategy to guide the nation in fighting *gomarara* (GOZ 2013)—this against a background of rising numbers of cases in recent years, with many of the cases being readily attributed to HIV and AIDS (Moyo 2012).

I am not even sure this is a question I am qualified to answer. Is there direct evidence linking OCPs and *gomarara* in Zimbabwe? Not yet. Is there enough evidence to raise concern? Absolutely. To begin this chapter, I first will explore the state of *gomarara* in Zimbabwe briefly, drawing out the incidence of those types of *gomarara* usually associated with OCPs. The statistics are quite staggering.

In the second section, I reconstruct debates about OCPs as environmental pollutants, an issue that was muted at the height of the spraying campaigns of the 1950s–1970s and is largely forgotten now. The 1964 protests among *vatema* were confined to loss of cattle, forced resettlement, and *nhimura* (the Native Land Husbandry Act). They did not extend to DDT and other OCPs as environmental pollutants with long-term health effects. *Vachena* were more concerned with "wildlife," and politico-military struggles to save Rhodesia from "communist terrorists," according to the Rhodesians, and the struggle for self-liberation (*kuzvisunungura*) from *hudzvanyiriri* (oppression, according to *vatema*). With the end of *Chimurenga*, Zimbabwe's war of independence, the issue was not remembered.

This silence is alarming given the global banning of the chemicals, along with other synthetic products like lead-based paint and asbestos that were once deemed very safe and that now turn out to be toxic. I will examine some of the investigations made into the environmental effects of OCPs elsewhere, marshalling that evidence to ask questions and to map and follow the itineraries of these pesticides in our bodies and those of our animal cousins.

Most of the historical literature on DDT and other OCPs focuses on the United States. I am not aware of any monographs on Africa. The carcinogenicity of OCPs is still generally unexplored from a humanities and social science perspective. This is despite the fact that scholarship has focused on the toxicology factor in the banning of pesticides and environmental regulation (Dunlap 2008). The histories of the development of DDT, caught between military and civilian uses against people and pests, has received detailed attention that enables those studying the chemical's deployments in Africa to understand where it is coming from technically and culturally (Russell 2001; Nash 2007; Dunlap 1981). Beyond the United States, the histories of industrial pollution–induced diseases (Walker 2010) betray a "how DDT changed the world" tone of technological determinism that ascribes transformative powers to the artifact (Kinkela 2011), contrary to preceding chapters. We see echoes of this with respect to cell phones in Africa: "How mobile technology is changing Africa." Yet these phones did not become technology, let alone mobile, alone. They are made such by people in Africa.

As noted, *varungu* in Rhodesia were concerned with OCP pollution because it affected *mhuka*, not because it endangered the lives of *vanhu vatema* or *vanhu* in general. In this, they have much in common with their fellow *vachena* in 1960s and 1970s America. The concern over the environmental effects of OCPs on birds and other forest animals in the United States in the wake of Rachel Carson's seminal *Silent Spring* was from *vachena*, not *vatema* (Davis 2014). Some of the environmental and public health texts about the United States deal with what Susan Bohme (2015) calls "toxic justice" (see also Tarr 1996; Stine and Tarr 1998). This chapter identifies more with recent works on chemical exposure, especially on the handling of toxic chemicals, uptake in plants, and contamination from wind drift, itself a form of aerosol mobility (Harrison 2011).

"The Statistics Are Classified"

To be fair, after the initial euphoria, the residual and aerial spraying of DDT, BHC, and dieldrin came under closer scrutiny among skeptics both within and outside the Tsetse Branch. In a detailed examination in 1951, entomologist Desmond Lovemore argued that simply because *mishonga* worked wonders "on economic farm crops growing on a few hundred acres," that did not mean automatically that they would work over "hundreds of square miles of remote and often worthless country with absolute success."[1]

Critics attacked the toxic footprint of these chemicals on the environment. The Tsetse Branch did not deny it. Nineteen years after Lovemore's criticism, Assistant Director of Veterinary Services (ADVS) Cockbill admitted: "It is a fact that we contribute to the pollution of the environment by using DDT." While defending the chemical for its effectiveness in killing *mhesvi* and its "cheaper cost than other effective insecticides," he also brought up another worrying fact: "it remains lethal ... for three or four months."[2] The Tsetse Branch had carried out tests on different *mishonga* with different chemical compositions, and found them "more lethal to the tsetse fly or more persistent in their action than the organochlorine compounds, dieldrin, telodrin, thiodan, DDT and BHC." Said Cockbill:

> The toxicity of telodrin to mammals is far too high for it to be used as a general insecticide. Dieldrin and thiodan, while effective and persistent, are costlier, and more toxic to mammals than DDT. BHC is less persistent than DDT, and therefore less effective under our conditions of application. In all our tests DDT was found to be superior in use to the organophosphorus and the carbamide insecticides. Of the range of commercial insecticides available at the present time there is no satisfactory alternative to DDT for the control of *G. morsitans* by means of applications of persistent insecticide to tsetse habitat. There are undoubtedly substances used as insecticides which, in a few days, break down into products less harmful to living things than DDT, but it would be futile to undertake costly and arduous spraying operations against tsetse with insecticide that was ineffective.[3]

Cockbill was philosophical in his defense of DDT and the problem of pollution. Urging "perspective" and the need to be "realistic," he argued for a curb on "those that dispense it most liberally," who could use less toxic pesticides (or DDT doses) and still achieve their objectives—like cotton farmers, who applied it at the rate of one to one and a half pounds of DDT per acre, ten to fourteen times a season, or fruit farming, in which "much, if not most of the pesticide falls to the ground or is lost in drift."[4] By contrast, 166.5 tons of 75 percent DDT wettable powder was sprayed in the *mhesvi* habitat over 3,496 square miles of infested country at a rate of 1.8 ounces of DDT per acre.

Usually the spray was a one-off application, the second coat being applied only "in some years," especially when rainfall had diluted the effectiveness. The applications targeted tree bark, rot holes, fallen logs, and ground holes, but not foliage. Cockbill stressed that the risk of poisoning wildlife was "not great" and was in fact "negligible compared with general farming practice."[5] However, he also stressed something telling: "The statistics are classified and are not available to me. I have been assured by the Central Statistical Office that the quantity of DDT utilized annually by this Branch is trivial compared with that used in agriculture generally."[6] DDT was also used for controlling cutworms in tobacco and maize, bollworms in cotton, and stalkworms in maize.[7] To be fair, although DDT continued to be imported as late as 1978, it had generally been replaced with "the very much more readily degradable" organochlorine compound thiodan, even though trials had shown the latter to be "very much more costly" than DDT and weaker.[8]

In 1978, Lovemore sought to assuage fears that BTTC was not concerned with the pollution issues surrounding DDT and residual sprays in general:

> No obvious effects on animal, bird, reptile, fish and other insect life have been observed during the very large scale spraying operations which have been conducted with DDT in Rhodesia. Field staff have been instructed prior to each operation to pay particular attention to this important aspect over the years, but nothing of interest has been recorded. Similarly, as regards the more insidious effects of the chemical the work done by Phelps and others at the University of Rhodesia has shown that no serious problem has developed as yet from tsetse control operations in Rhodesia. ... This work continues. ... It is also noteworthy that in Nigeria where DDT has been in use very extensively in tsetse control operations, in fact, probably very much more so than in Rhodesia, no serious "side-effects" have been noted.[9]

There was no mention of effects on *vanhu*—just *mhuka*.

Lovemore was also urging *vachena* to come to terms with an inconvenient fact: there simply was no other insecticide available that was "comparably as effective and sufficiently cheap to permit large scale operations." Research was ongoing on the sterilization of wild flies after attracting them in large numbers using "attractive odours" and the ultralow volume (ULV) application of thiodan by fixed-wing *ndege* operating at night. However, he cautioned: "Even in the event of one or both of these techniques being perfected tomorrow, it would be some years before one or other, or both, could be adequately applied in the field to replace ... DDT."[10]

Cockbill, however, would concede later that these organochlorines had residual effect *zvipfuyo* cattle of *vatema* resettled in the reclaimed areas and that the fear of poisoning was what made the government settle *vatema*, but not *vachena*, there:

> We have no knowledge of any record that damage to wild life had occurred from a concentration of DDT, or its breakdown products, in animal fats as a result of our

> insecticide spraying activities. We believe that no such records exist. However, during 1970 some donkeys belonging to this Branch died of unknown causes. They had been grazing for months within the Sebungwe spraying operations area, which had been treated with DDT. Specimens of liver and other organs taken at the postmortem examination were submitted for analysis for DDT. The Public Analyst reported that "traces of DDT were insignificant," despite the fact that modern gas chromatograph analytical methods are capable of detecting DDT in such minute quantities as one part in ten thousand million. There was thus no evidence of any accumulation of DDT in these donkeys.[11]

Even then, the reference was to *vatema*'s *zvipfuyo*, not *vatema* themselves.

Mafrayi were exposed to *chepfu*. This is clear from the "Safety Precautions" outlined in Farrell's report on the Chiredzi River spraying operation, wherein only the operator of the *mushini* wore overalls, but even he had no gloves, goggles, dust mask, headgear, or shoes. The overalls were supposed to be washed daily; each workman received daily soap allowances for the laundry and a personal bath. *Chimugondiya* (brown bar soap) was a detergent manufactured for hand-washing clothing, but it was the only soap *vatema* got, for laundry and bathing combined. By contrast, *vachena* received scented bath soap and had their clothes washed and ironed by the batmen (aides).[12] The official reports say that "no one came into contact with the neat Dieldrex 15" and that the chemical was "pumped from drum to drum with a semi-rotary petrol pump."[13]

However, the same report also says that "pouring of the 3.6% mixture was facilitated by buckets, which avoided the difficulty of pouring the insecticide from a five-gallon drum into a six inch [container]." By the end of Farrell's report, the statement that no one came into contact with the chemical is already exposed as a farrago of nonsense. "The operators frequently received spray drift in their faces. This is, of course, harmful," he begins to concede, but quickly plays this contamination down: "However, the maximum spraying time in any one month was 33 hours, and in any day 6 hours. This was divided between two operators. The WHO recommended maxima are 40 hours per week, and 8 hours per day."[14]

It was not until the late 1950s that investigations were undertaken to determine the toxicity of OCPs in Southern Rhodesia. This was to assuage the growing but nevertheless muted criticism on the possibility of poisoning the environment. The earliest tests involved the resistance of *zvipukanana* to *mishonga*. Every time that persistent (residual) pesticides were used in controlling bloodsucking *zvipukanana*, all but a few *zvipukanana* developed resistance. By 1959, no such resistance had been encountered in *mhesvi*, yet it was deemed essential to anticipate the likelihood of this *chipukanana* following most others in developing resistance according to the chemical composition of the *mushonga* used. Whenever such resistance was noticed, the *mushonga* was immediately replaced with another that differed profoundly in chemical structure.[15]

Another factor pushing the research was the fear of the *mhesvi* itself developing resistance to DDT, BHC, and dieldrin in future. Therefore, BTTC undertook preliminary tests comparing the toxicity of the organophosphate Bayer S1752 with that of dieldrin in Shell Dieldrex 15 form. After sixteen weeks of repeated experiments, it became clear that dieldrin retained toxicity sixteen weeks after spraying and that the organophosphate, though toxic, was slower in its action than the organochlorine as a pesticide. There was no indication that OCPs induced resistance in *mhesvi*.[16]

Investigators looking into the toxicity of OCPs were more interested in the effects of the chemicals on *mhesvi* and in boosting the efficacy of spraying operations than the environmental effects of pollution. They were clearly not concerned with the toxicity of the chemicals on *mombe*, let alone the fact that any *zvipukanana*, *shiri*, or *mhuka* coming into skin contact with or ingesting them would die or become sick. When they thought about the effects of rain on the deposits, they were not thinking about the diminishing effectiveness against *mhesvi*, not the downstream effects of the *mushonga* being washed away and, as runoff, entering streams, large rivers, and oceans.

After *Silent Spring*: Echoes from US Debates on DDT

For purposes of the discussion in the rest of this chapter, it helps to recap the contours of the debate on OCPs in the United States that was under way as the spraying campaigns against *mhesvi* and related experiments escalated in Southern Rhodesia. Rachel Carson's book *Silent Spring* provided the lightning rod for a campaign to ban DDT use in the United States. Published in 1962, the book detailed the negative environmental impacts of chemical pesticides like DDT. Carson accused chemical corporations of lying to the public and attacked politicians for their uncritical and even complicit stance toward the "facts" the industry provided about the safety of OCPs.

The wave of criticism from environmentalists—Carson being one among many—stung the US government into action. In 1964, most OCPs were banned from use in some forests and lands. Five years later, all nonessential DDT use was ended. In 1970, the Department of Agriculture banned DDT use on fifty food crops; on wood and lumber materials; on dogs, cats, and other domesticated animals; and in factories and on commercial and institutional premises. DDT was also outlawed in marshlands, forests, and plains of national parks and preserves of the United States. The crowning moment, that same year, came with the establishment of the Environmental Protection Agency (EPA) to oversee the enforcement of these and future environmental regulations. Two years later (in 1972), DDT and other chemical pesticides were banned from use on US soil. In 1978, the US Occupational Safety and Health Administration (OSHA) acknowledged the possibility that DDT was an occupational carcinogen (Coulston 1985, 373).

The emotions that environmental debates about OCPs in the United States raised among empowered citizens and chemical companies are instructive. To examine this, look no further than the words of biochemist and chemical industry representative Robert White-Stevens: "If man were to follow the teachings of Miss Carson, we would return to the Dark Ages, and the insects and diseases and vermin would once again inherit the earth" (McLaughlin 1998). There were those who still believed many decades later that Carson was "simply ignorant of the facts" (Bailey 2002) and must be held responsible "for almost as many deaths as some of the worst dictators of the last century," all in the name of small birds (Taverne 2005). There was big money riding on this; if DDT was banned in the United States, a global ban was next—such a huge market!

Carson's defenders pointed to an issue completely dismissed in Southern Rhodesia: that some of these besieged *zvipukanana* like *hutunga* had developed resistance to DDT (Quiggin and Lambert 2008). That is exactly what Carson had argued: that too many pesticides in the environment bred resistant types, and the less the spray the better (Carson 1962, chap. 16). In fact, DDT usage was banned neither in the United States nor internationally, whether by the US government or under the 2001 Stockholm Convention on Persistent Organic Pollutants on the basis of its connection to resistance. As we now know, within seven to ten years the pest's building genetic resistance to pesticides like DDT will be complete, and the chemicals become virtually useless (Oreskes and Conway 2010).

Defenders of DDT and other OCPs continued to argue that it was necessary and not as evil as its critics painted it to be. They argued, for example, that it degraded rapidly, its strength weakening the more it was exposed to sunlight (ultraviolet radiation) to a point at which it became a bunch of harmless chemicals. As one scientist said in 1985, that was why "levels of DDT worldwide have more or less remained constant and they have not accumulated" (Coulston 1985, 333).

Studies in the 1960s and 1970s showed that when pesticides were sprayed, besides the targeted animal the pesticide also killed land insects, aquatic insects, birds, shrews, mice, reptiles, and ants (Herman and Bulger 1979). So long as the pesticide remained, actively killing the targeted insects, other kinds of animals were also dying (Koeman et al. 1978, 55). Studies of helicopter spraying had shown it to kill predators of flies as well, such as flycatchers, spiders, ants, scorpions, grasshoppers, butterflies, hornets, wasps, and monkeys (55).

In particular, DDT and its principal metabolites, DDD and DDE, were found in nearly all *mhuka* in areas where the compound had been sprayed. Specifically, within these *mhuka* it was discovered in brain tissue, the blood stream, liver, and kidneys. It biodegraded very slowly into metabolites, at the same time also accumulating in adipose (fatty) tissue of birds and fish, making them mobile pollutants (i.e., as food for their predators). Researchers found DDT present in house sparrows, cowbirds, predatory birds, scavengers, and migratory birds. In times of extreme famine when the body needed to burn more fat to survive, DDT was found to move from its fatty deposits into the blood stream and thence into the brain, leading to death (Hill, Dale, and Miles 1971, 502).

This evidence began accumulating from the 1960s on. The concentration of DDT in the adipose tissue of egrets was discovered in Audubon Canyon Ranch (in San Francisco) as early as 1961 (Faber, Risebrough, and Pratt 1972, 111). In addition to neurological poisoning, increasing evidence was also pointing to reproductive poisoning. In Rhodesia, researchers could zone the effects of DDT poisoning according to immediate, intermediate, and secondary bioaccumulation. In the first category were *mhuka* in direct contact with spray—like small *mhuka* (including *zvipukanana*) in the air, plants, soil, and water. The second was composed of predator *shiri*, *mhuka*, and *hové* (fish) that fed directly on either contaminated water or *zvipukanana*. The third tier included predators of these predators.

Tests in the United States had shown that eggs of birds of prey like *twukodzi* (hawks), *makondo* (eagles; singular *gondo*), and *mazizi* (owls) had thinner than normal shells due to OCP poisoning (Faber, Risebrough, and Pratt 1972, 111–112; Elliott 1994). In Florida, eggs of gray herons were breaking under the birds' weight due to thin shells. In San Francisco the eggs of the peregrine falcon were observed to be disappearing mysteriously during incubation season, while among those birds whose eggs hatched, nestlings were dying in large numbers (Faber, Risebrough, and Pratt 1972, 112). The effects of DDT on populations of *hungwe* (fish eagle) and *rukodzi* (falcon) were first observed in 1983 (Tannock, Howells, and Phelps 1983).

Meanwhile, investigations in Norway from 1965 to 1983 discovered OCPs (including BHC and dieldrin) in the livers of dead *makondo* and *mazizi* (Frøslie, Holt, and Norheim 1986).

The evidence was not limited to *shiri* but extended to *mhuka* as well. Experiments with *mbeva* led to conclusions that DDT, dieldrin, and BHC also had neurotoxic, reproductive, hormonal, immunological, cardiac, renal, carcinogenic, and mutagenic effects in *mhuka* (Allen et al. 1979a, 514–518). The male and younger *mhuka* were found to be more susceptible than females and older ones, and higher dosages were more toxic. The OCPs mostly affected reproductive organs, the central nervous system, and metabolic pathways in liver and kidney. As neurotoxins, OCPs were found not serious enough to cause death. Reproductively, however, they were known to cause hormonal (estrogen and testosterone) changes that resulted in immature births in rats. In the same animal, they were also found to cause vascular congestion, tubular degeneration, and changes in kidneys. Livers meanwhile succumbed to mitochondria and fatty infiltration. In rabbits, the chemicals were discovered to limit the body's capacity to defend itself. Chronically exposed *mhuka* (especially *mbeva*) also exhibited a high risk of liver-cell tumor development. Carcinogenicity aside, the OCPs also damaged chromosomes and resulted in negative genetic (mutagenic) effects (Allen et al. 1979a, 514–521).

The Journey of OCPs in Our Bodies

OCPs are passengers and travelers in bodies. Metabolites like DDA, DDD, and DDE are formed as DDT moves through the body (or many bodies) and through food chains and ecosystems. *Vanhu* tend to excrete more DDT in their urine than do *mbeva* (mice), *imbwa* (dogs), and *tsoko* (monkeys). In order of rapidity of excretion overall, DDA was the most rapidly excreted, DDD was next, then DDT, with DDE staying longest in the body. These differences were due to each metabolite's water solubility; the higher the water solubility, the faster the ejection rate (Morgan and Roan 1971, 1972; Wallcave, Bronczyk, and Gingell 1974). The higher amount of DDE found in older women versus younger women was attributed to intake of DDT over longer periods of time (Coulston 1985, 364–365). Dietary ingestion was considered the primary source of DDT in adipose tissue, with fatty foods (poultry, milk) contributing to 95 percent of DDT intake. In the adipose tissue, DDT accumulated to as much as one hundred parts per million (ppm) and biodegraded at 4.1 mg/day; in two years, a body's DDT burden could shrink from 100 ppm to 40 ppm. It took approximately ten to twenty years for DDT to disappear from adipose tissue, while DDE persisted many years longer, even for an entire adult life (Keifer and Mahurin 1997; Le Couteur et al. 1999).

How do DDT and other OCPs—principally dieldrin and BHC—get into the bodies of *vanhu*? Here, the first port of call is *mufrayi* himself. Workers spraying without protective clothing and entire communities are told that the pesticide is harmless and for their own good. The villagers in the countryside are exposed to spray mist, especially from airborne spraying, through wind drift, evaporation, dusty air, and rains (Gil and Sinfort 2005). This is where an appreciation of ecosystems, atmospheres, climates, rainfall, and temperature as processes and as mobilities, as first advanced in the book introduction, is helpful (Coulston 1985, 341). These OCPs concentrate in the adipose tissue of the body—in the brain, kidney, liver, and heart and, in fish, in the gill and underbelly and muscles (Zhou, Zhu, and Kong 2007; Zhou et al. 2008). When *vanhu*, *shiri*, and other living things eat other *mhuka*, like fish and smaller *shiri*, or suckle their young (as the case may be), they ingest the primary sources of OCPs. The chemicals move through the trophic web, accumulating in larger quantities all the time, so that the higher we move up the food chain, the higher the concentration (Semenza et al. 1997, 1030).

If we think of the body as moving through time and of chemicals moving and being present in the body through time, then attention to mobilities opens a new vista. The older the body gets, the more these toxic chemicals accumulate in the adipose (fatty) tissues of the body (Coulston 1985, 364–5). The journey of OCPs and their metabolites (*musvo* in *chidzimbahwe*) also disables the body's mobility (or ability to be agile), not just through death or sickness, but also by inducing neurological, behavioral, morphological, and many other abnormalities (Rowan and Rasmussen 1992; Keifer and Mahurin 1997).

Of course, the body does not retain all DDT-based chemical in its adipose tissue; some of it exits with excretions. Liver, kidney, and breast tumors have been identified among people in DDT-concentrated areas. In *vanhu*, DDT was still considered to have only minor toxic effects in the 1980s, with acute cases of poisoning resulting in extreme muscular weakness, joint pain, extreme nervous tension, anxiety, confusion, inability to concentrate, and depression. Very high doses of DDT were associated with convulsions and even death, especially when containers were not disposed of permanently and safely (Allen et al. 1979a, 513).

Studies from the 1960s and 1970s had found no signs and symptoms of DDT among workers exposed to it, even after chest X-rays (Laws et al.

1967, 766; Deichmann and McDonald 1977). By the 2000s, however, this was no longer the case. The question was how high DDT dosages could lead to convulsions and death. A study had found "a significant association ... between DNA migration or percentage of damaged cells and blood concentrations of p,p´-DDT, p,p´-∂∂∂, and p,p´-DDE" (Yáñez et al. 2004, 22). This was especially the case among spraymen in Mexico, who worked in conditions like the ones in Southern Rhodesia, with parts of their bodies exposed (18). Other scientists found DDT not to be genotoxic, even while others associated it with Type 2 diabetes.

Workers that were exposed directly to chronic contamination with OCPs experienced jerks, seizures, hearing and visual problems, anemia, leukemia, fatty infiltration of the liver, degeneration of cardiac muscle, and damage to lungs, kidneys, and brain. They then descended into convulsions, leading to death due to asphyxiation and cardiac arrest. Other symptoms of poisoning also included chronic liver damage (cirrhosis), chronic hepatitis, endocrine and reproductive disorders, allergic dermatitis, breast cancer, non-Hodgkin's lymphoma, and polyneuritis. As early as 1989, scientists made connections between benzene-containing substances like BHC and various types of leukemia (Lee, Johnson, and Garner 1989; Tompa, Major, and Jakab 1994, 159; Kumar and Kumar 2007, 2). They also noticed marked rises in frequencies of chromosomal aberrations (CAs) in peripheral blood lymphocytes of workers involved in loading, packing, and transporting OCPs (Tompa, Major, and Jakab 1994, 160)—the work *mafrayi* were doing.

During the 1990s, research made a clear connection between parental exposure to pesticides and risk of *gomarara* in children, linked mostly to the exposure of the father in the preconception phase. Specifically, it suggested links with the two commonest *gomarara* among children, acute lymphocytic leukemia and central nervous system tumors. Links with Wilms' tumor, Ewing's sarcoma, and soft-tissue sarcomas were still tenuous (Flower et al. 2004). This was not limited to *mafrayi*; farmworkers and mineworkers were similarly exposed without knowing it (Gladen et al. 1998; Fenske 1997; Gomes, Lloyd, and Revitt 1999).

The *mishonga* usually entered *mafrayi*'s bodies through their skin and noses as they sprayed. Dieldrin in particular was easily absorbed through skin contact. By the early 1970s, some scientists were warning that dieldrin, even more than DDT, was "the greatest hazard" to spraymen. Elsewhere, impotence had been noticed in four out of five farm workers constantly exposed to OCPs (Espir et al. 1973). DDT-exposed males experienced reductions in testosterone levels, sperm count, and the semen volume of ejaculate. Meanwhile, abnormal sperm and sperm motility increased (Bush,

Bennett, and Snow 1986; Ayotte et al. 2001; Martin et al. 2002). In 2012, OCPs were identified as synthetic hormone-disrupting chemicals that led to a gradual increase in two male sexual-development disorders (cryptorchidism and hypospadias), testicular dysfunction, and testicular cancer (Bergman et al. 2013. It was speculated that occupational exposure could alter the sperm genetically prior to conception, damaging fetal development (Salazar-Garcia 2004). Some scientists, however, quibbled over the absence of any experimental studies to back up the birth defect claims. High doses of DDT, up to almost 1000 mg/man/hr., were recorded in places of sustained chemical use (Wolfe and Armstrong 1971, 169). The research indicated that DDT had little chance of skin absorption unless applied to the skin in fats or oils (Rivero-Rodriguez 1997). Yet, as we have discussed, in Southern Rhodesia DDT was *mixed with oil* to achieve penetration on leaves and on bark surfaces.

The chemical could be spread in any number of ways. This often occurred during application when the mist blew into the trees and the wind direction changed, blowing right back into the sprayer's face; or the *mushonga* might spill onto the body or clothing, making skin contact inevitable. It occurred due to lack of protective clothing like boots; overalls that were not torn and that covered the legs, torso, and arms completely; and face masks and helmets. Rubber gloves certainly insulated the sprayer from contact; leather or fabric ones were like sponges, absorbing and retaining the pesticide in contact with them so that it slowly seeped into the skin. Few *vatema* operating the *mushini*, carrying *mushonga*, or walking through sprayed swaths wore any shoes, and they stepped directly on the sprayed ground.[17] However, any such information about contamination was "classified."

Understanding the routes of OCPs into the environment and then into *mhuka* is critical for a comprehension of the full impact of DDT spraying in Southern Rhodesia. When it rained, the Tsetse Branch was far more worried about limits on flying and spraying time, diminished toxicity against *mhesvi*, and the inevitable leafing of the vegetation than about pollution and public human and animal health. By 1967, researchers had discovered that drinking water and eating food were the two major ways through which *mhuka* and *shiri* ingested DDT (Nash and Woolson 1967, 924; Menzie 1972, 199). OCPs were also inhaled through aerosols and dust from dry deposits or contaminated soils in the quantity of 5 mg/year. Although some scientists urged the use of respirators (Coulston 1985, 339), these were certainly never used in Southern Rhodesia.

During spraying, about 30–50 percent of the chemical was lost to the air and resulted in atmospheric contamination. Aerial spraying in particular deposited solid, gaseous, and liquid forms of OCP into the atmosphere through wind drift and evaporation (Gil and Sinfort 2005). Pesticides also entered the atmosphere after application when they evaporated out of crops and soil where they would have dissolved (volatilization), through degradation pathways (e.g., hydrolysis and photolysis), and through sheer wind erosion. In the air, the distribution of OCPs varied according to their chemical and physical characteristics and meteorological conditions. They exited the atmosphere as acid rain or as solid and gas injections or through intake by living things (Gil and Sinfort 2005). In Rhodesia, windy days provided optimal spraying conditions, the drift enabling the aerosol droplets to move sideways into the vegetation, touching *mhesvi* hiding on the underside of the leaf or log and in between cracks in bark.

Workmen returning with OCP-contaminated clothing brought the chemicals into the home; *tsika/chivanhu* (both then as now) made women responsible for washing their husbands' clothing. This exposed pregnant women—and their unborn children, through the placenta (Salazar-Garcia et al. 2004). Comet assay tests performed on pregnant women in 2004 demonstrated the correlation between DDT, DDD, and DDE levels in the blood and DNA damage in women in *hutunga* and agricultural spraying (Yáñez et al. 2004, 18).[18] Connections were found between DDT and DDE and altered menstrual cycles, fetal loss, and earlier or delayed menopause. In 2012, the hormone-disruptive activity of OCPs—especially the development of new substances mimicking the behavior of estrogen—was noted to be increasing (Bergman et al. 2013; Garcia-Rodriguez et al. 2004).

Mothers also contracted OCP poisoning by eating contaminated food or drinking contaminated water. For the latter, people and other *mhuka* ingested OCPs when they drank water from wells and rivers, where the runoff from sprayed areas accumulated. In winter, when most rivers stopped flowing and evaporation was at its peak, the ratio of animals to waterholes also increased (Castilhos et al. 2000; Zhou, Zhu, and Kong 2007). Each rainy season, the flooded rivers gathered the contaminated water and deposited it downstream and thence into the ocean. A 2010 study of dolphins on the Zanzibar coastline indicated the presence of lindane and DDT metabolites, the health effects of which were not established, but they were deeply troubling from an environmental health perspective (Mwevura et al. 2010). Studies of persistent organochlorine pesticides in Lake Tanganyika showed negative consequences to fish and fish eaters like *shiri* and *vanhu* (Manirakiza et al. 2002).

Therefore, the polluted environment becomes one source of many routes through which chemicals enter us, travel within us, lodge inside our organs, and exit us into the food chain or the environment (Menzie 1972, 199; Nash and Woolson 1967, 924; Coulston 1985, 337). The route from the spring or well to the mouth, through the stomach(s) to the exit points (urine, fecal matter, sweat, and mucus) becomes the transportation infrastructure and transient workspace the chemical passes through, acts and is acted upon, and moves on by dint of the body's own biologically mobile workshop (Coulston 1985, 341).

(How) Do OCPs Cause *Gomarara*?

The itineraries of OCPs in the body began to raise alarm when some scientists began to associate them with *gomarara*—itself a cellular mobility of sorts, if we think of mutagenicity that way. The argument began to be made in the 1970s that, on entering bodies, certain chemical compounds trigger the growth of tumors that are simply abnormal masses of tissue that are functionless but not inflammatory, especially where preexisting health issues exist. A chemical that does this is called a *carcinogen*; the tumor does not have to be painful or harmful (if it is a benign instead of malignant tumor); just its sheer presence is enough to qualify a chemical that causes it as a carcinogen (Coulston 1985, 348).

The next question, then, is whether chemicals like OCPs cause, promote, or escalate *gomarara*. Inevitably, the first link drawn between DDT and *gomarara* was that the former promoted rather than caused the latter. DDT was not a "true" carcinogen (Coulston 1985, 349). By 2004, DDT, DDD, and DDE were confirmed to cause DNA damage in women; to alter their menstrual cycles; to cause fetal loss, early menopause, retarded childhood or pubertal growth; to induce significant structural and functional neurodevelopmental changes, testicular disorders and tumors, breast cancer, chronic liver damage, Parkinson's and Alzheimer's diseases, kidney diseases, and immunosuppression; and to cause a decrease in semen quality, increase in testicular and prostate cancer, increase in defects in male sex organs, and increased incidence of breast cancer (Yáñez et al. 2004, 18; García-Rodríguez 1996, 1093; Allen et al. 1979b, 679–680; Wolff and Toniolo 1995; Kumar and Kumar 2007; Pan et al. 2009).

As Freya Kamel and Jane Hoppin noted in 2004, the tragedy of pesticide poisoning is that it usually affects the poor—especially farm workers, who cannot afford medical consultations. For that reason, it usually goes undiagnosed; hence "workers who have never been diagnosed with pesticide poisoning may still have sustained high exposures or experienced pesticide-related illness; therefore, using diagnosed poisoning as a criterion for inclusion in an exposed group or exclusion from a comparison group may incorrectly classify individuals. … Farm owners who employ others to apply pesticides may have limited personal exposure to pesticides" (Kamel and Hoppin 2004, 950; see also Moses et al. 1993).

OCPs are a global problem. Countries where *hutunga* and *mhesvi* are found have high temperatures and heavy rainfall that enable the rapid movement of these pesticide residues through air and water into the global environment (atmosphere, sea; Ramesh et al. 1990, 290). DDT and HCH (hexachlorocyclohexane; not to be confused with BHC—that is, benzene hexachloride), for example, are the major *mishonga* in Indian foodstuffs such as grains and vegetables. Milk and milk products are the major sources of dietary exposure to DDT and HCH in India, above FAO/WHO-approved levels. Untreated water is a potential source of DDT and HCH (Kannan et al. 1997).

It is striking—but hardly surprising—that research into chemical exposures and breast cancer effects was not a high priority in the United States. To start with, it only affects women in a predominantly male-dominated scientific field. Moreover, research into the carcinogenicity of cancer, just like research into guns and violence, tobacco and its harmful effects, or that of genetically modified crops, has met with resistance and downright suppression from the industries that stand to benefit from continued use of these drugs (pun intended). The connections between breast cancer and reproductive hormones must at the very least have justified an investigation into environmental chemicals and their effects. Yet more than four decades after the banning of DDT use in the United States in the wake of proven adverse reproductive effects in wildlife, the agenda for a rigorous debate on the carcinogenic effects of OCPs remains elusive.

There is a serious contradiction in responses to chemicals and other goods that are later found to be either toxic or defective in the United States and Africa; I specifically mean companies headquarters in the former with subsidiaries in or selling chemicals to the latter. In the United States, it is common for goods made locally or abroad to be "recalled" or banned entirely after successful class-action lawsuits are brought against them in court: goods such as lead-based paints, asbestos, pharmaceutical drugs, faulty automobiles, and so on. Adverts are floated everyday inviting claims from potential victims of mesothelioma who worked in the navy, shipyards, mills, heating, construction, or automotive industries who were exposed to asbestos to submit claims. By contrast, those same companies are never required to extend this same compensation to victims who did the actual mining, transportation, shipping, and handling in Africa, where many of these companies extracted these ores. Gabrielle Hecht's (2012) extensive study of these commodity chains demonstrates this in the case of uranium that played such a critical role in both the military and energy projects of the United States and South Africa. There is also usually an assumption that simply banning something marks the end of the problem. Lead-based paint is a perfect example: it remains on the walls, as does the asbestos, poisoning away *vatema*, with neither consequences for those that have poisoned others nor compensation for the victims, many of whom just die at home, unable to afford a visit to the clinic or to get tested.

Worse, a ban is a door that closes a chapter of inconvenient questions investigators might ask. Put differently, it seems that investigators look for answers so long as a drug or chemical is being produced and used. The moment it is withdrawn or its production is terminated, they look for another cause célèbre. Perhaps that is exactly why science, technology, and society (STS) are essential in Africa or anywhere else. That is, to investigate and raise questions when goods are no longer here to do what is *good* for people and the environment—bar a few who get rich, consequences be damned—but have instead become the problem. Once DDT was banned, there seemed no further incentive to conduct detailed research into and evaluate its effects in the long term, especially in places remote from the United States where the *chepfu* came from.

In fact, DDT and dieldrin, and then endosulfan, continue to be used in Africa, Latin America, and Asia long after 1985, when their production in the United States for export was restricted. Translated, this means that US companies were exporting *chepfu* that the country had deemed too dangerous on its own soil, in its own atmosphere, to its own people. Are we all humans, then? Do we bleed blood too?

The problem with OCPs is that they do not kill people en masse or instantly, their gestation period being more drawn out over almost a person's lifetime. Thus, the incentive to investigate the long-term effects of OCPs did not exist, because the generation affected by them was still healthy and much alive—and in fact bought into the idea that DDT was very safe for *vanhu* (Wolff and Toniolo 1995). The Rhodesians were more worried about the chemicals' effects on *mhuka*. The way a society treats its animals is a good indicator of how it treats other people. We may assume since they called and treated *bobjaan* (baboons), they included *vatema* in their compassionate worry for *mhuka*.

The deafening silence on the possible carcinogenicity of OCPs was neither localized to Rhodesia nor limited to the mid- to late-twentieth century. Research from the 1970s and 1980s seemed to indicate that DDT, DDD, and DDE accumulations in the human body had a fifty-fifty chance of causing any health problems, let alone *gomarara*. The same researchers, however, also admitted to the lifelong bodily burden of DDT and its metabolites once ingested, especially DDE (Coulston 1985, 366). Nearly half a century later, silence!

All of this makes an inquiry into the environmental history of DDT and other OCPs in Zimbabwe, Africa at large, and the Global South more broadly important. As Epstein (1994) noted, the definition of "environmental" causes of *gomarara* research has almost exclusively focused on viruses, vitamins, diet, tamoxifen, smoking, occupational exposures and radiation, while ignoring pollution or industrial chemicals. Almost to the end of the twentieth century, studies made no attempt to examine the obvious link between carcinogens and breast cancer, choosing instead to explore dietary factors (Wolff and Toniolo 1995). Even as recently as 2009, scientists have lamented "the lack of knowledge about human exposure and health effects in communities where DDT is currently being sprayed" (Eskenazi et al. 2009).

By 2000, breast cancer incidence rates were on the increase, and it was no longer enough to say that this was purely a result of more efficient screening and detection of *gomarara*. What accounted for the cases revealed efficiently to begin with? Could it be that environmental chemicals were responsible, just as they had already proven to be endocrine disruptors? How could it be ignored that *gomarara* figures had risen at the same time as the rise in the use of OCPs (Allen et al. 1979b, 679)? Already, epidemiological studies were showing that women who had the highest DDE levels in their blood also had a fourfold risk of breast cancer (Allen et al. 1979b).

Therefore, at the very least, Zimbabwe's *gomarara* crisis deserves notice. In the limited space of this final section, and in the absence of a systematic ethnographic study on carcinogenicity, I now seek only to raise awareness of the issue and ask questions. This is not a conclusion, but a start—not necessarily my own start, but one for all those who care for humanity. Although studies of *gomarara* in oncological wards in Africa take us to the scene of encounter between cancer and medicine (Livingston 2012), we also need to step out of the ward and explain the cases under chemotherapy. To what causes are they bearing witness?

Opening a Discussion: Environmental Pollutants and *Gomarara* in Zimbabwe

The record-keeping on *gomarara* in Zimbabwe has been erratic at best, making the availability of up-to-date statistics on the countryside difficult, especially for cases originating from tsetse-related OCP-sprayed areas. Add to that the smallholder farmers and farmworkers that have extensively used OCPs like Gammatox (BHC) since at least the 1960s as pesticides when growing vegetables and grain, the tons of OCPs sprayed to guard against *hutunga*, and the large tracts of *mapurazi* on which fruit trees, tobacco, cotton, and maize that require pesticides grow!

Only one government agent, the Zimbabwe National Cancer Registry (ZNCR), currently collects and collates statistics on *gomarara*. The first ever cancer registry in the country was established in the second-largest city, Bulawayo, in 1963, but it closed in 1977, after which *gomarara* registration lapsed, with information confined to data from histopathology series. The registry reopened in 1985, this time in the capital, Harare (formerly Salisbury), but it only covered statistics for the city. Its numbers are drawn from routine weekly visits to the wards of the nation's two central referral hospitals, medical records with a discharge diagnosis of *gomarara*, government laboratory pathology reports, completed notifications from the Radiotherapy Department, monthly reports from the private pathology laboratory, and death certificates for Harare residents. The major limitation is that these figures are drawn only from Harare residents or patients admitted to Harare hospitals (Kadzatsa and Chokunonga 2016; Chokunonga et al. 2013).

The elderly and the poor usually retire to, or never leave, the rural countryside. They live far from Bulawayo or Harare—in Hwange, Binga, Lupane, Sanyati, Gokwe, Hurungwe, Guruve, Centenary, Muzarabani, Rushinga, Mutoko, Mudzi, Nyanga, Chipinge, or Chiredzi East. These outskirts of Zimbabwe are the same places where massive amounts of OCPs were dumped in the environment exactly when (1950s–1970s) the groups being struck down by *gomarara* now were children or young adults, including those employed in tsetse control operations as *magocha* and *mafrayi*. If any in this group go to the hospital, they go to the one nearest their village, or to a general hospital at which only painkillers such as Panado, Disprin, or paracetamol are dispensed and where no doctor, let alone testing equipment, is to be found. There are also those that go to a *n'anga* (traditional healer) or to *maprofita* (prophets) instead—not because they are ignorant of the efficacy of clinical medicine, but because they are too poor to afford a radiological test, bus fare, or hospital fees.

They die from what society commonly calls *kuroyiwa* (witchcraft), *chirwere chegomarara* (disease of the mistletoe), *ronda rinongobva pasi risingapori* (a wound that emerges from the ground and never heals), or *kutemwa nemusoro* (headache). Statements like "*Akangoti mudumbu mudumbu, ndiye sarai*" ("He started complaining of a headache, then said goodbye"), "*Gumbo rake rakabva pasi*" ("Her leg came from the ground"), "*Rakangotanga, hapana anoziva kuti chii*" ("It just started, nobody knows what it is"), and "*Ane chitsinga*" ("He has an implant inserted by a witch") are common. These people die quietly, without even an autopsy, carrying their secrets to the grave.

Gomarara is becoming a serious policy issue in Zimbabwe, a nation of fourteen million people. This is especially true after prominent politicians Morgan Tsvangirai and Thokozani Khupe, president and deputy president of the opposition Movement for Democratic Change (MDC), respectively, admitted to colon and breast cancer in 2016 and 2011, respectively. They can count themselves lucky to have lived longer (Tsvangirai succumbed to the disease on February 14, 2018), because they are powerful people of means; up to 1,300 people die of *gomarara* in Zimbabwe every year without ever getting tested or treated. Since 2007, the number of new cases of *gomarara* per year has more than doubled from 3,349 to over 7,000 and rising; only 700–1,500 are treated. There are only two radiotherapy treatment, chemotherapy, and surgery centers for *gomarara* in a country of fourteen million citizens. The machinery at the two centers, the Mpilo and Parirenyatwa referral hospitals, is constantly broken.

The World Health Organization (WHO) has emphasized sex, tobacco use, and alcohol as contributing to 40 percent of the *gomarara* case load (Jemal et al. 2012; Shafey, Eriksen, and Mackay 2009; Glynn et al. 2010). Indeed, it has urged *vatema* to make "lifestyle changes"—to stop "risky sexual behavior," eat a healthy diet, perform regular physical activity, limit alcohol intake, avoid or reduce smoking, and have regular health checkups ("Fighting the Cancer Scourge" 2012). Completely excluded from this approach is an environmental intervention, in a country and continent whose history and landscape has been so thoroughly drenched in OCPs. The most common *gomarara* in Zimbabwe are cervical, breast, prostate, and skin cancer, but *gomarara* of the digestive system are on the rise. The focus of the Cancer Association of Zimbabwe (CAZ) is to establish the relationship between *gomarara*, HIV and AIDS, and other conditions, like diabetes and hypertension (Moyo 2012; "Call to Decentralize Cancer Treatment Services" 2012). OCPs are not on the agenda.

In a groundbreaking article on the social aspects of cervical cancer in Zimbabwe in 2006, J. F. Mangoma and colleagues said something that is as true for this specific *gomarara* as it is for breast, bladder, or prostate cancer. According to the authors, "Cancer of the cervix is a gender-sensitive condition in that only women suffer from it. Thus, its importance may easily be marginalized. Often, *gomarara* has to compete for meager resources with more dramatic diseases like HIV and AIDS, malaria and tuberculosis"

(Mangoma et al. 2006, 93). Thus, though preventable and curable, its morbidity and mortality is increasing in Zimbabwe (Chirenje et al. 1998, 1999). Most studies of *gomarara* focus on the scientific situation, not on sociological, historical, and environmental aspects. Mangoma and others thus sought "to give a sociological and anthropological insight into rural black women's understanding of cervical cancer, its symptoms and the importance of screening" (Mangoma et al. 2006, 93). These scholars used Mutoko as a case study, but limited their studies to the social environment (patriarchal societies, kinship, marital and sexual factors), perhaps oblivious to the fact that Mutoko experienced sustained anti-*mhesvi* spraying operations in the 1960s and 1970s.

The funding priorities for combating *gomarara* in Zimbabwe are decided by donor countries and organizations in the United States and Europe, as well as WHO—not by Zimbabweans, let alone the Zimbabwean government, which is broke. These priorities are heavily skewed toward communicable disease–related *gomarara*. Hence, Kaposi's sarcoma and cervical *gomarara* receive the most funding because of their association with communicable diseases like HIV/AIDS and HPV (papillomavirus), whereas those associated with noncommunicable carcinogenic agents receive less (Kachala 2010). Some of the money, just as with HIV/AIDS funding, goes toward advocacy work to convince people that *gomarara* is not "a curse unleashed by angry ancestral spirits on errant individuals" but a matter of tumors (*gomarara*; Moyo 2012).

Studies of *gomarara* among Zimbabwe's white population over thirty years found the pattern of *gomarara* roughly typical of populations with high socioeconomic status living in Europe or North America, with elevated incidence rates of breast cancer, large-bowel cancer, and, in women, lung cancer. However, there were also some rather unusual features, like higher skin cancer (including melanoma) rates and liver and bladder cancer rates than were often seen in white populations. The statistics for 1990 to 1992 found that 551 suffered from *gomarara* of the skin other than melanoma (just nineteen), with the forty-five and over age group the most vulnerable; 318 suffered *gomarara* in all other sites except the skin, again with the forty-five-plus age group most susceptible; and sixty-seven suffered prostate cancer, mostly affecting those over fifty-five. Among white women, 362 suffered *gomarara* of the skin, mostly from age thirty-five and over; 135 suffered breast cancer, again in those thirty-five and over; 339 suffered "all sites but skin" cancer, also in those thirty-five and over. The skin cancers were mostly skin tumors. Figures for 1995 showed that breast cancer in Zimbabwe was "virtually the highest in the world," and it was put down to people's diet, along with the very high incidence of ovarian cancer (Bassett et al. 1995, 24–28).

Among black people in Harare, the 1995 statistics showed high rates of liver, prostate, and cervix cancer and low rates of large-bowel and breast cancer. Incidences of *gomarara* of the esophagus, bladder, and (in men) lungs were also high; the increase in Kaposi's sarcoma was attributed to AIDS. At that point, no evidence yet existed of non-Hodgkin's lymphoma or cervical cancer. *Gomarara* occurrence in Africa was readily attributed to "urbanization, with its accompanying changes in diet and lifestyle, as well as the recent AIDS epidemic" (Bassett et al. 1995, 29). Again, there was no discussion of environmental factors.

Another instance of this bias can be seen in research on primary carcinoma of the liver, one of the most common *gomarara* in sub-Saharan Africa. It is attributed to aflatoxin exposure, but even though scientists acknowledge the aflatoxin's contribution to geographic variations in liver cancer in the region, they make no mention of its sources or promoters—like DDT. Rural men have also tested positive for iron overload, blamed on the high iron content of home-brewed beer, but there is no attention to OCP content in such foods, especially grain or crops treated with gammatox, malathion, or DDT. Tobacco, for example, is blamed for esophageal cancer, but the high rates observed in certain regions defy this simple explanation in the absence of environmental studies (Bradshaw and Schonland 1974; van Rensburg et al. 1985; Segal, Reinach, and de Beer 1988; van Rensburg 1981; Marasas, van Rensburg, and Mirocha 1979; Sydenham et al. 1990).

There is thus far no satisfactory explanation for why *gomarara* incidence rates in Zimbabwe have been higher than anywhere else in Africa. In 1995, esophageal cancer was high, lung cancer moderately high, and liver cancer also high. Cervical cancer was the highest recorded in Africa. Breast cancer was previously not common but increasing in incidence, as was melanoma, 90 percent of it affecting the legs. There were also age-specific incidences of leukemia, non-Hodgkin's lymphoma, and myeloma (Bassett et al. 1995, 29–30). The incidence of stomach and lung cancer was moderate, with large-bowel cancer relatively rare. Kaposi's sarcoma is now the most common *gomarara* in men (23.3 percent), accounting for two-thirds of *gomarara* in men aged twenty-five to thirty-four. However, the increase in women is striking; it is ranked third in frequency after cervical and breast cancer, the most common *gomarara* among young women (29 percent at ages fifteen to thirty-four). Kaposi's sarcoma was long endemic in Zimbabwe in elderly men—62 percent of them aged forty-five or more, a quarter of them sixty-five and over, and almost all of them skin tumors. The pattern of cases was credited to HIV and AIDS; none of them to environmental pollution factors. Non-Hodgkin's lymphoma incidence was not very high, but is increasing—again attributed to HIV (Bassett et al. 1995, 34; Gordon 1973).

The question I end with is this: How would an environmental approach—moreover, one that examines OCP use in mass campaigns against *mhesvi*, *hutunga*, *zvimokoto* (quelea birds), *hwiza* (locusts), and *mhunduru* (armyworms), as well as field uses in crop and fruit tree protection—aid understanding of the *gomarara* crisis in Zimbabwe today? I do not ask as a scholar—just as a son whose father died of it and a son-in-law whose father-in-law has it.

14 Chemoprophylactics

Chemoprophylaxis refers to the administration of medication to prevent disease or infection. This method was based on the conviction that *hutachiwana hwen'gana,* which *vachena* called *Trypanosoma congolense* (literally, "the trypanosome of Congo," because that is where it was first "discovered" by bench scientists), occurred only within the blood of infected *mombe* (cattle), not outside the vascular system. Although most strains produced low concentrations of *nyongororo* (parasite) in the jugular vein, large numbers of *hutachiwana* could be observed in microcirculation, especially in the brain, heart, and ear veins (Maxie and Loses 1977). In Rhodesia (from 1963 on), chemotherapy was almost as old as white rule itself. By the time that drug prophylactics and other trypanicides were extended to deal with *n'gana* outbreaks in resettled areas cleared of *mhesvi* (but nonetheless prone to reinvasion) in the 1950s, they had been in use since the early twentieth century.

One of the biggest problems that Rhodesia's Branch of Tsetse and Trypanosomiasis Control (BTTC) faced throughout the Rhodesia period was *hutachiwana*'s habit of developing a degree of resistance to the drug deployed to destroy it. The condition that made resistance possible was one in which the "dosage or concentration of drugs [was] too low to kill outright, so that the microbe [had] time to organize its resistance."[1] To develop that capacity, large numbers of *hutachiwana* had to be constantly exposed to a drug circulating within the animal's body. To deal with such resistance, vets across *n'gana*-infected Africa developed "stoppers"—combinations of drugs used when resistant *hutachiwana* were encountered and one drug alone proved inadequate to conquer them. The task for which stoppers were deployed—overcoming drug-resistant trypanosome strains—was called a "challenge."[2]

As early as 1944, skeptics were already warning that the new chemical weapons against pests were "turning out to be double-edged weapons" that "may at the same time destroy both useful and harmful agricultural insects." Indeed, as Jane Stafford cautioned, "They may rid your dog of fleas, but insidiously ... damage his liver or paralyze him through nerve damage. They will rid your home of mosquitoes, flies and vermin, but the price may turn out to be high in human health and life" (Stafford 1944, 90). This is what happened in Rhodesia.

This chapter first gives a historical overview of chemoprophylaxis in Southern Rhodesia, then turns to the problem of drug resistance and photosensitization, which, as noted earlier, is a clinical condition in which the skin's negative exposure and reaction to sunlight is heightened due to phototoxic drugs and chemicals. This photosensitivity occurs when these substances absorb sunlight (ultraviolet radiation), triggering a burning sensation, redness, and swelling. Within twenty-four hours, the skin becomes pigmented and starts peeling off. The animal dies a slow, painful death; the owner endures the pain of seeing his wealth, his cherished cow, ox, or bull, losing its skin piece by piece. *Hutachiwana* attached itself to capillary walls, and trypanicidal drugs worked by disturbing and forcing them off the walls into the general blood circulation system. Any treatment regimen had to work around that to prevent drug resistance; few drugs could.

The chapter ends with a case study of chemoprophylaxis operations in Southern Rhodesia, exploring how the early promises of chemoprophylaxis ended with unforeseen complications that poisoned instead of cured animals of *n'gana*. The argument made is one about pollution of the most intimate kind: within the body of both the animal and *hutachiwana* itself. This chapter will show a general pattern among all the drugs; they worked well initially before *zvipfuyo* (livestock) either relapsed or exhibited signs of drug resistance (Whiteside 1962), prompting the deployment of one drug to cure another.

This chapter must be read in the context of what kinds of prophylactic regimes these new materials were now replacing—specifically, two forms of inoculation already referred to that *vatema* practiced. The difference lies in the toxicity of the new methods. These methods of dealing with pestiferous micromobilities inside their bodies and those of their *zvipfuyo* was discussed earlier in chapter 2. The glossary of all *chidzimbahwe* and other local keywords, found at the back of the book, will aid the reader.

Chemoprophylactics in Southern Rhodesia: A Historical Overview, 1909–1973

In 1909, entomologist Llewellyn Bevan visited the Pasteur Institute in France and returned with a method of injecting *mombe* with alternating doses of antimony salt and arsenic to kill *hutachiwana*. Bevan later found that the antimony was the key effective therapeutic agent, so antimony potassium tartrate, rather than arsenic, became the preferred chemotherapy against *n'gana*-struck *mombe* in Southern Rhodesia, alongside antimosan.[3] In 1928, the Division of Entomology even claimed that since 1909, antimony had "saved the lives of thousands of animals,"[4] a conviction that continued to govern *n'gana* policy until 1938[5]—the year that phenanthridine, the compound that eventually replaced antimony potassium tartrate and antimosan, began to be used. One such phenanthridine, dimidium bromide, was being used widely to treat *n'gana* elsewhere in Africa. Early on, homidium bromide, a methyl-substituted analog of dimidium, was shown to be an effective curative against *hutachiwana* that *varungu* called *T. congolense* and *T. vivax*. However, although it was much less toxic to *zvipfuyo* than dimidium, homidium bromide turned out to have limited use as a prophylactic (Dolan 1990). In 1946, the transition was made to intravenous application of dimidium bromide after a rather ineffectual homidium regimen—still with little improvement.[6] This marked the first time that *T. congolense* resistant to drugs had been noticed in the country.

The name *homidium* was changed to *ethidium* in the 1950s, and its derivative, prothidium, was hailed as "one of the more hopeful ... newer drugs" on the market. In East Africa, when applied in *mombe* at 2 mg/kg, the drug had a six-month effectiveness.[7] Compare this with studies in Northern Nigeria that had found prothidium to have unusually short periods of protection in *mombe* (Williamson and Desowitz 1956; Stephen 1958; Stephen and Williamson 1958; Williamson and Stephen 1960). However, these negative results were set aside in Southern Rhodesia because the Nigeria studies had used "defective batches containing impurities or too high a dosage rate."[8] By 1960, however, homidium-resistant *hutachiwana* had become a "distinct possibility" (Williamson and Stephen 1960, 366).

By then, antrycide had become the most widely used drug throughout Africa and "the first piece of real progress in the chemical control of trypanosomiasis."[9] Antrycide came in two forms: (1) demethyl sulfate (DMS), which was very soluble, acted rapidly, and passed out of the body equally rapidly (in three to four weeks); and (2) chloride salt, relatively insoluble and forming a deposit in tissues, from which it slowly passed into the body over two to three months. The chloride salt was therefore prophylactic. The methyl sulfate was not; it was viable only for treating or curative purposes. These two drugs were then combined into a "pro-salt" to achieve both a curative and a prophylactic effect. The much cheaper chloride salt could still be used as a prophylactic in situations of sporadic infection by carried *mhesvi*; pro-salt was, of course, always on standby should the challenge become too difficult.[10]

Antrycide offered a wide range of actions that few other drugs could match against the various *hutachiwana*. The DMS was absorbed rapidly once it entered the body, whereas the pro-salt was slower, the drug "lumping" or "banking" in a kind of "bag" at the injection site from which it was released over three months. The chloride in the "lump" of pro-salt was absorbed into the body system through the intramuscular or subcutaneous route. That way—the vets said—the chemical traveled virtually nontoxically, whereas intravenously inoculated drugs might poison the animal. The chloride was preferred to provide a "bank" of the drug at the site of injection for release over a period of months.[11] Treated *mombe* were easily identified by the orange-sized lump that developed at the site of injection, which often led to huge muscle abscesses. This problem had already led to falling *mombe* prices in Kenya when, in 1955, the meat inspector at Mutare "condemned quite a large proportion of rump steak from carcasses showing lumps" for that reason.[12] It turned out that the risk of a lump could be lessened by simply massaging the site of injection.

The complications of pro-salt also led to a revised formula (RF) for antrycide treatment, reducing the chloride element that did not dissolve while maintaining the sulfate radical element that did. This two parts chloride and three parts sulfate solution immunized *mombe* for over one hundred days (among exotic breeds on *mapurazi*), or even longer (over five months among indigenous breeds). Not only did it offer better prophylaxis, but RF also was cheaper than ordinary formula (OF) and induced smaller local reactions that disappeared quickly, unlike OF-induced lumps that remained even after three years. To avoid lumps and reduce incidents of "condemned meat" even further, *mombe* requiring a few inoculations could be injected on the caudal fold site rather than the neck.[13]

However, antrycide turned out to be poisonous to *mombe*. Among the toxic symptoms were increased salivation, sweating tremors, and collapse—even death—due to overstimulation of the parasympathetic nervous system (PSNS).[14] The PSNS is one of two key divisions of the autonomic nervous system (ANS) that regulates internal organs and glands unconsciously. When the body is in repose, it stimulates rest-and-digest activities, such as salivation, lacrimation, sexual arousal, defecation, digestion, and urination. The PSNS complements the sympathetic nervous system (SNS), which stimulates fight-or-flight responses. A photosensitized animal exhibited poor health, exhaustion, overheating, fright, undue exertion, and dehydration before succumbing to death. Postmortems normally revealed hemorrhagic gastroenteritis with or without zebra markings of the large intestine, pulmonary edema, excessive fluid in the thoracic and cardiac cavities, and evidence of serious kidney damage.[15]

With antrycide's problems of photosensitization mounting, the Veterinary Department switched to samorin (isometamidium), a product of Samorin in the United Kingdom. It went on to become one of the most widely used drugs of its time against *n'gana* in *mhesvi*-infested areas of Africa. The beauty of samorin was its flexibility of deployment, as both a prophylactic and a curative. By contrast, drugs like homidium and berenil (diminazene aceturate) were solely curative. Samorin and berenil would be used as a sanative combination successfully in *n'gana* control in the 1960s (Moloo and Kutuza 1990).

By 1966, reports of drug-resistant trypanosome populations were increasing, casting serious doubts on the hopes of a drug-based solution to the *n'gana* problem. The first response to the samorin-resistant *hutachiwana* was to search for a better method of administering treatment. Vets decided to inoculate the drug intravenously instead of intramuscularly, thereby achieving an initial concentration of the drug in the blood that was significantly higher than that achieved through intramuscular administration. Under experiment, all the *hutachiwana* present in the host were eliminated (Whitelaw et al. 1991; Kinabo and Bogan 1988; Dowler, Schillinger, and Connor 1989).

A second problem was the relapse of all *zvipfuyo* treated with samorin, which demonstrated that *hutachiwana* infections had not been eliminated, nor had the demonstrable presence of *nyongororo* in the blood (parasitemia) been delayed to any significant degree. The experiments had shown that only a dosage of 2.0 mg/kg or higher offered the best opportunity for eliminating infections with *hutachiwana* resistant to samorin. Finally, it was found that the trypanicidal efficacy of samorin depended not only on the concentration of the drug to which *hutachiwana* were exposed but also on the length of exposure (Sutherland et al. 1991; Moloo and Kutuza 1990.

Berenil (diminazene) entered the Rhodesian chemotherapeutic scene as a stopper with an entirely different chemical composition from its predecessors: a diamidine used in the form of an aceturate (N-acetylglycinate). Berenil was a product of German pharmaceutical company Farbwerke Hoechst AG and by far the most commonly used treatment for *n'gana* throughout Africa. In the 1968–1969 agricultural season alone, 47,577 doses were given in Rhodesia, with not a single case of toxicity reported. Years earlier, only one beast had suffered an allergic reaction. It was discovered, however, that berenil fatalities occurred only when it was dispensed right after samorin, as a result of changes in liver fat.[16]

The problem with berenil early on was that it was rapidly eliminated from blood through urinary excretion, even though biologically active quantities remained in *mombe* for up to three weeks after treatment. In 1977, it was discovered that berenil did not kill *hutachiwana* directly, but could instead make them available for destruction by the "big eaters" of the immune system, the white blood cells called *macrophages* (Maxie and Losos 1977, 280–281).

Interestingly, *vachena*'s experimentation with these drugs did not end with *zvipfuyo*; in East Africa, white doctors injected black patients suffering from *gopé* with a dose of berenil as a primer before treatment with melarsoprol. It was believed that the berenil would reduce the incidence of the reactive encephalopathy that often followed the use of melarsoprol alone. The physicians considered oral berenil a nonirritant, with no significant side effects even after prolonged administration. Hopes were high in 1968 that the drug could be used as chemotherapy for *gopé* as well. An experiment duly conducted on *vatema* infected with *gopé* confirmed its effectiveness (Bailey 1968, 122). I have not yet found evidence of berenil use on *vatema* in Rhodesia, but that does not mean it was not used that way.

Curiously, berenil was applied to cure the photosensitization and drug-resistant *hutachiwana* that samorin had created, only for berenil itself to induce diminazene-sensitive and diminazene-resistant trypanosome infections in domestic *zvipfuyo* later (Aliu, Mamman, and Peregrine 1993). A study in 1981 then found berenil to be highly toxic to camels, with the main signs of poisoning being hyperesthesia, salivation, intermittent convulsions, frequent urination and defecation, itching, and sweating. It also found damage to the liver and concentrations of ammonia and decreases in the concentrations of calcium and magnesium in serum (Homeida et al. 1981). Indeed, berenil summarizes the double role of the trypanicidal drugs as *mushonga* (medicine) and *chepfu* (poison) all at once—even more so when *mhesvi* drew blood containing such drug-resistant *hutachiwana* and spread them!

Experiments in Drug Resistance

The experiments that determined the "facts" about these drugs were conducted at three stations. The first research started at Lusulu Field Research Station after the discovery of antrycide- and berenil-resistant *hutachiwana* in 1963. The early research tested for effective trypanicidal drugs and the mechanical modes of transmission of *hutachiwana*, including and excluding an animal medium.[17] It focused on stock responses to drug treatment "under conditions of high trypanosomiasis risk"[18] and the development of drug resistance.

Observations during 1965 confirmed that administering samorin to all stock for one year "successfully eliminated" the antrycide-fast strain of *hutachiwana* that was now firmly established in *mombe* "and which was evidently being transmitted by the tsetse fly." Further trials showed that the drug, administered at 1 mg/kg dosage, could "completely control" *n'gana*. Even a 0.5 mg/kg dose at the start of the disease had been shown to offer a two-month protection.[19]

The early optimism evaporated in 1966 when further trials showed samorin to have "unpleasant side effects," described as follows in the director's 1966 annual report:

> It produces extensive muscle destruction with replacement fibrosis. Injections are given into the neck of the animal on each side. The contraction of the scar tissue has produced in some animals a condition resembling opisthotonus. Two animals have to kneel to crop short grass, but despite the disfigurement they are in satisfactory condition. In the animals that have received a prolonged treatment the replacement fibrosis is rendering it increasingly difficult to find suitable sites for the intra-muscular injections. After two or three treatments into the neck muscles, the extent of the replacement tissue renders the neck valueless as meat. Apart from this drawback, the drug at 1.0 mpk (milligrams per kilogram) and 2.0 mpk at two and three monthly intervals has successfully maintained cattle at Lusulu, where they have been exposed to an estimated 4.5 infected bites per hour.[20]

The experiment also showed that 56 percent of the infections happened one hundred or more days after treatment—"an unexpectedly prolonged period of protection."[21] Very heavy doses of 2.0 mpk at three monthly intervals in *zvipfuyo* carrying *hutachiwana* highly resistant to antrycide had proved "quite successful": 223 days of protection from 33 to 506 days after dosing. However, the trials were stopped rather abruptly when samorin-fast strains of *hutachiwana* developed. Twenty-two *mombe* died of samorin treatment, displaying distinctive lesions.[22]

By contrast, berenil at 3.5 mpk at two-month intervals kept *zvipfuyo* healthy for three years and was declared successful, albeit causing parasitaemia in the fourth to fifth week after routine treatment, which then vanished for a while after the next treatment. The 1967 experiments were conducted to determine the toxicity of samorin and berenil following illness and mortalities at the station among experimental *mombe* undergoing long-term trials with the two trypanicidal drugs.[23] As we will discuss later, the drug-resistance problem would persist in the two chemotherapies, leading to a search for alternatives.

By the end of 1964, veterinary staff at Rekomichi Field Station had started conducting trials with three groups of *zvipfuyo*, divided according to the drug they were being administered—that is, antrycide, ethidium, and berenil groups. The first group was on antrycide pro-salt RF and berenil, monitored for drug resistance. The second group bore a homidium-resistant (i.e., ethidium-resistant) *hutachiwana* strain that broke out in June that year, leading to the immediate termination of ethidium and the switch to berenil 7 mpk every fourteen days. The third group was divided into two, for immediate and delayed treatment with berenil, both administered to adult *mombe* and calves. Besides *mombe*, *nguruve*, *mibhemhe*, and *makwayi* were also experimented on.[24]

The 1965 trial sought to prove whether "the extended period of aparasitaemia which develops after prolonged treatment of cattle with a prophylactic drug is due to immunity and not to an accumulation of the drug in the tissues or body fluids."[25] In February, the Food and Agriculture Organization (FAO) immunologist M. A. Bolton examined the immune response in *mombe* to a number of drug treatments, including ethidium and berenil.[26] The trial was slated for eighteen months to allow immunity to develop in *mombe* under drug administration; afterward, the treatment would be stopped and the *mombe* removed to join others. Their differential behavior relative to continued exposure to high *mhesvi* risk would prove whether immunity had developed. Ethidium was chosen because berenil was the sanative (restorative drug) in the event ethidium-resistant *hutachiwana* developed. Because berenil was already being used, there was no risk of interfering with results in other ongoing experiments at Rekomichi, as there might be with a new drug.

The following year, the experiment was continued with two groups of five *mombe* each maintained at Gwebi Research Station and two similar groups at Rekomichi. At both stations, one group was kept on berenil, the other on ethidium, each at similar dosage rates, and each subjected to repeated infections. The results from the berenil group were consistent with the hypothesis that "repeated infections confer a form of immunity." The ethidium group showed that "the prolonged use of ethidium under conditions of repeated infection" did not result in any sustained immunity after withdrawal of treatment as antrycide did.[27]

Like at Lusulu, Rekomichi Research Station also conducted tests on the toxicity of samorin when used in combination with or after berenil. Another herd of bulls and slaughter cattle was also maintained under ethidium prophylaxis and berenil. At Rekomichi, the experiments also extended to the pathology and immunology of *makwayi* and *mbudzi*, the latter an indigenous flock maintained since 1965. It was divided into two: one receiving no prophylactic treatment, another berenil therapy. Just four of the ten untreated survived, forcing the staff to dose them with berenil.[28] Donkeys, by contrast, were maintained in perfect condition under samorin. *Imbwa* responded adversely to berenil, with many succumbing to side effects; the use of the drug became "to say the least a somewhat controversial subject," even though the work continued because of the ethics of subjecting *imbwa* to certain death.[29]

Case Study: The Eastern Districts and Drug-Resistant Chemoprophylactics

The Mkota communal land is in the Mudzi district, right on Zimbabwe's northeastern border with Mozambique. It was in that area that the first major trial of dimidium bromide, involving block inoculation of 4,723 head of *mombe*, was conducted in 1948. After four months, just five head were discovered to have died, and just two had contracted *n'gana*. The conclusion was obvious: a huge success without adverse effects.

The following year, the drug was extended to Chikwizo, also along the border but south, almost on the boundary of Nyanga District. Here, 11,300 *mombe* were block inoculated, "mortality being immediately arrested and no ill effects ... observed." Further outbreaks in 1951 were dealt with decisively using dimidium bromide, with 20,000 *mombe* block inoculated, even though "doubt was being expressed as to the future efficacy of the block inoculation in view of so little being done to prevent the tsetse fly itself becoming fully established in the areas concerned."[30] In both cases, *mhesvi* was coming from Mozambique, constituting the western edge of a long *mhesvi* belt that extended into Catandika, Gorongosa, and all the way to Mtarara.

The infections resurfaced in Mkota in 1951. This time, when vets treated *mombe* with dimidium bromide as before, the *mombe* did not respond; in fact, one hundred *zvipfuyo* were dying every month.[31] A new problem arose. Twenty cases of photosensitization were recorded in 1952, but inoculations of dimidium bromide continued. In March 1953, as the casualties mounted, the government decided that the remaining 250 *mombe* should be slaughtered and arbitrary compensation paid. By the end of 1953, only 254 out of 4,723 *mombe* inoculated in Mkota had survived. Meanwhile, the repeated inoculations—twenty-five to thirty times at two-week intervals—had created a super-resistant *hutachiwana*. Test after test after such inoculations was positive, with relapses the order of the day.

To solve the problem, the veterinarians increased the dosage strength. The relapses continued.[32]

Desperate, the veterinarians switched to antrycide, but it was just as bad in terms of drug resistance. Attention was redirected to the trial of ethidium bromide; it too failed to cure *n'gana* and had the same relapse rate as dimidium bromide. Further trials showed that antrycide was effective in curing cases resistant to ethidium and dimidium bromide. In 1954, twelve thousand dimidium inoculations were carried out, but they only resulted in a high number of photosensitization cases (10 percent of the herd).[33] In 1955, thirty-six thousand inoculations were undertaken with dimidium, resulting in 2,326 deaths and two hundred cases of photosensitization.

By 1956, it had become customary for *varungu* to praise themselves when everything went well but blame *vatema* when things went wrong. The shortage of white staff meant that trained *vatema* were deployed to inoculate *mombe*; they became the scapegoat for the "toxic symptoms [of] overdosage ... or too frequent dosing" in Mkota. Said one veterinary officer that year: "Admittedly a professional officer did supervise the block treatments, but the native inoculators had access to further supplies of dimidium, which no doubt they used (and sold) on the same animals many times after the block inoculation." The officer also found that "resistant strains ... negatived [*sic*] any therapeutic effect of the drug and in fact probably speeded up its toxic effect by overdosage."[34] The "native inoculators" were relieved of their duties, but the mortalities still continued.

By 1958, some seventy thousand head of *mombe* were under prophylactic and curative drugs throughout Southern Rhodesia. The inoculation campaign was operated at *madhibhi* (cattle dip tanks) (see figure 14.1), with each stockowner issued a stock card to be presented to the dipping attendant every dipping week. The attendant was supposed to tally the numbers of *mombe* physically presented for dipping with those indicated on the cards. Then, the beasts were driven into the plunge area and, as they exited, were injected with the drug using a syringe. Alternatively, a proper dose was poured through the mouth using a Coca-Cola or Fanta bottle. The dipping attendant (or vet) then marked the card, and off the stockowner went. In the absence of dip tanks, operated by *vatema* on the Native Department's payroll, it was virtually impossible for the vets to dispense drugs.[35]

In some cases, the treatments took place at each homestead's *danga* (kraal) full of *mombe*. The animal to be inoculated had to be cast (brought down by tying and pulling its legs in different directions to unbalance it) amid "a permanent cloud of dust." The white vets' characterization of the

Figure 14.1
A typical dipping tank in Zimbabwe, as it was during the days of Rhodesia.
Source: Author 2016.

vatema as "usually inexperienced, frightened or frankly lazy" is inconsistent with the depth of *ruzivo rwemombe* (cattle knowledge and practice) we discussed in chapter 1.[36]

As the number of antrycide-fast strains of *T. congolense* increased, the Veterinary Department decided to switch to berenil as a sanative and ethidium as a routine curative. Ethidium turned out to be effective; for three months, there were no infections.[37] Without discounting the drugs' efficacy, it is possible that the improvement was also a result of insecticidal sprays, which drastically reduced *mhesvi* catches at Makaha fly chamber in Chikwizo from thirty-four in 1963–1964 to just seven in 1964–1965. Nonetheless, the dramatic reversal of explosive *n'gana* situations at Chikwizo Centre and Zano owed much to the use of berenil.[38] At the latter center, berenil curative treatment slashed the cases from 170 in 1967–1968 to a mere fifty-six in 1968–1969. By 1970, the *n'gana* situation had stabilized in Chikwizo and Mkota.

Just south of Mudzi, the Nyanga North chemotherapeutic campaign got under way in 1958, essentially in response to the spillover of *mhesvi* and *n'gana* from Chikwizo. By chance, a Tsetse Department employee coming back from leave in his village captured one *mhesvirutondo* as it tried to bite him. Even after this early warning, the Tsetse Branch did not post regular patrols. Two years passed. Then in January 1960, in a community called Fombe, several outbreaks of *n'gana* took place barely a month after antrycide treatment.

In response, the vets dispensed sanative treatment with berenil and followed it up with antrycide pro-salt prophylaxis. However, twenty days later, twenty-four *mombe* tested positive to *n'gana* again. Based on the Mkota and Chikwizo experiences, *hutachiwana* were now known to be resistant to antrycide, so the switch was made to inoculation with samorin. It worried the authorities that the *mhesvirupani* density in Ruenya was so low, and yet the risk of *hutachiwana*—never mind one that was becoming drug-fast—was so high.[39]

The position deteriorated even further in 1961—not just at Fombe, but also at Mandeya, Chimusasa, Chifambe, and Matisi. In the first two centers, all *mombe* were subjected to a bimonthly prophylactic regimen of antrycide pro-salt. Chifambe and Matisi were placed under review due to outbreaks of civil disobedience among *vatema*, who rejected any sort of diagnostic or therapeutic measures, which they blamed for the miserable deaths of their *mombe*. Two hundred died untreated. In Mandeya, where twelve cases had been diagnosed, *vatema*'s civil disobedience was so serious that police had to be called in and arrests and prosecutions made. In Nyanga North and Holdenby, 13,032 *mombe* were treated, 1,353 of them with prophylactics and 11,679 with curative drugs. Out of the 10,527 inoculations, 5,036 were antrycide DMS, 954 antrycide pro-salt, and 4,537 berenil.[40]

Vatema's refusal to cooperate in producing their *mombe* for treatment at *madhibhi* continued throughout Nyanga North, barring Fombe, Chimusasa, and Chapferuka. The political temperature in the urban and rural areas of Rhodesia was rising. The Federation of Rhodesia and Nyasaland had ended in 1963, with Northern Rhodesia (Zambia) and Nyasaland (Malawi) being re-Africanized as independent states. However, the *vadzvanyiriri* (downpressors, or white settlers) in Southern Rhodesia refused to hand over power to black people and moved instead toward a Unilateral Declaration of Independence (UDI) from Britain to form the Republic of Rhodesia. Already in 1964, "African nationalist" leaders had been rounded up and dumped in remote areas, where they were held in detention camps, but with freedom to hold rallies within a fifty-mile radius. They were accused—not without truth—of inciting the rural population to civil disobedience. In particular, they urged *vatema* in rural areas to boycott dipping their *mombe*, which was the most important mechanism through which vets could inspect and inoculate *mombe*.[41] It was such threats that led the government to unleash the Law and Order Maintenance Act (LOMA) on the rural countryside in 1965, which coerced *vatema* to capitulate to chemotherapy.

Even on veterinary grounds, it was considered unbeneficial to carry out immediate treatment, because the drug resistance had become so widespread. Two months were needed to choose and field-test drug regimens to control the various strains of trypanosome not responding to treatment. Ultimately, the vets turned to samorin.[42] It seemed for a brief period that the drug had solved the problem.

Unbeknown to the vets, samorin was about to become a problem as well. In June 1966, the animal health inspector for Nyanga reported at least forty deaths in *mombe* at the Chifambe, Samakande, Manwere, Nyamasara, Chipatarongo, Ruvangwe, Fombe, Nani, and Matisi inoculation centers. His postmortem indicated extensive liver damage strongly suspected to stem from the drug samorin. The *mombe* had been treated in the week starting May 9, with berenil inoculation ten days later. After three weeks, the deaths started. By June, the death toll had risen to 152 *mombe*, with 154 more infected and "likely to die." At Chipatarongo, one owner lost five or six *mombe*, others none—a reflection of the tendency of the strain to be concentrated at certain cattle kraals.[43]

Along with a "capricious appetite," diarrhea was perhaps the most immediately recognizable sign of infection, with the beast suffering a running stomach for four to seven days before dying. In its final hour, the bovine "appeared tucked up, dehydrated, weak and thin," its temperatures subnormal. Its digestive system would also be in "ruminal stasis," the feces "hard dry and mucus-covered."[44] Many of the *mombe* had a "crusting around the eyes," evidence of excessive lacrimation. A few of them showed increased nasal discharge and excessive salivation. Some that were sick had recovered; others had not. The postmortem confirmed the lesions to be "a massive acute fatty degeneration of the liver" and the kidney.[45]

The laboratory tests took months to run. By the time prophylaxis resumed in April 1967, the reinfections had drastically fallen due to two factors. One was the prolonged protective effect of a series of samorin treatments; the other was the drastic reduction of *mhesvi* due to the 1966 spraying operation. Although the department could not distinguish the exact contribution of each factor, the cumulative effect was that no significant *n'gana* outbreak occurred until January 1967. Thereafter, bimonthly sanative berenil treatments were religiously applied until April, when a reduced 0.25 mg/kg dosage of samorin treatment commenced, which did not cause photosensitization.[46]

The weakened regimen covered Nyanga, Matisi, and St. Swithun's. First the Fombe, Chimsasa, and Chapferuka centers and then, from December 1967, Chifambe, Samakande, Manwere and Mangezi were treated with this drug. From March on, berenil was incrementally deployed as a sanative. The *n'gana* situation was brought under control temporarily, only for the infection to resurface in September.[47] Berenil was also effective in eliminating the mild outbreak at the Matisi and Nyamasara centers. In Sawunyama, the use of berenil was so effective that only thirty-four cases were recorded in 1967–1968, compared to a staggering 230 in 1966–1967. Meanwhile, at Matisi, Mangezi, Chifambe, and Samakande, just seven cases were recorded.[48]

Mombe in Nyanga District had maintained their condition well. "Cooperation from the local stock owners was of a very high standard," BTTC Assistant Director Gerald Cockbill commended. Prophylaxis with samorin was maintained at Chimsasa and Fombe until July, when the last treatment was given. Drug cover was then withheld from *mombe* at these centers so that incidence of *n'gana* could be used as an indication of the effectiveness of the 1969 spraying operation. Elsewhere in the central area, satisfactory control had been maintained by use of berenil on the individual *zvipfuyo* or on infected herds in areas such as Samakande.[49]

When *n'gana* broke out in Musikavanhu in Chipinge in 1954 and antrycide and dimidium bromide were deployed as the prophylactic and therapeutic, respectively, to contain it, everything seemed to be going smoothly for the Veterinary Department.[50]

To avoid the "human error" that it blamed for the buildup of drug resistance in *hutachiwana*, the department stripped all "native assistants" of the role of dispensing the medicines and closely supervised their activities. However, the deaths continued, with the blame now being shifted to the drug itself and the *mombe* that had grazed on toxic lantana types.[51] The director of Veterinary Services found that "in the course of time one or both jugular veins became practically occluded due to phlebitis as a result of injections being badly placed, or leakage of the irritant drug from the vein after withdrawal of the needle."[52] Some "experts" warned that the real nature of dimidium's toxicity would probably never be known. It was not clear if fresh grass prevented or worsened the drug's toxic effects, or whether in fact it was the lack of grass in November that caused heavy mortality in Chipinge in 1955. Rift Valley fever was known to exist in those *mombe*, which may have worsened the liver damage—but then, similar liver damage in other areas had nothing to do with Rift Valley fever.

The postmortem findings at Salisbury Central Laboratory revealed extensive peeling of teats and udders, enlargement of the liver and kidneys, and increased blood flow and accumulation of fluid in the lungs.[53] At the end of the day, dimidium bromide was considered to have so many unpredictable characteristics when specifically applied to Rhodesia that it had to be discontinued except in areas of small groups of *mombe*, such as *mapurazi*; ethidium bromide took its place. By 1958, dimidium bromide generally had fallen into disuse in Africa "due to a type of photo-sensitisation leading to liver and other tissue damage."[54]

After antrycide pro-salt was authorized for use in Chipinge, 13,517 head of *mombe* were inoculated. The 177 deaths that followed were blamed on the excessive dosage, which had caused emaciation, extreme dehydration, and internal bleeding. The lymph nodes were enlarged, edematous, and hemorrhagic. One animal had bled extensively into the duodenum and small intestine, where large clots were found during postmortem. To solve the problem, the dosage was reduced.[55]

Subsequently, several recommendations were made: First, a rest period of two hours should always be observed before and after inoculation. This involved watering *mombe* before the commencement of the inoculation and shade afterward. Second, *mombe* must not be beaten or chased into the inoculation arena, but walked in gently. Should some beast run away, it was not to be chased and brought in for inoculation kicking and mooing, but must be returned under calm conditions, after the meek ones had been treated. Third, all inoculators were supposed to be familiar with the live weight estimation of the beasts they were working with.[56]

Conclusion

By 1973, only ethidium, berenil, and samorin were listed as treatments. Hurungwe had become ethidium and berenil country; Lupane, Hwange, Kadoma, Hurungwe, Nemakonde, Rushinga, Centenary West, Mutoko, Nyanga, and Chipinge had become exclusively berenil country after the resistance to antrycide and photosensitization problems with samorin; and Binga, Gokwe, and Guruve were berenil-samorin combo country.[57] All told, Rhodesia's *mhesvi*-prone districts were almost equally split between berenil and samorin, with ethidium used only in token quantities.[58]

Meanwhile, investigations were under way to determine whether a 1:1 mixture of berenil and pyrroli-dinomethyl-tetracycline (reverin) was efficacious against *n'gana*. A 1973 experiment found that this mixture could cure *hutachiwana* infections in *mombe* if administered at a dosage rate of

2.5 mg/kg of the berenil fraction. The experiment was simultaneously undertaken at Central Laboratory in Salisbury and at the three BTTC field stations—namely, Lusulu (Binga), Rekomichi (Hurungwe), and Gwebi (outside Salisbury).[59]

At this point, the thinking in the Branch of Tsetse and Trypanosomiasis was that chemotherapy was only an interim measure to "maintain the health of cattle until such time as the vector is eliminated."[60] The primary enemy of both the glossinologist and the veterinarian was *hutachiwana*, especially what *vachena* called *T. congolense*, the chief *nyongororo* of cattle. This microbe had thus far, after seven decades of combat, exhibited "a remarkable facility for antigenic change that makes it resistant to the host's antibody and its ability to develop drug resistance."[61] As an enemy, it was a nifty, ever-shifting target, "capable of altering its sensitivity to hazardous substances or detrimental material introduced into its milieu to such an extent that it becomes insusceptible." Consequently, the *nyongororo* shifted the matter of success or failure of *vanhu*'s assault upon it away from the strengths of *hurumende yevadzvanyiriri* (the government of the downpressor) and blamed it on two factors: "the lack of a vaccine and the frequent appearance of drug fast strains of trypanosomes."[62] In both, the Rhodesian state was vulnerable.

No other country in *mhesvi*-occupied Africa had executed chemotherapy and chemoprophylaxis to the extent and with the efficiency that Rhodesia had. The stoic persistence in maintaining *mombe* in areas with the greatest *mhesvi* challenge had astonished its critics who cautioned about drugs that started with much promise, only for *hutachiwana* to get used to them and resist their effects. The Rhodesians beat their chests, crowing that they had succeeded in using *mishonga* against *hutachiwana* since 1950. Yet that view did not make anything of the suffering of those that lost entire herds in the name of experiments the outcomes of which, they dreaded, would only bring social, spiritual, and economic ruin as their *mombe* breathed their last. *Hurumende* did not care, because *mombe* and the lives it was experimenting with were of *vanhu vatema*, not *vachena*. In its approach, *vatema* and their *mombe* were not only *midziyo* (instruments) for clearing lands of *mhesvi* and keeping *hutachiwana* from getting in. They were also experiments in a vast laboratory that was *ruzevha* (the Native Reserve). Indeed, in experimenting with the one thing that was so central to *vatema*'s livelihoods, wealth, and social being, *hurumende yevachena* was experimenting with *vatema* themselves.

15 Unleashed: *Mhesvi* in a Time of War

Starting around 1963, two black political parties—the Zimbabwe African People's Union (ZAPU) and the Zimbabwe African National Union (ZANU)—turned to armed struggle to demand independence for *vatema*. From 1966 to 1968, *mhesvi* efforts in Sebungwe, Binga, Hurungwe, and Guruve had experienced setbacks as ZAPU's armed wing (later Zimbabwe People's Revolutionary Army, or ZPRA) skirmished with Rhodesian Security Forces. Meanwhile, further east, ZANU's army, the Zimbabwe African National Liberation Army (ZANLA), carried out operations in Nemakonde, Kadoma, and surrounding areas that culminated in the Battle of Chinhoyi in 1966. These operations were dramatic—for publicity purposes for both the guerrillas and Rhodesia—but were failed first attempts.

But in 1972, ZANLA's resumption of operations from Portuguese-ruled Mozambique with aid from that country's national liberation movement, the Frente de Libertação de Moçambique (FRELIMO), threatened the conduct of tsetse control operations. The insurgents were infiltrating via the Dande tsetse control front, their mobilities also producing "a steadily decreasing ability ... to carry out tsetse control measures, ... with a concomitant massive increase in the incidence of *n'gana*."[1]

Field operations to administer chemotherapy and drug prophylactics to *mombe* in the rural countryside were now severely curtailed. *Mombe* were decimated. BTTC had pushed *mhesvi* back into Mozambique, beyond the border into Tete Province. Now *mhesvi* had not only returned but was spreading throughout Chiswiti, Chimanda, and Masoso Tribal Trust Lands. By the end of 1972, it had crossed the Mazowe River south into the Mutoko and Murehwa districts. Here and in Binga, guerrillas foiled repeated attempts to treat *mombe*; over twelve thousand head died.[2]

The war of self-liberation paralyzed the entire tsetse control operation on the northeastern border. It was no longer possible to maintain the four hundred to six hundred miles of game and cattle fences without being shot at. The branch acknowledged that "wire and even standards are being stolen in a wholesale manner and ... not a single conviction has been obtained."[3] Restrictions of *mombe* from specific areas to prevent them from "seriously embarrassing control procedure" were no longer possible because of the security situation. There were only two solutions possible: either political settlement or flood the countryside with the Security Forces and thereby protect personnel on tsetse control duties. Only then could the operations resume.[4]

Hurumende was faced with two kinds of pests, *mhesvi* and *magandanga* ("terrorists," as *vachena* called them, who were *varwi verusununguko*, or freedom fighters, to *vatema*). An internal memo in 1972 is very clear that *mhesvi* was catching a ride on the "terrorists," putting fifty percent of Rhodesia's national herd at risk.[5] "Carried fly" caught a ride on the backs of "terrorists," whose mobilities were too irregular for, and indeed totally disrupted, the cleansing chambers and tsetse gates. The antitsetse operations had relied on the unchallenged access of *vachena* to means of coercion such as guns. Now the tables had turned: the rural people became very stubborn, their resistance to continued arbitrary tsetse policies increased by the knowledge that freedom fighters were in the bush, their fingers curled around the triggers of AK-47 rifles and bazookas. *Vatema*'s mood was now anything but submissive:

> In the interim the veterinary organization would be called upon to attempt by chemotherapy the protection of large numbers of cattle, an operation regrettably not based upon so firm a foundation as previously. Freedom of choice will be the popular order of the day and suggestions of coercion will at least at first be frowned upon because unpopular as such measures inevitably are, they might be turned to good account by unscrupulous opportunists. ... In the past the considerable success we have enjoyed in controlling trypanosomiasis and its vector has been due in no small degree to our ability to enforce the appropriate regulations framed specifically for those purposes but often unpopular and seldom completely understood. ... It is now suggested that this will, at least at first, no longer be possible and some compromise solution must be sought without delay. In the past there have been instances where the stock owners have been allowed to refuse treatment in the belief that the ensuing deaths would convince the owners of the correctness of the official policy. In nearly every case this has proved a singularly potent weapon and it promises to be even more important in the future.[6]

The arbitrariness that had accounted for the success of the prophylactic settlement and inoculation campaigns of the 1950s and 1960s was gone. The tone had changed. (The glossary at the back of the book will aid the reader in understanding *chidzimbahwe* registers of this change.)

Places Rendered Inaccessible by War

The BTTC memo quoted previously illustrates clearly how war (mobilities of freedom fighters) rendered the borderlands inaccessible and virtually opened the floodgates to the mobilities of *mhesvi*. The freedom fighter and *mhesvi*—what a pestiferous combination of mobile workshops!—were both terrorists in the eyes of *hurumende*. The freedom fighters used guns, landmines, and mortar tubes, firing bullets, killing through detonations and bombs. *Mhesvi* used its long mouth, depositing *hutachiwana* and death in the herds. Both terrorized *hurumende* and *vachena* with their mobilities. The terrorist was a freedom fighter, fighting for freedom from Rhodesian rule. *Mhesvi* was exercising its freedom to move where it pleased. We have heard this story of *zvipukanana* and insurgents (i.e., problem *vanhu* or *mhuka*) before from Hugh Raffles (2010), but here we are talking about *hwiza* (locusts) and freedom fighters terrorizing *hurumende yevadzvanyiriri*. Not separately—together!

It was not just a matter of *mhesvi* originating from and because of the existence of a conflict zone. ZANLA and ZPRA freedom fighters were waging war from *mhesvi*-infested terrain that Rhodesia's mounted unit, the Grey Scouts, infamous for its tracking in roadless terrain, could not reach. The freedom fighters had mined the roads, and *ndege* was always fair game for SAM-7 (Strela) missiles, antiaircraft guns, bazookas, and, for choppers flying just above treetop cover, AK-47s aimed at pilots. That included places like Sebungwe, Binga, Hurungwe, Guruve, Rushinga, and later Nyanga and the Chipinge to Savé-Runde junction stretch, where battles had raged for over half a century between the *chipukanana* insurgent and *hurumende yevadzvanyiriri*. In the Rushinga area, for example, ZANLA attacked just when Rhodesia's Vet Department was finalizing plans for a major spraying campaign in the Rio Luia (Ruya), "the stronghold of the persistence of tsetse in that area." It had "put in tracks, arranged camps and had the full permission of the Portuguese," but military considerations forced the cancellation of the mission ("Tsetse Fly the Winner in Terror War" 1977).

The tsetse control situation ebbed and flowed with the plateaus and valleys of war. After the commotion of 1972 involving ZANLA infiltrations into Rushinga, political talks began. Détente followed. The guns fell silent, freedom fighter mobilities vanished, the tsetse control teams returned, and *mhesvi* was on the retreat. In 1974, Assistant Director Desmond Lovemore declared the situation "very satisfactory," barring areas along the international border between the Ruya and Mazowe rivers, where fear of attack from inside Mozambique kept the tsetse control teams away. In areas in which no "terrorists" were active, residual spraying and selective game elimination continued, the inoculations with ethidium, berenil, and samorin resumed, and the pneumatic sprayers dumped DDT over six thousand square miles of *mhesvi* habitat.[7]

Détente delivered a crippling blow to ZANLA and ZPRA. Only in 1976 did serious missions from Mozambique and Zambia resume. Mozambique had become independent in 1975, and the new FRELIMO government of Samora Machel wasted no time in inviting the new unified ZANLA-ZIPRA army, now named the Zimbabwe People's Army (ZIPA), to establish staging bases inside Mozambique along the entire eastern border with Rhodesia. The unity was short-lived; as ZANLA consolidated its operations from Mozambique, ZPRA continued infiltrating from Zambia through the entire frontage of Rhodesia's Zambezi River boundary. ZPRA's campaigns into Lupane, Gwaai, Shangani, Binga, and Guruve, and ZANLA's into Rushinga, Mutoko, northern Nyanga, and south into Honde and Savé-Runde, were (barring a few) conducted through *mhesvi* country.

The war took place in exactly the areas infested by or recently cleared of *mpukane*. ZPRA freedom fighters operated from Zambia and its *gopé*-prone Zambezi and Luangwa valleys, from Sebungwe (intermittently) to Guruve, and ZANLA from Dande to eastern Nyanga and from Chipinge to the Savé-Runde river junction. Fortunately, the freedom fighter experience with *mhesvi*—especially its excruciating bite—is beginning to be told ("Jane Ndlovu" 2015). ZPRA crossing points passed through thick *mpukane* country astride the Zambezi on either side of the Kariba Dam—that is, between Kazungula and Victoria Falls (into Hwange), Batoka and Victoria Falls (into Lupane, Nkayi, and Binga), and from the Kariba Dam to Chirundu (into Hurungwe, Binga-Kariba, and Gokwe). These were the areas where prophylactic settlement of Gwai, Shangani, Gokwe, and Hurungwe had generated immense ill feeling against the settler government.

In the last half of Mugabe's ill-fated personalized rule, state propaganda took pains to emphasize how top politicians were "stung by mosquitoes, tsetse flies, spiders, dangerous snakes and other harmful bugs" to show their own sacrifices at the expense of the two constituents that really fought the war: the freedom fighter and the *povo* (the masses) inside the war zone (Mhombera 2012; Bwititi 2015). The anthrohistorian David Lan's account of Dande, scene of ZANLA's highly effective covert infiltration starting in 1970, has shown that, because of *mhesvi* (and *mhuka*) presence, the area had no *mombe*; any that were brought in died within months. All cultivation was undertaken *chibhakera* (by hand), not with plows (Lan 1985). These material conditions forced ZANLA to rely on head porterage and the *mhesvi*-resistant donkey to move ammunition and supplies from their Mozambican rear bases and cache them in local forests and hills. The routes to and from the ZANLA training bases in Chimoio (Praça Adriano), Nyadzonia, and Tembwe passed through the *mhesvi*-infested stretches in northern Nyanga, Mudzi, Mutoko, Dande, and Muzarabani.

Whereas the *mhesvi* advance created serious veterinary risks for *mombe*, with *vatema* being resettled as human shields to make the land impassable and unlivable to *mhesvi* the freedom fighter incursions turned this entire philosophy on its head. The prophylactic against pests had become, through political mobilization, the water in which the fish (freedom fighters) now swam; resettled people became a logistic- and intelligence-support infrastructure for a successful war of self-liberation. In an interview in 1978, Robert Mugabe, leader of ZANU, the parent party of ZANLA, described this reversed role of *vatema* from prophylactics against *mhesvi* into the rock upon which to build a strong push for independence:

> You must rely on the people above all. Without them, the use of arms is of no value ... win a base in the heart of the people and later operate out of the people, with the people, against the enemy. ... It is not enough to talk about destroying the enemy, transforming certain zones into areas that are safe for the people. The people have to be organized into bases of support for their own struggle, so that it will be the people themselves that will continue the struggle. ... The role of the guerrilla is to act as a vanguard, to guide the revolution. This is why what must come first after the building of support is the establishment of administrative structures for the people. The people have to govern themselves. They must create the conditions for survival based on their own resources through the developments of projects in agriculture, education, and health. (Mugabe 1978, 28)

Mhesvi struck its covenant with *mhuka*; the freedom fighters struck theirs with the people. Through their bite, *mhesvi* had protected *mhuka* by limiting the mobilities of *vanhu* riding on horseback or ox-wagons; now the freedom fighter protected the people by lining roads with landmines and harassing fire, thereby denying the enemy of the people freedom of movement. It was not altogether benevolent; the sole reason for *mhesvi*'s proximity to *nyati* the buffalo and other big game was to feed upon their blood. By contrast, the freedom fighter was very much interested in the people as logistic support. To the degree that both were so reliant on their hosts, *mhuka* and *vatema*, their risks could be eliminated using similar methods: starving them by eliminating their "hosts,"' creating buffer zones between them and their hosts, and, if it came to it, chemical warfare to destroy them en masse.

Mugabe was talking about more than just sucking blood out of the people to gain the energy necessary to attack and conquer *hurumende yevadzvanyiriri*. He was saying that freedom fighters must build the capacity to enable the people to establish an administration parallel to *hurumende yevadzvanyiriri*, not within the space that *hurumende yevadzvanyiriri* still governed but that was now a liberated zone—in other words, rolling away the carpet that was Rhodesia and spreading over it the carpet that was the new Zimbabwe. Then, like a *mhesvi* catching a ride on game—like carried fly— the freedom fighters and the nationalists would ride to power, to take over *hurumende* and insert Zimbabwe while deleting Rhodesia as a geophysical expression:

> From the rural zones we can now expand into the urban zones that are the strong points of the regime. They have many military bases in the highlands where the cities are located and where the main railroads, roads, and other lines of communication are. Our next stage will be to surround the enemy in those areas. As we advance out of the rural zones, the people will have a very important role to play. The war turns into a people's war with the people struggling, placing the mines and attacking the enemy. Our army will attack the most difficult targets and the people the easier ones. (Mugabe 1978, 28)

Rhodesia's *mhesvi* operations had been so successful not only because *hurumende* had a monopoly over the means of violence, but also because people in the borderlands had lent their *ruzivo* and labor to *hurumende*'s projects to extract some form of livelihood. The extension of age-old *hunyanzvi hwekuvhima* or *huhombarume* (hunting expertise) and practices to *hugocha* is a perfect example of this. Those days were now gone. In Chibwedziva, for example, seasoned hunters abandoned their *misha* in fear of *hurumende*'s retribution against *vanhu* for supporting "terrorists" and extended their skills to hunting for the freedom fighters with a new kind of weapon: the Soviet-made SKS rifle. Many became the eyes and ears of the freedom fighters.[8] As Mugabe said in 1978, most of the *mhesvi*-infested border areas had been lost to the freedom fighters, who now declared them liberated zones.

The situation as viewed from the capital, Salisbury, was dire. On March 3, 1977, Chief Veterinary Officer (Trypanosomiasis) Bill Boyt told the *Rhodesia Herald* (see figure 15.1) that the war had "put back Rhodesia's battle against the tsetse fly 20 years." The situation had become "very serious" as *mhesvi* advanced "at a rate of knots." In the northeast, *mhesvi* had reached Bindura and Shamva and was strongly anchored in Mutoko. In the southern areas, it had reached as far inland as Buhera, Gutu, and Bikita, spread no doubt by freedom fighters deploying into combat. Further east and southeast, the scenario already seen in the north was becoming inevitable by the day. Boyt was a very dejected man:

Rhodesia Herald 3/3/77

Tsetse fly the winner in terror war

Herald Reporter

THE terror war has put back Rhodesia's battle against the tsetse fly 20 years. It will cost millions to regain the ground lost to the fly.

The situation is very serious. Tsetse is advancing "at a rate of knots", said Dr Bill Boyt, Chief Veterinary Officer (trypanosomiasis) in the Ministry of Agriculture.

"In the Mount Darwin area, we are faced with a major advance of tsetse," he said. Already several thousand head of cattle have died of nagana, and the advance threatens an area of 19 000 km² and 164 000 head of cattle.

He said that because of the terror war, tsetse and trypanosomiasis control teams were unable to operate. "We have been put back to a position we were in 20 or more years ago. It will take millions of dollars to put it right."

In 1973 Rhodesians were about to embark on a major spraying operation in Mozambique it-

DR BOYT

trees and bushes," said Dr Boyt.

Gona-re-Zhou, we have had a very successful campaign in conjunction with the Portuguese in Mozambique and the South African authorities from 1963 to 1974 we pushed the tsetse completely out of Rhodesia and we were working 80 km into Mozambique.

"Had we been allowed to carry on we would now be within smell of the sea. And that would be the problem removed from that area. The threat to the Limpopo and Kruger National Park would have been removed.

HERDS

"Now, we haven't even got any sentinel herds along the border. These were herds we kept as alarm systems to check the presence of fly and of the trypanosome parasite

Figure 15.1

Linking *mhesvi* and *magandanga*.

Source: Rhodesia Herald, March 3, 1977.

East of the Sabi and south of Chipinga (sic, Chipinge) in the Lowveld we pushed the tsetse back into Mozambique, we eliminated them from Rhodesia, and we worked about nine or 10 km into Mozambique. Now this facility has been denied to us. The cattle in that area could be infected. … It could cross the Sabi flood plain, it could cross the Sabi and could be in areas like the [Sangwe] Reserve, Ndanga and so on. … And further south, in Gona-re-Zhou, we have had a very successful campaign in conjunction with the Portuguese in Mozambique and the South African authorities from 1963 to 1974. We pushed the tsetse completely out of Rhodesia and we were working 80 km into Mozambique. … Had we been allowed to carry on we would now be within smell of the sea. And that would be the problem removed from that area. The threat to the Limpopo and Kruger National Park would have been removed. ("Tsetse Fly the Winner in Terror War" 1977)

To say that Boyt's interview illustrates the curtailment of the freedom of mobility on which the tsetse control operation had relied is to understate the role of mobility in denying *mhesvi* freedom of movement. When the hunters, bush-clearing teams, spraying "spans,"' *zvikopokopo*, and light *ndege* had enjoyed freedom of movement, *mhesvi* was on the run—or had nowhere to run. Now that freedom was gone and *mhesvi* was advancing, especially as carried fly, on the backs of freedom fighters infiltrating through hitherto restricted spaces, sometimes driving *mombe* through minefields to breach them, as well as on special and ordinary forces returning from "hot pursuit" operations in neighboring Mozambique, now an independent country and the rear base of ZANLA.

The entire infrastructure—"the intelligence system of tsetse," game elimination, and prophylactic settlement—was coming apart. *Vatema* were no longer anybody's human shields or machines—nor their *mombe*:

> Now, we haven't even got any sentinel herds along the border. These were herds we kept as alarm systems to check the presence of fly and of the trypanosome parasite. So the first intimation we would have for the return of tsetse would be infection of cattle in Matibi No. 2 Tribal Trust Land, north of Gona-re-Zhou Game Reserve. … Because we can't work along the border for security reasons, we can't even monitor the tsetse in that area. The whole of Gona-re-Zhou is closed to everyone except the security forces. From Mount Darwin [Dande] down to Vila Salazar [Sango] very little work could be done. … A quite serious position might develop anywhere, or everywhere, along that border.[9]

It was a sign of the times that District Commissioner (Gokwe) P. G. Dix wrote a letter to the senior tsetse field officer expressing utter helplessness in the face of collapsing *mombe-mhesvi* boundaries in Gokwe, now a ZPRA war zone, in 1978. He was calling for an exercise to be "mounted to restore some discipline in the matter," stressing he could no longer "continue 'fobbing off' people." In the angry letter, Dix makes clear the pressure building from below and exposes the effects the ban on cattle-keeping in reclaimed lands had on people whose livelihood was based on farming:

> I am continually receiving requests to be allowed livestock in the Huchu and Masuka Areas. … The people now point out that no control is maintained; and why should the people who flout the law be "allowed" livestock and those law-abiding citizens have to "plough" with *badzas*. If control is to be relaxed then all should be allowed animals. If control is to be maintained then no animals should be allowed. The present situation is to say the least embarrassing as no action is taken against the offenders. … Headman Nembudzia also asks if he may be allowed cattle as he says he is "surrounded" by cattle.[10]

Boyt found Dix's letter to be "unreal" but clearly reflecting "the feelings, frustrations and difficulties" the sort of which, he said, starts with "a couple of obvious truths, passes through a phase of tub thumping and ends on a plaintive note" while failing to "admit the truth": that the government was powerless to enforce *n'gana* regulations and the disorderly mobility of *mombe* was "inevitable." Dix's was "a masterpiece of rhetoric echoing the vain hopes of all those who yearn for the return of order and discipline, factors which can no longer be imposed other than by the people themselves. ... Self-discipline is a rare virtue arising only as a result of privation and necessity."[11] Still, Dix's letter highlighted the need for "a major change of policy demanded by a changed and changing political situation." Boyt recommended a review of policy governing restrictions upon *zvipfuyo* in "tribal areas," because the legislation was unenforceable in any case and the deployment of staff hampered by the risks to their own lives at the hands of the freedom fighters.[12]

Rhodesia's tsetse control strategy was coming apart. The mobilities of the freedom fighters with AK-47s and bazookas in hand were cutting off the ability of *hurumende* to put a finger on the pulse of *mhesvi*. The freedom fighters were establishing a militaristic order in the countryside and blasting their way to Salisbury. *Mhesvi* was no longer in the hands of the Rhodesian state. By April 1980, even Rhodesia was no longer on the map. In place of the green and white flag, the green, yellow, red, and black, with the *hungwe* (*shiri yedzimbahwe*/Zimbabwe bird) occupying a red star perch on a white scalene triangle, was flying high over the land.

After seven decades of battling *mhesvi*, *vachena* were utterly helpless as the little *chipukanana* not only returned to its old haunts, but broke new ground. Therefore, the Rhodesians dumped the chemicals instead on the cause of their failure: the nationalist freedom fighters. The extension of weapons intended for animal pests to human pests of *hurumende* occurred exactly at the time (from 1974 on) that spraying and other operations against *mhesvi* became impossible to execute. The excess ordnance had found a new use: as a force multiplier to attenuate the superiority that rapid mobility on foot through roadless terrain gave ZPRA and ZANLA freedom fighters. That story awaits its own monograph.

The New Cordons Sanitaires

As I was about to close this book out, I came across a quite informative article from one Rhodesian combat veteran, Mark Craig, paying homage to the "fly men" and their role in the Rhodesian Security Forces' strategy of attenuating the "terrorists'" numerical strength through the construction of cordons sanitaires. The article explored, among other things, the

Rhodesian use of mechanical and chemical defoliation in both cordons and non-*cordons sanitaire* operations. After briefly outlining the US military's use of Agents White, Purple, Pink, Green, Blue, and Orange in Vietnam in an operation codenamed *Ranch Hand*, as well as British use of Agent Orange in Malaya, Craig then turns to Rhodesia's "apparently not squeaky clean" war.

Craig's account is unusually self-reflexive, which is a rarity among Rhodesian war writers nostalgic about their ninety-year "moment" in the country's history. He acknowledges that very little is known on the topic, that "no objective evidence ... shows what if any residual effect there was on the local population and indeed our own troops. Perhaps this is an aspect that no one wants to talk about or perhaps it was just one of those activities no one knows much about."[13] His account is also quite well researched; where he cannot find actual operational orders and debriefs of specific infrastructural assignments, he has deferred to a vibrant Rhodesian online community of ex-soldiers that have kept in touch, often reconstituting the entire chain of command involved in these covert operations.

From his own research and based on his war experience, Craig first confirms that in the intensive phase of the war (1976–1979), the BTTC was an integral part of Rhodesia's chemical warfare project. The branch's personnel could be found, for example, carrying "back-pack hand-operated sprayers containing HIVAR-X," a wettable powder herbicide mixed in water and applied to destroy brush. However, the *chepfu* proved very expensive to use for the envisaged expansion of vegetation-free spaces along cordon fences from 25 to 150 meters wide. The switch was made to TORDON 225—"a costly mistake as this product was ineffective and resulted in Rhodesia instituting court action against the South African manufacturers."[14] The chemical defoliants were apparently deployed to clear cordon frontage on the Musengezi, Mukumbura, and Nyamapanda to Ruenya minefield.

Rhodesia's Army Corps of Engineers provided *mabhurudhoza* and graders for the mechanical clearing of all vegetation on areas demarcated for the cordons sanitaires, and BTTC installed the fence. The branch also did the defoliating—for example, on Chete Island on Lake Kariba, where they sprayed HIVAR-X as one would fertilizer, along the banks. Says Terry Griffin, the officer in charge of the operation on the ground: "This would (as it did) clear that sector of all foliage and thereby (hopefully) deny natural cover. After the first rains it was evident all was dying off and it did clear all fairly quickly creating a rather bare scar along that section of the island. Some 10 years later it was still very visible but on my last fishing trip there +/- 4 years ago all had now regrown."[15]

The tsetse personnel and army engineers worked together on fence construction or maintenance on the North Eastern Border Game Fence (NEBGF) in Mukumbura. The former group handled the construction and maintenance work on fences on both the "home" and "enemy" sides of the minefield. The personnel were usually composed of the TFO and numerous *mafrayi* under him. Without them erecting fences further ahead, the sappers could not lay any mines. Keeping the fly men within the "protective boundary" was critical lest they move too far off, become isolated, and be captured or killed by the "terrorists"—which occasionally happened. They would be severely outgunned; only the TFO was armed with a Fabrique Nationale (FN) rifle.[16]

How did we move from tsetse game fences to *cordons sanitaires* against freedom fighters? Here again, Craig's account is important as a primary source: "The main idea of the fences being constructed by the Tsetse Fly crews was to stop the migration of host animals from one area to the next. It was quite a clever idea to use these fences as minefield perimeters as well."[17] As *varungu*, TFOs had the power to shoot *mhuka* for the pot, which made them popular with soldiers fed up with corned beef and other rations. After all, they camped together, at the very same tsetse camps we discussed throughout the book. Only this time, their Land Rovers were no longer soft-skinned, but retrofitted with mine-resistant armor plate. We may not like it that defoliants and Land Rovers were extended to kill black people fighting for the right to be free in their own land.

However, what we cannot do is ignore that innovation, even though it was intended for purposes we may see as evil. A significant body of literature addresses Rhodesia's use of chemical and biological weapons (CBWs), with *vatema* as victims. Writing on science, technology, and innovation is very scarce, and where it exists in R&D and academic form, it focuses on Rhodesian CBWs and the renovation and innovation of military vehicles (Nass 1992, 1992–1993; Lawrence, Foggin, and Norval 1980; Sterne 1967). Craig's reconstruction of the tsetse men's participation in the counterinsurgency operations is a small slice of ex-Rhodesian military personnel and their biographers beginning to publish texts about their wartime technological innovations, including the retrofitting of light-skinned vehicles into landmine-proof troop carrier vehicles (see also Lester 1996; Wood 2005). Other countries have looked past the moral repugnance of these regimes and have long since noticed this innovative, very applied and pragmatic past; by contrast, Zimbabwe has looked past the innovation and seen only the evil deeds of the downpressor.

Some examples: A decade after the war ended in 1980, the US Army commissioned the RAND Corporation to undertake a study into the valuable lessons to be learned from the Rhodesian counterinsurgency experience (Hoffman, Taw, and Arnold 1991). In 1994, Australia's Department of Defense began plans to refurbish its fleet of armored vehicles. It subsequently commissioned a study of Rhodesian innovations in the protection of light-skinned vehicles against landmines. It is hard to imagine that the US counterinsurgency strategy since the 1990s and Australia's turn to the Bushmaster Protected Mobility Vehicle after the year 2000 did not benefit from these lessons (ACIL Tasman 2009).

Those things we do not see as important, others see their value. They come, study them carefully, take them, and make money with them.

Conclusion: *Vatema* as Intellectual Agents

This book is a project of tracing the itineraries of *ruzivo* in a historical moment of encroachment and dominance by an incoming one: *vachena*'s "science." It was concerned with talking about *vatema* and their knowledge (*ruzivo*) of and practices toward *mhesvi, ndedzi, mpukane,* and *tsetse* and what happened to that *ruzivo* when *vachena* arrived in 1890, imposed themselves through fraud and force, and during their reign which ended in 1980. Mobility helped to stipulate the conditions of possibility within which *mhesvi* influenced such encounters (or not) between *vanhu* (people). Only an examination of the regimes or conditions—forces, presences, absences, movements, stasis, affordances, and preclusions—can help us parse the relations between *vatema* and *mhesvi* and the incoming *vachena*'s encounters with both (see the glossary for *chidzimbahwe* and other local keywords).

Instead of using race relations as a central analytic in the encounter between *vatema* and *vachena,* I see the mobility of multiple actors (*ruzivo* in particular) as more of a vantage point from which different entities reveal their presence, impact, indefatigability, and unignorability and become key factors affecting relations between *vatema* and *vachena,* between *vanhu* and other *zvisikwa,* between breathing beings and nonbreathing ones. *Mhesvi* and *ruzivo* of it serve as transient sites from which to analyze and understand the reasoning behind and stratagems *vanhu* mobilized against this mobile, deadly *chipukanana.* I do not see the intellect and *ruzivo* of *vatema* as "subordinate" to that of *vachena,* nor their *ruzivo* as "subaltern"; to the contrary, the book has argued and shown the opposite. I seek not merely a historical method, or a mere critique, but rather to open new analytical platforms from which *ruzivo* that seems to have been disappeared might be retraced. As a body of ideas and practices that travels through time and is adopted, shared, or appropriated, *ruzivo rwemhesvi* (knowledge of *mhesvi*) constitutes an entity or body in motion that can be followed into and within the knowledge *vachena* claim to have discovered—in search not of some kind of *afterimage of earlier ruzivo and practices*, but the trajectories of *ruzivo*.

The point I make is not the same as the one that Africanist anthropologists like Melville Herskovits (1940) made in the early to mid-twentieth century—namely, that before *vachena* first robbed *vatema*'s land and wealth (*kupamba ivhu nehupfumi hwavatema*), there existed survivals, vestigial, fossil forms of cultural practice of *vatema* that are only awaiting our extraction, pure, autochthonous representations of *tsika/chivanhu* (culture) of *vatema* preserved in the amber of contemporary practice. As *munhu mutema* who is trained in an enlightenment education tradition (but wary of its historical associations with a project of imperialism that dehumanized my own existence), I am sensitive to change over time, while also insisting that such change does not obliterate *ruzivo rwevatema* overnight in the same way, at the same pace, or always, anywhere in Africa.

Herskovits was writing as an Africanist scholar, but I am writing as *mudzimbahwe* with a stake in this *ruzivo*, cautious about the dangers of letting deconstruction run away from the essentialisms that ordinary people insist on and strategically deploy. Somebody once quipped, when I was giving a lecture on *Transient Workspaces* at Harvard, that speaking as an insider makes it impossible for the outsider to say anything. I take this view of the *insider* not to silence those who come from *outside*, but to access an experiential analytical location from which to incite and enrich conversation so that those coming from elsewhere can also be free to use their own analytical locations. Imagine being a professor with a class in which each student brings his or her experiences to bear on a specific subject; one common issue, yes, but many different eyes, minds, hearts, and hands.

The challenge of reintellectualizing, dethingifying, and rehumanizing *munhu mutema* as an insider is this double placement. On the one hand is the self-writer (*munyori ari kuzvinyora*), declaring independence of thought and intellect, finally exhaling: "We write about ourselves, *at last*!" The keywords must be legible among the people, hence the extensive deference to *chidzimbahwe*. On the other hand is a scholar who answers to a larger academic audience that has certain expectations. *The Mobile Workshop* has its feet firmly planted in *chidzimbahwe* idioms and addresses the academic from there, emphasizing that our writing as *vanhu vatema* must be relevant to *vanhu vatema*, just as most Africanist writing is relevant to the countries such scholars hail from first and foremost. To be meaningful to the people of Africa, these narratives must be grounded in a language that the people are articulating, as opposed to simply mobilizing local knowledge as empirical evidence in a thesis that simply seeks to advance an idea originated from and addressing debates in US or European academies. To do that, these narratives must start seeing the people of Africa as more than just their "informants" (which in Zimbabwe explicitly means "sellouts," given our history of *vatema* who sold out their fellow *vatema* to the Rhodesians during the war of self-liberation)—as people engaged in intellectual conversation through everyday engagement with things and challenges.

To return to Herskovits, *fossil* and *vestige* are words of *vachena* with no meaning in *chidzimbahwe*—the analytical location I write from—and both fall far short of the meanings and uses of *ruzivo* in the context of *tsika/chivanhu*. On the contrary, the irony explored in this book is precisely the ruse of *varungu* like Duerden, Rheinallt-Jones, Hoernlé, Junod, Fantham, and Bieshevel, busy fossilizing and turning *vanhu vatema* into unthinking objects of study while their colleagues like Watt and Brandwijk, Lloyd, van Warmelo, Maingard, and Laydevant were besieging the homes of every *inyanga* (healer) to make them show and tell them about every medicinal and poisonous plant they knew. We are talking of Wits and UCT medical schools and anthropology departments, local and overseas laboratories, missionaries, and Native Commissioners—a vast meshwork of *vachena* plundering the knowledge of *vatema*.

The temptation for most "indigenous knowledge" scholars is to want to restore the golden age and mourn its death when *vachena* arrived. The displacement of *vatema* and the forcible partition and occupation of their land is turned into a sudden rupture between the regime of the fossil or savage (*vatema*) and the civilizing mission (*vachena*). The challenge is to breach this epistemic boundary between "precolonial," "colonial," and "postcolonial" and to write from within the everyday lives and terminologies of people, which are paced and temporalized in their own complicated ways (Karp and Masolo 2000, 3). Nothing like a wholesale clearing out of ideas and practices marked *vachena*'s occupation, as the "savage's 'fables'" died and fossilized into "prehistory." The clock of *ruzivo* certainly did not automatically rewind to zero, no void in *ruzivo* was ever created, and no wholesale "transfer" of revolutionary ideas from Europe filled a void that, after all, did not exist.

The process of writing for and as *munhu mutema* is one of continuing rehumanization, and we should be clear that the Africanist—the one for whom Africa is an object and subject of study—is justified in seeing Africa from a different place. I only insist that the Africanist understands that we all write from a particular analytical location, that no one viewing platform is more legitimate than another. We just need to understand its reasons for articulation and, as people of Africa variously located, to judge for ourselves whether it expresses us or something else entirely.

I write from where I write because of what history forced me to write. Some write as beneficiaries of Europe's Enlightenment and its intellectual aftermaths; I write in refusal of being a mental victim of its thingifying impulses and actual aftermaths. I did not choose to be dehumanized, but I inherited the consequences of the Enlightenment tradition, which placed upon me the burden of first having to justify my intellect based on a standard set by the same system that thingified me and then having to self-educate myself on *vatema*'s modes of knowledge, because not only were they never taught in school and university, but they were deemed *fabulous* (i.e., fable matter).

Instead of seeing the coming of *vachena* and the moment they ruled Africa as a rupture, I see *ruzivo rwevatema* as the bridge—or the drift—that enabled *vachena* to cross, occupy, partition, settle, exploit, and make a home out of Africa. Whichever part of Africa one may be in, the story of *varungu*'s arrival was not only a journey guided by *vatema* who knew the land and its bounties, but also one in which the travelers tapped into *ruzivo rwevatema* to survive. Indeed, one could make the same case for *vachena*'s disruptive encounter with the Maoris in New Zealand, Aboriginal Australians, and the first nations of North and South America.

This is where the historical and "indigenous knowledge" scholarship on Africa thus far is weak, in that it has not defined such *ruzivo* from the names and meanings its producers give it. One way to do so is for *vatema* to tell their stories, to do so by revisiting their own vocabularies and experiences, to go back and relive the activities and spaces they inhabited and still inhabit, and use them as equally powerful sources and filters for reading these libraries compiled by outsiders. Methodologically, that is what this book aimed to do.

Glossary

Terms are in *chidzimbahwe,* except where specifically indicated.

Abantu, bantu; sing. ***muntu***: People (isindebele)

Abazingeli: Hunters (isindebele)

Amatsheomnyama: Black water (isindebele)

Bani, rupani; pl. ***mapani***: Valley

Basa, mushando: A task or work to be done, employment

Bhurudhoza; pl. ***mabhurudhoza***: Corruption of bulldozer

Butalifi: Special expertise (silozi)

Chakasviba: Dark

Chepfu: Poison

Chibhakera; pl. ***zvibhakera***: Fist; plowing with the fist; zero tillage

Chibharo: Forced labor

Chidzimbahwe: Language, culture, and things of *dzimbahwe*

Chiguraura; pl. ***zviguraura***: Literally, "the one that has cut off its intestines," what *vachena* called larva

Chikopokopo; pl. ***zvikopokopo***: Helicopter

Chikukwa; pl. ***zvikukwa***: The insect at its worm or pupa stage, what *vachena* called puparia

Chimokoto; pl. ***zvimokoto***: Quelea bird

Chimugondiya: Brown laundry bar soap

Chimukuyu: Dried meat

Chimumbure: Net

Chinanga: The hooked-thorn tree *vachena* called *Acacia nigrescens*

Chinhu; pl. ***zvinhu***: Thing

Chinosvosvoma; pl. ***zvinosvosvoma***: Crawler, reptiles

Chipfuyo; pl. ***zvipfuyo***: Livestock

Chipukanana; pl. ***zvipukanana***: Small animals, insects

Chirwere chegomarara: Disease of the mistletoe

Chisikwa; pl. ***zvisikwa***: [God's] creation

Chitatarimbo; pl. *zvitatarimbo*: The birdlime tree

Chivanhu: Culture; custom; ways of the people

Chokwadi; pl. ***zvokwadi***: Truth

Danga: Kraal

Démó: Blade

Dewulana: Bat (xitsonga)

Dikgomo, ***kgomo***: Cattle (setswana)

Diphôlôgôlô: Wild animal (setswana)

Diruiwa: Livestock (setswana)

Dondo, ***rutondo***; pl. ***matondo***: Forest

Dzimbahwe: Houses of stone; headquarters of kings; graves

Dzoma: Bushbuck

Ekhayeni, ***ekhaya***: Village (isindebele)

Fanakaló or ***fanagaló***: Hybrid language composed of some English, Nyanja, and Shona; also known derogatorily as *Kitchen-Kafir*, *Mine-Kafir*, *Pidgin Bantu*, *Isikula* (*Coolie* or Indian), *Chirooroo*, *Fanikaroo*, and *Chiraparapa* (*Silapalapa*)

Fenzi yemhuka: Game fence

Fenzi yemombe: Cattle fence, locally

Fenzi yetsetse: Tsetse fence

Frayi chemba: Fly chambers

Gareta: Bush cart

Gedhi retsetse: Tsetse gate

Goji: Holes (xitsonga)

Gomarara: Cancer; derived from a plant that usually grows on other plants, deposited in the fecal droppings of birds roosting or stopping over

Gondo; pl. ***makondo***: Eagle

Gopé, hotsikotsi: Sleeping sickness

Gora, pl. ***magora***: Vulture

Gudo, pl. ***makudo***: Baboons

Gumba: White stork

Guta, pl. ***maguta***: City

Hambautare: Iron tortoise (bicycle)

Hembe dzemapisi: Clothes from pieces

Honci-nhova: Warthog (xitsonga)

Hové: Fish

Hudzvanyiriri hwevachena: The downpression of white people

Hufrayi: The work of flycatching

Hufuro, mafuro: Pastures

Huku: Chicken

Hugocha: Game-destruction work

Hungwe: Fish eagle

Hunhapwa, hutapwa: Slavery, bondage

Hunhu: Behavior, being

Hunyanzvi: Expertise

Hunyanzvi hwekuvhima or ***huhombarume***: Hunting expertise

Hupfu: Meal, ground grain

Hurimbo, urimbo: Birdlime

Hurumende: Government

Hurumende yehudzvanyiriri: The oppressive government

Hurumende yevachena: White government

Hutachiwana: Literally, "we have found it, the invisible sickness causer"

Hutachiwana hwegopé: Used when referring to sleeping sickness

Hutachiwana hwen'gana: *N'gana*; animal trypanosomiasis

Hutongi hwavachena: White rule

Hutsi: Smoke

Hutsiny'e hwemabhunu: The cruelty of the Boer

Hutunga: "The one that gores"; *varungu*'s mosquito

Hwayi: Sheep

Hwema: Odor, olfactory

Hwiza, mhashu, ndongwe, domboji, dzviti: Swarming locust, grasshopper

Idlelo: Pasture (isindebele)

Imbuzi: Goat (isindebele)

Imbwa: Dog

Imvu: Sheep (isindebele)

Inda: Louse

Inja: Dog

Inkomo, izinkomo: Cattle (isindebele)

Intesi: Tsetse, corruption of *ndedzi* (xitshangana)

Inyamazana: Animal (isindebele)

Inyanga: Healer, diviner, doctor, physician

Inyongo: Fever (isindebele)

Isibungu: Insect (isindebele)

Izinkomo, inkomo: Cattle (isindebele)

Kabhunu: Thin, emaciated Boer/white man

Kasiri: Cattle breed indigenous to *vedzimbahwe*

Khumba: Bushpig (xitsonga)

Kitsô: Knowledge, knowing (setswana)

Kubika: To cook, cooking

Kuchenesa, kuchenura: Cleaning

Kudheruka: To come from nowhere

Kuenda kusango: Going to the bush (urinating or defecating)

Kufamba: Mobility, walking, traveling, moving

Kufamba kweruzivo: The mobility of knowledge

Kufamba kwevanhu nemhuka: Human and animal traffic

Kugadzira: Making

Kugarisika kwevanhu: Human settlement

Kugobora: Stumping

Kugumha hurimbo: Tapping birdlime

Kuhugocha: At *magocha* work (at the work of tsetse control hunting)

Kumafrayi: The place of the fly people

Kumusha: In/at the village

Kupa: Giving

Kupamba ivhu nehupfumi hwavatema: Abducting the land and wealth of blacks

Kupesvedzera, kupesva: Inciting

Kupihwa: Receiving, being given

Kupisa sora: Burning weed/grass

Kurodza: Sharpening

Kuroyiwa: To be bewitched

Kusangana: Encountering, coming into contact

Kusangana kweruzivo: Encounter between knowledges

Kushanda, kuita basa: Working, doing work, efficacy

Kutemwa nemusoro: Headache; literally, "being chopped by the head"

Kuziva: Knowing

Kuziva kwevachena: Knowledge of the white people

Kuzvisunungura: Untying, freeing oneself, self-liberation

Likomu; sing. ***komu***: Cattle (silozi)

Limbweletele: Human excrement (silozi)

Lisu: Sun-dried cattle dung (silozi)

Mabisi: Milk (silozi)

Mabwidi: Those from Northern Rhodesia in *chidzimbahwe* (colloq.)

Madheruka magochamiti: Newcomers who roast trees (colloq.)

Madhibhi, sing. ***dhibhi***: Dip tanks

Madzviti, sing. ***dzviti***: Pillagers

Mafrayi, sing. ***mufrayi***: Fly men, fly catchers

Mafulô: Pastures (setswana)

Mafuro, sing. ***hufuro***: Pastures

Magandanga, sing. ***gandanga***: Terrorist (colloq.)

Magirosa or ***zvitoro***: Grocery store

Magobo: Stumping

Magocha, sing. ***mugocha***: Those who are always roasting meat; black hunters employed by the government to kill wild animals to starve *mhesvi*

Magora, sing. ***gora***: Vultures

Magwa, sing. ***igwa***: Canoes

Mahlomalavisi: Pupa, chrysalis (xitsonga)

Maitiro: Ways of doing

Majerenyenje, sing. ***jerenyenje***: Anteaters

Majuru: Termite

Makandiwa, ***madhunduru***: Contours

Makarwe, sing. ***garwe***: Crocodiles

Makonzo, sing. ***gonzo***: Rats

Makorekore: Those *vedzimbahwe* who live or hail from Gorekore, who taboo *gudo* the baboon or *soko* the monkey

Makwayi, sing. ***gwai***: Sheep

Mangwa: Zebra (xitsonga)

Manhuwe: Repellents

Manyasarandi: Those from Nyasaland in *chidzimbahwe* (colloq.)

Many'atera: Sandals

Manyepo: Falsehoods

Mapete, sing. ***bete***: Cockroaches

Maprofita: Prophets

Mapundu: Boils

Mapurazi, sing. ***purazi***: Farms, ranches; from the Portuguese word *prazo*

Marengeny'a: Tattered clothes

Mari: Money

Marukisheni, sing. ***rukisheni***: Locations

Mashuramurove, sing. ***shuramurove***: The bird that portends waterlogging; stork

Matanga emombe, sing. ***danga remombe***: Cattle stockades

Matore, sing. ***dore***: Walking carcasses

Mazizi: Owls

Mbavala: Bushbuck (xitsonga)

Mbeva: Wild (therefore edible) rodents, mice

Mbudzi: Goat

Mbumburu: Bullet

Mbuwe, ***mbuwo***, ***uo***: Blind fly

Mfenhe: Baboon (xitsonga)

Mhangele: Guinea fowl (xitsonga)

Mhara: Impala

Mhelembe: Rhinoceros (xitsonga)

Mhene: Duiker

Mhesvamukono: The one that drives the bull (*mukono*) crazy

Mhesvi: From the word *kupesva*; tsetse

Mhesvirupani: *Mhesvi* that lives in the valley; *varungu*'s *Glossina pallidipes* and *Glossina palpalis*

Mhesvirutondo: *Mhesvi* that lives in the forest; *varungu*'s *Glossina morsitans*

Mhopfu: The blind one; eland (xitsonga)

Mhuka: Animal of the forest

Mhuka yomurukova, pl. ***mhuka dzomurukova***: Animal that lives in water

Mhunduru: Armyworm

Mhunga: Millet

Mhunti: Little blue duiker (xitsonga)

Mhuri: Family

Mibhemhe, sing. ***mubhemhe***: Donkeys

Midziyo, sing. ***mudziyo***: Instruments

Migodhi, sing ***mugodhi***: Mines

Mikono, sing. ***mukono***: Bulls

Mirhi, sing. ***murhi***: Tree (xitsonga)

Misengwa, sing. ***musengwa***: Luggages, loads

Misha, sing ***musha***: Villages, homes, homesteads

Mishini, sing. ***mushini***: Machines

Mishini yekupureya: Spraying machines

Mishonga, sing. ***mushonga***: Poisons, medicines

Misika yemabhazi, sing. ***musika wemabhazi***: Marketplaces for buses (bus stations)

Mlilo: Fire (isindebele)

Mogare: Pathogens (setswana)

Mombe: Cattle

Monang: Mosquito (setswana)

Moto: Fire

Motokari: Motorcar

Motse: Village (setswana)

Mpukane: Tsetse (isindebele)

Mpukane zegangeni: Tsetse of the forest (isindebele)

Mubhemhe: Donkey

Muchena, pl. ***vachena***: White person

Muchetura, pl. ***michetura***: Poison

Mudziti: Stick around which birdlime is wound, set to catch birds

Mudzviti: Magistrate

Mufambi, pl. ***vafambi***: Traveler, walker

Mufrayi, pl. ***mafrayi***: Fly man

Mugocha, pl. ***magocha***: The ones who are always barbequing

Mugwagwa, pl. ***migwagwa***: Road

Mukonde: ***Euphorbia ingens***

Mulaha: Liquid cow dung (silozi)

Mulilo: Fire (silozi)

Mulumbeti: The devil (xitsonga)

Mumusha: In the village

Munhu**, pl. **vanhu: People, those who are familiar, kin, or known

Munhu mutema, pl. ***vanhu vatema***: Black person

Munhuwi wemunhu: Human odor

Munyori ari kuzvinyora: A writer writing about him/herself

Mupani, ***musharu***: Mopane (setswana)

Mupembere, ***mubondoroko***: *Vachena*'s *Combretum* bushland

Mupfuti: All *vachena*'s *Brachystegia*

Murungu, pl. ***varungu***: White person

Musha: Village, homestead

Mushini, pl. ***mishini***: Machine

Mushini wekupureya, pl. ***mishini yekupureya***: Knapsack sprayer

Mushonga, pl. ***mishonga***: Poison, medicine

Mushonga wezvipukanana (shortened to ***mushonga***), pl. ***mishonga yezvipukanana***: Pesticide

Musvo: Metabolites; literally, "residue"

Muswe: Tail

Mutema, pl. ***vatema***: A black person

Mutemo, pl. ***mitemo***: Law

Muteyo, pl. ***miteyo***: Trap

Mutsetse: In the Tsetse Department

Muvatli: Carpenter or sculptor (xitsonga)

Muvengahonye: Literally, "the maggot hater"

Mvubu: Hippo (isindebele/isizulu)

Mvura: Water, rain, rainfall

Mvuu: Hippo

N'anga: Traditional healer

Nare: Buffalo (setswana)

Ndedzi: Tsetse (xitsonga)

Ndege: Fixed-wing aircraft

Ndlopfu, ***tindlopfu***: Elephant (xitsonga)

Ndlovu: Elephant (isindebele)

Ndunguza: Antelope

Ndzilo: Fire (xitsonga)

Ngamboto: A matter of common knowledge (kilozi)

N'gana: Animal trypanosomiasis

Ngano: Fable

Nghala: Lions (xitsonga)

Nghwari: Crested francolin (xitsonga)

Ngongoni: Wildebeest

Ngoroyemoto, pl. ***ngorodzemoto***: Carriage of fire (motorcar)

Nguruve: Bushpig

Nguruve yemusango: Bushpig

Ngwahle: Monitor lizards, iguana (xitsonga)

Ngwarati: Sable

Ngwenya, ***garwe***, ***gambinga***: Crocodile

Nhéma, ***chipembere***, ***mhembere***: Rhinoceros

Nhimura: Literally, "the slashing"; the term *vedzimbahwe* gave to the Native Land Husbandry Act

Nhoro: Kudu

Nhoveni: Forest (xitsonga)

Nhungu: Kudu (xitsonga)

Nhunzi: Housefly

Njiri: Warthog

Nkawu: Monkey (xitsonga)

Ntsandza: Poles (xitsonga)

Ntsi, pl. ***lintsi***: Housefly (setswana)

Ntsutsu: Egret (xitsonga)

Nyama: Meat

Nyarhi: Buffalo (xitsonga)

Nyati: Buffalo

Nyaudzosingwi: Ideophone

Nyemba: Beans

Nyimo: Round nuts, Bambara nuts

Nyoka: Snake

Nyong'o: Fever, malaria

Nyongororo: Parasite

Nyongorosi: Earthworm

Nzira: Footpath, way

Nzira nemaitiro: Ways and means

Nzou, zhou: Elephant

Nzungu: Peanuts

Pabasa, pamushando: The workplace, being at work

Pesva: The action of plunging a sharp object like a needle or thorn into flesh (ideophone of *kupesva*, to probe; hence, proboscis)

Pfungwa dzavatema: Ideas, idioms of black people

Purazi: See *mapurazi*

Ronda rinongobva pasi risingapori: A wound that emerges from the ground and never heals; a wound with no explicable origin

Rukodzi: Falcon

Rusosa, ruzhowa: Fence

Ruzevha: Native reserve (transliteration)

Ruzivo: Knowledge

Ruzivo chairwo: True knowledge

Ruzivo rwemhesvi: Knowledge of *mhesvi*

Ruzivo rwemombe: Knowledge of cattle

Ruzivo rwevachena: Knowledge of whites

Ruzivo rwevarungu: Knowledge of whites

Ruzivo rwevatema: Knowledge of black people

Ruzivo rwezvehurimbo: Birdlime knowledge

Rwendo: Journey

Sango, pl. ***masango***: Forest

Setshidinyana: Insect (setswana)

Shayishayi: Monarch butterfly

Shiri: Bird

Shumba: Lion

Sinuka: Odorous grasses (silozi)

Sondela enkosini: Draw near to the king (xitshangana)

Stupa, ***chitupa***: An identity document (Fanakalo, Chiraparapa, chidzimbahwe)

Sutelezi: Repellent (silozi)

Svexilungwini: The things of the white people (xitsonga)

Svidvelo, sing. ***xidvelo***: Pastures (xitsonga)

Svifufunhunhu, sing. ***xifufunhunhu***: Insects (xitsonga)

Svifuri, sing. ***xifuri***: smiths (xitsonga)

Svigalanyundzu, sing. ***xigalanyundzu***: Hammers (xitsonga)

Sviharhi, sing. ***xihloka***: Animal (xitsonga)

Svihloka, sing. ***xihloka***: Axe (xitsonga)

Svikokovi, sing. ***xikokovi***: Reptiles

Svimun'wana, sing. ***ximun'wana***: Springs (xitsonga)

Svindlu svasvifuri, sing. ***xindlu xaxifuri***: Forge (xitsonga)

Svinyenyana, sing. ***xinyenyana***: Birds

Tau: Lion

Tawana: Cub, from the word *tau* (adult lion) (setswana)

Tihomu, ***homu***: Cattle (xitsonga)

Tiko: Land, village, country

Timbuti: Goats (xitsonga)

Timbyana, sing. ***mbyana***: Dogs (xitsonga)

Timhala: Impala (xitsonga)

Tlou: Elephant (setswana)

Tsangadzi: Runner grass, which *vachena* called *Loudetia superba*

Tsetse: *Mhesvi* (setswana)

Tsika: Culture, custom

Tsika, chivanhu: Culture, custom

Tsika dzevachena: White culture

Tsika dzevatema, chivanhu: Ways or culture of black

Tsoko, shoko: Monkey

Tsvina: Dirt

Twiza: Giraffe

Twukodzi, sing. ***rukodzi, hodzi, kakodzi***: Hawks

Ukulagisa: Whereby *izinkomo* are lend to relatives to look after in exchange for milking, manure, and draft power (isindebele)

Vachena, sing. ***muchena***: Whites

Vadzimu, sing. ***mudzimu***: Ancestors

Vadzvanyiriri, sing. ***mudzvanyiriri***: Downpressor, oppressor

Vafambi, sing. ***mufambi***: Travelers, walkers, itinerants

Valungu, sing. ***mulungu***: Whites (xitsonga)

Vanhu, sing. ***munhu***: People, person

Vanhu vachena: White people

Vanhu vatema, sing. ***munhu mutema***: Black people

Vanhu vetsetse: Tsetse people, people who work for the Tsetse Department/Branch

Vantima, sing. ***muntima***: Black people (xitsonga)

Vapambepfumi, sing. ***mupambepfumi***: Abductors of wealth

Vapambevhu, sing. ***mupambevhu***: Abductor of the land/soil

Varungu, sing. ***murungu***: Whites

Varwi verusununguko, murwi werusununguko: Freedom fighters

Vashandi vatema: Black workers

Vatema, sing. ***mutema***: Blacks

Vazungu, sing. ***muzungu***: Whites

Vechishangwe: colloquial *shangwe* people, actual name *vakorekore*

Vedzimbahwe: Those of *dzimbahwe*

Vekupureya: Spraymen

Vutivi: Knowledge (xitsonga)

Vutivi bya valungu: The knowledge of whites (xitsonga)

Waya nemakurundundu: Crude wire fence made of felled tree poles

Waya yehurumende: Government fence

Xekolodi: Millipede (setswana)

Xifufunhunhu, pl. ***svifufunhunhu***: Insects

Xiharhi: *Mhuka* (xitsonga)

Xikanyakanya: Bicycle, named from the sound of pedaling, *kanya-kanya-kanya* (xitsonga)

Xindedzi xa nhoveni, pl. ***svindedzi sva nhoveni***: *Mhesvi* of the forest (xitsonga)

Xindedzi xa nkova, pl. ***svindedzi sva nkova***: *Mhesvi* of the valley (xitsonga)

Xinyenyana, pl. ***svinyenyana***: Bird (xitsonga)

Xitivi, pl. ***svitivi***: Expert (xitsonga)

Xitsongwatsongwana, pl. ***svitsongwatsongwana***: The microorganism that caused it (xitsonga)

Zambuko, pl. ***mazambuko***: Drift

Zengeni, ***shengezhu***: Tall sweet grass, what *vachena* called *Hyparrhenia spp.*

Zeze: Kokolo name for *mhesvi*

Zimbabwe: Big house of stone

Zimunhuwenhuwe: Smelly plant

Zumbani, ***sumba***: Both indigenous mint varieties

Zuva: Day, sun

Zvakatikomberedza: Surroundings, vachena's "environment"

Zvechirungu: Literally, "the things of the white people"

Zvenhando: Trivia

Zviguraura, sing. ***chiguraura***: Literally, "the ones that cut off their intestines"; larva

Zvikepe, sing. ***chikepe***: Boats

Zvikopokopo, sing. ***chikopokopo***: Helicopter

Zvikukwa, sing. ***chikukwa***: Chrysalises, puparia

Zvikwekwe: Ticks

Zvimokoto: Quelea birds

Zvinhu, sing. ***chinhu***: Things

Zvipfukuto, sing. ***chipfukuto***: Weevil

Zvipfuyo, sing. ***chipfuyo***: Livestock

Zvirimwa, sing. ***chirimwa***: Crops

Zvokwadi, sing. ***chokwadi***: Truths, facts

Notes

Preface

1. Notes on my visit to speak with RAW Material Company fellows in Dakar, Senegal, May 20, 2017. Here was this wonderful taxi brother, pointing out very exciting things outside, and I sat there in the back seat, so many questions on my lips, neither of us able to say anything to the other.

2. Peter Tosh, one of the pioneers of reggae music, popularized this term in his protest song "Downpressor Man" (1977).

3. A glossary is included at the back of the book.

Introducing *Mhesvi* and *Ruzivo Rwemhesvi*

1. Interview, Fema Ngonda, Nembudziya, March 15, 2016. ZVI/G/N/6; Interview, Solomon Bvekenya, Makuleke, November 1, 2010. MOV04A, November 1, 2010. All archival sources are listed in endnotes, but not as references.

2. A. D. Marks, "Preliminary Notes on the Relationship of the Behaviour of Tsetse Flies and Impala," 1967. South African Centre for Epidemiological Modelling and Analysis/Tsetse Archives [hereafter SACEMA/TA]; A. D. Marks, "Notes on the Behaviour of Tsetse Fly (*G. morsitans* and *G. pallidipes*) with Regard to Impala," September 1, 1966. SACEMA/TA.

3. "Report of the Director of Tsetse and Trypanosomiasis Control and Reclamation for the Year Ended 30th September 1960," 233. SACEMA/TA.

4. R. W. Jack, "The Life Economy of a Tsetse Fly," January–February 1944, 7. SACEMA/TA.

5. B. M. Leggate, "The Diurnal Feeding Activity, Resting Sites and Trypanosome Challenge of *G. Pallidipes* at Rekomitjie," 1960, 250. SACEMA/TA.

6. Jack, "The Life Economy," 3.

7. Jack, "The Life Economy."

8. R. W. Jack, "The Tsetse Fly Problem in Southern Rhodesia," May 1933, 1. SACEMA.

9. Annual Report, Chief Entomologist 1923, 1. SACEMA/TA.

10. J. A. Whellan, "Tsetse Fly in S. Rhodesia, 1949," 2. SACEMA/TA.

11. R. W. Jack, "Tsetse Fly: A Four Years' Experiment in Game Elimination," June 1923. SACEMA/TA; Jack, "The Tsetse Fly Problem"; R. W. Jack, "The Control of Tsetse Fly in Southern Rhodesia," March 13, 1935. SACEMA/TA.

12. E. Bursell, "Tsetse Control in Relation to Wild Life Conservation," University College of Rhodesia and Nyasaland, May 19–26, 1961, 4. SACEMA/TA.

13. E. Bursell, "Tsetse Control in Relation to Wild Life Conservation," Wild Life Conservation Course, University College of Rhodesia and Nyasaland, May 19–26, 1961, 4. SACEMA/TA.

14. Bursell, "Tsetse Control," 4–5. SACEMA/TA.

1 How *Vanhu* Managed Tsetse

1. The author wishes to thank Lerato Motshwarakgole for these nuanced translations. E-mail exchange, Cambridge, MA, September 19, 2017.

2. Interview, Willias Chabata.

3. The predominant names used in southeastern Zimbabwe and adjacent parts of Mozambique are in *xitsonga*; I shall use *xitsonga* concepts in all references to the southeast.

4. J. K. Chorley, Director Tsetse Fly Operations, "Destruction of Game in Native Reserves and Native Areas Situated on the Edge of a Fly Belt, or in Areas Covered by Game Elimination Operations" [n.d., likely January–March 1953], 2. SACEMA/TA.

2 Translation into Science and Policy

1. E. Bursell, "Tsetse Control in Relation to Wild Life Conservation," Wild Life Conservation Course, University College of Rhodesia and Nyasaland, May 19–26, 1961, 4–5. SACEMA/TA.

3 Knowing a Fly

1. E. Bursell, "Tsetse Control in Relation to Wild Life Conservation," Wild Life Conservation Course, University College of Rhodesia and Nyasaland, May 19–26, 1961, 4. SACEMA/TA.

2. Bursell, "Tsetse Control," 4–5. SACEMA/TA.

3. Report of the Commission of Inquiry into the Cold Storage Commission of Southern Rhodesia, CSR 29-1952; CSC, 7th Annual Report and Accounts, 1954; Director of Agriculture to Treasurer, December 20, 1921. NAZ S1193/M5.

4. CSC, Expropriation of the Rhodesian Export and Cold Storage Company, 1937–1951. NAZ S2704/3/1.

5. Memo from the Minister of Agriculture and Lands. NAZ S2704/3/1; Cold Storage Commission, General, 1948–1950, Minister of Agriculture to Cabinet. NAZ S2704/1/1340/2.

6. Cold Storage Commission, General, 1948–1950, Minister of Agriculture to Cabinet. NAZ S2704/1/1340/2.

7. Southern Rhodesia, *Legislative Assembly Debates* (1st Nov. 1937, 3rd Session, 4th Parliament), XVII, 2447.

8. Government Policy toward Assisting the Development of Industries in Southern Rhodesia. NAZ S482/134/17/48.

9. CSC, Annual Report and Accounts, 1938–1947.

10. Cold Storage Commission: Agreement for Supply of Meat to Britain, September 1939–November 1942. NAZ S1215/1342/1.

11. Southern Rhodesia, Committee to Enquire into the Costs of Distribution of Imported and Local Products in Southern Rhodesia, Interim Report on Livestock and Meat, with Special Reference to Cattle and Beef, 1936 (A. R. Burnett-Hurst—Chairman), 14.

12. Cattle Commission, Evidence1962–1963 and Cattle Inquiry Commission, Evidence Received 1963, Janneck and Swart, "The Place of the Cold Storage Commission." NAZ F259/1131/l-2.

13. Southern Rhodesia, Report of the Commission of Enquiry into Certain Sales of Native Cattle in Areas Occupied by Natives, 1939 (R. J. Hudson—Chairman, hereafter Hudson Report, Presented to the Legislative Assembly, 1939); Southern Rhodesia, Report of the Commission Appointed to Inquire into the Marketing of Slaughter Cattle and the Products Thereof, 1942 (W. E. Thomas—Chairman).

14. Cattle Commission, Evidence 1962–1963, S. R. Ministry of Internal Affairs, Evidence for Submission to the Commission of Enquiry into the Cattle Industry (June 1963), NAZ F259/1131.

15. Slaughter Cattle Prices Order—Government Notice No. 18 of 1942. NAZ F259/1131.

16. Hudson Report.

17. Phimister 1978, 397.

18. CSC, 19th Annual Report and Accounts, 1956.

19. Rupert Jack, "The Control of Tsetse Fly in Southern Rhodesia," Salisbury, 1935, 1. SACEMA/TA; G. Cockbill, "The Branch of Tsetse Fly Operations," Professional Officer's Conference, March 28–29, 1955, 1. SACEMA/TA1; Annual Report, Chief Entomologist, 1921. SACEMA/TA; Annual Report, 1923, 3; Annual Report, Chief Entomologist, 1925, 2. SACEMA/TA.

20. Cockbill, "The Branch," 1.

21. Report of the Chief Entomologist 1938, 18. SACEMA/TA.

22. Annual Report, Chief Entomologist, 1930, 7. SACEMA/TA.

23. Annual Report, Chief Entomologist 1931, 2. SACEMA/TA.; Report, Chief Entomologist, 1932, 6. SACEMA/TA.

24. Report 1932, 6.

25. J. K. Chorley, "Tsetse Fly Operations, 1942," 1. SACEMA/TA.

26. Annual Report: Director Tsetse Fly Operations, 1951, 5.

27. Ibid.; Annual Report: Director Tsetse Fly Operations, 1953, 4. SACEMA/TA.

28. "Retirement of Dr. Gerald F. Cockbill, 1972," 1. SACEMA/TA.

29. Annual Report 1955, 5.

30. Report, Acting Director, DTTCR, 1956, 137. SACEMA/TA.

31. "Retirement," 2.

32. Report 1956, 137; "Retirement," 2.

33. Report, Director: Tsetse and Trypanosomiasis Control and Reclamation (TTCR), 1958, 202. SACEMA/TA; John Ford, "A Programme for Tsetse and Trypanosomiasis Control in Southern Rhodesia" 1958, 1. SACEMA/TA.

34. Ibid.

35. Ibid., 2.

36. Report, Director: DTTCR, 1959, 196. SACEMA/TA.

37. Report, Director: TTCR, 1960, 224. SACEMA/TA.

38. Report 1959, 196; "Retirement," 2.

39. Ibid., 3.

40. Ibid.

41. W. P. Boyt, "Trypanosomiasis Southern Rhodesia 1964: A Review," February 4, 1964, 2. SACEMA/TA.

42. Ibid.

43. Report: ADVS (TTC), 1964," 11. SACEMA/TA.

44. "Trypanosomiasis Southern Rhodesia 1964: A Review," 1. SACEMA/TA.

45. G. F. Cockbill, "Report: ADVS (TTC), 1965," 1. SACEMA/TA.

46. Ibid., 2.

47. Ibid.

48. Annual Report: Branch of Tsetse and Trypanosomiasis Control (BTTC), 1970, 21. SACEMA/TA.

49. Annual Report: BTTC, 1972, 1. SACEMA/TA.

50. Ibid., 1–2.

51. Annual Report: BTTC, 1973, 1. SACEMA/TA.

52. Ibid.

53. Annual Report: BTTC, 1982, 2. SACEMA/TA.

54. Annual Report: BTTC, 1984, 3. SACEMA/TA.

55. Interview, Chabata.

56. Interview, Fema Ngonda.

57. Report 1958, 206.

58. Ibid.

59. Interview, Reuben Mavenge.

60. Interview, Ngonda.

61. Report 1965, 25.

62. Ibid., 55.

63. Ibid., 55.

64. Report NPB 1968, 35.

65. Ibid., 34.

66. Report NPB 1969, 23.

67. NPB 1970, 26.

68. Annual Report 1969, 27.

4 How to Trap a Fly

1. Annual Report, Chief Entomologist, 1929, 6. SACEMA/TA.

2. Annual Report 1931, 8.

3. Report 1932, 9.

4. "J. K. Chorley, "The Tsetse Fly Problem in Southern Rhodesia, 1937," 13. SACEMA/TA.

5. R. W. Jack, "The Tsetse Fly Problem in Southern Rhodesia," May 1933, 17. SACEMA/TA.

6. Annual Report 1970, 48. SACEMA/TA. The Manitoba trap was named after its origin, Manitoba, Canada, where the entomologist A. J. Thorsteinson used it to trap horseflies effectively.

7. Annual Report 1982, 14.

8. Report, Acting Director: DTTCR 1957, 8. SACEMA/TA.

9. Annual Report 1982, 16; Report 1957, 8.

10. Annual Report 1967, 37.

11. Ibid., 38.

12. Ibid.

13. Ibid.

14. Ibid., 39.

15. Ibid.

16. Ibid.

17. Annual Report 1969, 57.

18. Ibid.

19. Ibid.

20. Annual Report: BTTC, 1972, 7. SACEMA/TA.

21. Annual Report 1973, 10.

22. Interview, Chabata.

23. Annual Report 1973, 10.

24. Annual Report 1973, 10–12.

25. Annual Report 1972, 9.

26. Annual Report 1982, 16.

27. Ibid., 17.

28. Annual Report: BTTC, 1985, 16–17.

29. Ibid.

30. Annual Report 1982, 19.

31. *Nhamo urimbo, inonamira.*

32. *Iri paurimbo ndeinoti tsviritsviri; iri muriva inoti denga rawa.*

33. C. Mossop, "Report: Acting Chief Entomologist, 1947," 16. SACEMA/TA.

34. Annual Report 1969, 56.

35. Report 1965, 26.

36. Ibid.

37. Ibid., 27.

38. Ibid.

39. Annual Report: ADVS (TTC) 1968, 40. SACEMA.

40. Ibid., 35.

41. Annual Report 1969, 37.

42. Ibid.

43. Ibid.

44. Ibid.

45. Ibid., 58.

46. Ibid., 60.

47. Ibid.

48. Annual Report 1970, 50.

49. Annual Report 1982, 17.

50. Annual Report 1985, 19.

51. Annual Report 1984, 18.

52. Annual Report 1982, 21.

53. Annual Report 1984, 18.

54. Annual Report 1985, 19.

55. Annual Report 1984, 19.

56. "Observations on the Application of Odour-Baited Targets in Large Scale Tsetse Control Operations," Technical Report of the Advisor, Zimbabwe Component of the RTTCP, June 7, 1995. SACEMA/TA.

57. "Observations," 9.

58. Ibid., 2.

59. Ibid., 11.

60. Ibid., 11.

61. Ibid., 11–12.

62. Annual Report 1973, 10.

63. Annual Report 1971, 35.

64. Ibid., 36.

65. Annual Report 1971, 35.

66. Ibid.

67. Ibid.

5 Attacking the Fly from Within

1. Annual Report 1923, 1.

2. Ibid., 2.

3. Ibid.

4. Annual Report 1982, 15.

5. No. 12-14-100-5453-(33), Cooperative Agreement between the Agricultural Research Council of Rhodesia and Nyasaland and the United States Department of Agriculture, Agricultural Research Service, Entomology Research Division, 1963, 5. Kenya Agricultural and Livestock Research Organization Tsetse Archives [hereafter KALRO/TA].

6. David A. Dame, "Research on the Sterility Method of Tsetse Fly Control or Eradication: First Progress Report," 1963, 1. KALRO/TA.

7. Ibid.

8. Annual Report 1982, 18.

6 Exposing the Fly to Its Enemies

1. Annual Report, Chief Entomologist, 1928, 2–3. SACEMA/TA.

2. Ibid.

3. Jack, "The Tsetse Fly Problem," 14.

4. J. K. Chorley, "Tsetse Fly Operations, Short Survey 1943," 3. SACEMA/TA.

5. Ibid., 4.

6. Ibid., 2.

7. Ibid., 3.

8. E. Bursell, "A Prospect of Tsetse Flies: A Prospect of Tsetse Flies," 1968, 12. SACEMA/TA.

9. Chorley, "The Control," 6.

10. Ibid.

11. Jack, "The Tsetse Fly Problem," 7; "A Program for Tsetse and Trypanosomiasis Control in Southern Rhodesia," 1958, 6. SACEMA/TA.

12. Ibid.

13. Annual Report, Chief Entomologist, 1933, 5. SACEMA/TA.

14. Bursell, "A Prospect," 16.

15. Ibid.

16. Ibid.

17. Jack, "The Control," 18–19.

18. Annual Report, Chief Entomologist, 1935, 14. SACEMA/TA.

19. J. K. Chorley, "Tsetse Fly Operations 1940," 7. SACEMA/TA.

20. J. K. Chorley, "Tsetse Fly Operations 1941," 3. SACEMA/TA.

21. Ibid., 7.

22. J. K. Chorley, "Tsetse Fly Operations 1944," 7. SACEMA/TA.

23. Chorley, "Tsetse Fly Operations 1943," 3.

24. Annual Report 1952, 11.

25. Report 1956, 145.

26. Ibid., 137.

27. Ibid., 159.

28. Ibid., 14.

29. Gerald Cockbill, "The Control of Regenerating Brachystegia Spiciformis Benth By Means of Tordon 22K, February 1968," 1–2. SACEMA/TA.

30. Report 1959, 209.

31. Cockbill, "The Control," 1–2.

32. Ibid., 1–2.

33. Ibid.

34. Chorley, "Tsetse Fly Operations 1943," 8.

35. Report 1956, 137.

36. Ibid., 147.

37. Report 1958, 210–211.

38. Report 1959, 203.

39. Ibid., 205.

40. Report 1958, 211.

41. Ibid., 201.

42. Ibid.

43. Ibid.

44. Ibid., 205.

45. Annual Report: Director BTTC, 1962, 208. SACEMA/TA.

46. Ibid.

47. Interview, Mugocha Julius Mavasa.

48. Ibid., 201; "Trypanosomiasis Southern Rhodesia 1964: A Review," 5–6. SACEMA/TA.

49. Bursell, "A Prospect," 12.

50. "A Program," 6.

51. Bursell, "A Prospect," 13.

52. Ibid., 14.

53. Ibid., 14.

54. Ibid., 15.

55. Ibid.

56. Ibid., 16.

57. Panning: Chipinda Pools Clips; Interview, Bvekenya, Makuleke.

58. Panning: Chipinda Pools Clips.

59. Interview, Watson Machiukele.

60. Ibid.

7 *Cordon Sanitaire*

1. "Land Settlement and Tsetse Reclamation," 4–5.

2. Ibid., 3.

3. Chorley, "Destruction," 2.

4. Interview, Evelyn Musengi.

5. Interview, Ngonda.

6. Report 1932, 13–15.

7. Ibid., 12–13.

8. Ibid.

9. Chorley, "Tsetse Fly Operations," 8.

10. Ibid., 8–9.

11. Chorley, "Tsetse Fly Operations," 7.

12. Report, Chief Entomologist, 1934," 10. SACEMA/TA.

13. Report 1959, 226.

14. Jack, "Annual Report," 10.

15. Annual Report 1935, 9.

16. Chorley, "Tsetse Fly Operations 1941," 3.

17. Interview, Musengi.

18. Report 1957, 18.

19. Ibid., 17.

20. Ibid.

21. Ibid.

22. Ibid.

23. Ibid., 18.

24. Report 1958, 210.

25. Interview, Mavasa.

26. Interview, Solomon Bvekenya, Nthlaveni; Interview, Gazeni Komundela; Interview, Hanyani Chauke.

27. Report: Director Veterinary Services, 1963, 214. SACEMA/TA; Report 1965, 3.

28. Annual Report 1973, 7; "A Note on Tsetse and Trypanosomiasis Control in Rhodesia … 1975," 3. SACEMA/TA.

29. Jack, "The Tsetse Fly Problem," 7.

30. Ibid., 18.

31. Ibid., 8.

32. Chorley, "Tsetse Fly Operations 1940," 3–6.

33. Annual Report 1953, 3–5.

34. Annual Report 1952, 6.

35. R. R. Gladdon (Land Development Officer), "Urungwe Native District: Outline Scheme for the Resettlement of Natives in a Tsetse Fly Area, 1956 Onwards," 4–5. SACEMA/TA.

36. Ibid., 3.

37. Annual Report 1952, 2–6.

38. 4629 Statement, Provincial Native Commissioner Hurungwe, August 10, 1952, 1–2. SACEMA/TA.

39. Ibid., 2.

40. Ibid.

41. Ibid.

42. Chorley, "The Control of Tsetse," 2.

43. Interview, Mavenge.

44. Interview, Chabata.

45. Interview, Raymond Muzanenhamo.

46. Ibid.

47. Interview, Mavenge.

48. Ibid.

49. Ibid.

50. Ibid.

51. Interview, Musengi.

52. Interview, Ambuya Misi.

53. Interview, Ngonda.

54. Interview, Chabata; interview, Musengi; interview, Muzanenhamo.

55. Interview, Musengi.

56. Ibid.

57. Interview, Mavenge.

58. Interview, Ngonda.

59. Interview, Misi.

60. Interview, Mavenge.

61. Interview, Ambuya Misi.

62. Interview, Anonymous.

63. Interview, Ambuya Misi.

64. Ibid.

65. Interview, Musengi.

66. Interview, Ambuya Misi.

67. Interview, Ngonda.

68. Report 1964, 11; Report: ADVS (TTC), 1966, 24. SACEMA/TA; Annual Report: ADVS (TTC), 1967, 41. SACEMA/TA; Annual Report 1970, 2; Annual Report 1972, 10.

69. Ibid.

70. Interview, Musengi.

71. Ibid.

72. Interview, Ambuya Misi; interview, Mavenge.

73. Interview, Anonymous.

74. Interview, Ambuya Misi.

75. Interview, Ngonda.

76. Interview, Chabata.

77. Interview, Cleto Zharare.

8 Traffic Control

1. Report of the Director of Tsetse and Trypanosomiasis Control and Reclamation for the Year Ended September 30, 1960, 225. SACEMA/TA.

2. Annual Report 1928, 4.

3. Annual Report 1929, 6.

4. Annual Report, Chief Entomologist, 1936, 17. SACEMA/TA.

5. Annual Report 1930, 5–6; Report 1960, 232.

6. Annual Report 1936, 16; Chorley, "Tsetse Fly Operations 1941," 6–9. SACEMA/TA; J. K. Chorley, Report: Chief Entomologist, 1946," 12, 13. SACEMA/TA.

7. Report 1960, 234.

8. Annual Report 1952, 7.

9. Ibid.

10. John Ford, "The Advance of the *Glossina morsitans* and *Glossina pallidipes* into the Sabi and Runde River Basins, Southern Rhodesia" [n.d., likely 1958], 6. SACEMA/TA.

11. "Meeting," 3.

12. Ibid.

13. Jack, "Report," 8.

9 Starving the Fly

1. Interview, Mavasa, Chibwedziva.

2. Interview, Muzanenhamo.

3. Interview, Chabata.

4. Interview, Muzanenhamo.

5. Secretary, Salisbury to Mining Commissioner, Lomagundi, September 6, 1894. NAZ M 1/1/1; Resident magistrate, Umtali, to Public Prosecutor, Salisbury, October 23, 1894. NAZ J 1/1/1.

6. For a discussion in science studies on the term *lay* versus *expert* (the former of which I shall not use, because it is a bad word that ignores *kuziva*/knowing by experience), see Wynne 1996.

7. Annual Report, Government Entomologist, 1919. SACEMA/TA.

8. Annual Report, Chief Entomologist, 1920, 2. SACEMA/TA.

9. Annual Report 1921.

10. Annual Report, Chief Entomologist, 1922. SACEMA/TA.

11. Annual Report 1923, 2. SACEMA/TA.

12. Interview with Fema Ngonda.

13. Ibid.

14. Ibid.

15. Ibid.

16. Sabi Programme: South of Runde and Sabi East, 1959. SACEMA/TA.

17. Jack to Director of Agriculture, March 18, 1919. NAZ G1/3/2/15.

18. "Tsetse Fly Experiment. Wankie, Draft Plan of Operations," April 17, 1919. NAZ G/3/2/14.

19. A. H. Holland to D. Souter Robertson Esq., January 5, 1911. NAZ T2/1/10.

20. Interview, Muzanenhamo.

21. Interview, Bvekenya, Makuleke.

22. Whellan, "Tsetse Fly," 3.

23. Interview, Muzanenhamo.

24. Ibid.

25. Interview, Ngonda.

26. Ibid.

27. J. K. Chorley to Secretary, Agriculture and Lands, August 15, 1945: "Motor Mileage Allowance: Tsetse Fly Tsetse Field Officers." NAZ S3106/11/1/6.

28. B. Du Plessis to Provincial Native Commissioner Umtali, November 30, 1945: "Tsetse Fly: Chipinga District." NAZ S3106/11/1/6-6.

29. Senior Entomologist to Game Officer, May 27, 1953: "Chiredzi ICA Committee." NAZ S3106/11/1/7.

30. Chief Entomologist to W.N. Odendaal, February 12, 1946: "Distribution of Native Hunters." NAZ S3106/11/1/6.

31. G. Davison, "Review of Hunting Operations, June 1971–March 1972," May 4, 1972," 6. SACEMA/TA.

32. D. F. Lovemore, "Report by Mr. Davison," July 14, 1971, 1. SACEMA/TA.

33. Ibid.

34. Ibid.

35. Davison, "Review of Hunting Operations," 1. SACEMA/TA.

36. Interview, Ngonda.

37. Ibid.

38. Ibid.

39. Ibid.

40. Ibid.

41. Ibid.

42. Ibid.

43. Ibid.

44. Ibid.

45. Ibid.

46. Ibid.

47. Ibid.

48. Interview, Bvekenya, Makuleke.

49. Ibid.

50. Ibid.

51. Ibid.

52. Interview, Mavenge.

53. Interview, Ngonda.

54. Child and Riney, "An Examination," 15.

55. Ibid., 18.

56. Interview, Bvekenya, Makuleke.

57. Interview, Mavenge.

58. Interview, Anonymous.

59. Interview, Anonymous.

60. Ibid.

61. R. Heathcote to ADVS (TTC), January 26, 1972, 1. SACEMA/TA.

62. Davison, "Review," 2.

63. G. Davison, "Annual Review of Hunting Operations 1973," 4. SACEMA/TA.

64. Davison, "Review," 1.

65. Ibid.

66. Davison, "Annual Review," 9.

67. Ibid., 8.

68. Ibid., 5.

69. Ibid., 8.

70. J. K. Chorley to Secretary Agriculture, July 26, 1955. SACEMA/TA.

71. Child and Riney, "An Examination," 15–16.

72. Davison, "Review," 3.

73. Lovemore, "Report," 2.

74. Davison, "Review," 4–5.

75. Davison, "Annual Review," 6.

76. Ibid., 6.

77. Ibid., 4.

10 The Coming of the Organochlorine Pesticide

1. D. F. Lovemore, "The Role of Modern Insecticides in the Control of Tsetse Flies," April 5, 1951, 1–2. SACEMA/TA.

2. Ibid., 1.

3. Ibid., 3.

4. Ibid.

5. D. F. Lovemore, "The Use of DDT for the Control of Tsetse Flies in Rhodesia" [n.d., likely 1978–1979], 1. SACEMA/TA.

6. J. A. K. Farrell, "Insecticide Spraying Operations, Chiredzi River, September-November 1959," November 21, 1959, 5. SACEMA/TA; J. A. K. Farrell, "The Control of a Tsetse Fly (*Glossina*) Advance by Use of Residual Insecticide, Chiredzi River, Southern Rhodesia," September–November 1959, 3. SACEMA/TA.

7. B. Leggate, "Experiment to Text the Reaction of Blow-Flies in the Laboratory to Residual Films of Dieldrin," February 6, 1958, 1–2. SACEMA/TA.

8. Lovemore, "The Use of DDT," 3.

9. Annual Report 1970, 34–35.

10. Annual Report 1973, 9.

11. Lovemore, "A Note," 3.

12. Lovemore, "The Use of DDT," 7–8.

13. Annual Report 1982, 12–13.

14. Ibid.

15. Lovemore, "A Note on Tsetse," 3.

16. Lovemore, "The Use of DDT," 6.

17. Ibid.

11 Bombing Flies

1. Annual Report 1952, 5.

2. Annual Report 1953, 8.

3. Ibid., 9.

4. Ibid., 9.

5. Ibid.

6. Annual Report: Director Tsetse Fly Operations, 1954, 3. SACEMA/TA; Interview, Fema Ngonda.

7. Ibid., 10.

8. Annual Report: Director Tsetse Fly Operations, 1955, 13. SACEMA/TA.

9. Annual Report 1954, 10.

10. Annual Report 1955, 13.

11. Report 1956, 153.

12. Report 1956, 153.

13. Report 1957, 9.

14. Ibid.

15. Ibid., 10.

16. Ibid.

17. Ibid.

18. Ibid., 17.

19. Report 1956, 153.

20. Ibid.

21. Report 1957, 17.

22. R. Goodier, "The Tsetse Position in the Kariba Area Subsequent to the 1956 Aerial Spraying Campaign," 1–4. SACEMA/TA.

23. Ibid., 4.

24. Annual Report 1982, 4.

25. N. G. Chapman, "Aerial Spraying of Tsetse Flies (*Glossina* spp.) in Rhodesia with Ultra Low Volumes of Endosulfan," 1976, 3. SACEMA/TA.

26. Ibid., 3.

27. Ibid., 8, 13.

28. Lovemore, "A Note," 7.

29. Annual Report 1982, 5.

30. Ibid., 6.

31. Ibid., 4.

32. Ibid., 6.

33. Annual Report 1984, 5.

34. Ibid., 6.

35. Annual Report 1985, 8.

36. "A History of Tsetse and Trypanosomiasis in Zimbabwe's Mid-Zambezi Valley North of 17° S and between 29° and 32°E for the Period 1980–1997," 32. SACEMA/TA.

37. Ibid., 33.

38. Ibid., 27.

39. Ibid., 28.

12 The Work of Ground Spraying

1. Farrell, "Insecticide Spraying Operations," 2; Farrell, "The Control," 2.

2. J. A. Farrell, "A Preliminary Report on the Humani-Devuli Spraying Operation of August–September, 1960," 10. SACEMA/TA.

3. Ibid., Appendix B.

4. D. F. Lovemore, "A Report on the Schefenacker Machines Used in the Maseme Spraying Operations," October 3, 1960, 2. SACEMA/TA.

5. Ibid.

6. Ibid.

7. Ibid.

8. Ibid.

9. Ibid.

10. Ibid.

11. Report 1963, 205.

12. Report 1964, 7.

13. Interview, Bvekenya, Makuleke; interview, Mavasa, Chibwedziva; interview, Timothy Sumbani; Report 1965, 19; Annual Report 1968, 21; Annual Report 1969, 24; Report 1970, 20.

14. Annual Report 1930, 7.

15. Report, Division of Entomology, 1943, 2. SACEMA/TA.

16. Chorley, "Tsetse Fly Operations, 1944, 3.

17. Report, Division of Entomology 1945, 7. SACEMA/TA.

18. Annual Report 1953, 5.

19. Chorley, "Tsetse Fly Operations 1942," 1.

20. Chorley, "Tsetse Fly Operations, 1944," 3.

21. Report 1945, 7.

22. "A Review of the Tsetse Fly Situation in S. Rhodesia, 1948," 4. SACEMA/TA.

23. Report 1958, 202.

24. Interview, Sumbani; interview, Mavasa.

25. Report 1960, 227.

26. Annual Report 1969, 4; Annual Report 1970; Annual Reports 1971; Annual Reports 1972; Annual Report 1973.

27. Report 1962, 210.

28. Ibid., 214.

29. Interview, Mavasa; interview, Sumbani.

30. Interview, Bvekenya, Makuleke.

31. Report 1962, 210.

32. Report 1965, 19.

33. Interview, Mavasa.

34. Ibid.

35. Ibid.

36. Annual Report 1969, 24.

37. Annual Report 1968, 21.

38. Annual Report 1969, 24.

39. Ibid.

40. Annual Report 1970, 18–21.

41. Ibid.

42. Report 1972, 5.

43. Annual Report 1973, 7.

44. Interview, Pastor Staff Masungwini.

45. Interview, Piet Barnard.

46. T. J. Casewell, "The Control of Tsetse and Trypanosomiasis in the North Eastern Districts of Rhodesia," 1968, 3–4. SACEMA/TA.

47. Interview, Sumbani.

48. Farrell, "Insecticide Spraying Operations," 2.

49. Ibid., 5.

50. Casewell, "The Control of Tsetse," 5.

51. Farrell, "A Preliminary Report," 13–14.

52. Ibid.

53. Casewell, "The Control," 4.

54. Farrell, "Preliminary Report," 14.

55. Ibid.

56. Casewell, "The Control," 5.

57. Farrell, "Preliminary Report," 14.

58. Ibid.

59. Farrell, "Insecticide Spraying Operations," 5.

60. Ibid.; interview, Sumbani; interview, Mavasa; interview, Ngonda.

61. Farrell, "Preliminary Report," 13–14.

62. Ibid.

63. Ibid.

13 DDT, Pollution, and *Gomarara*

1. Lovemore, "The Role of Modern Insecticides," 1.

2. G. F. Cockbill to Secretary, Ministry of Agriculture, October 21, 1970, 1. SACEMA/TA.

3. Ibid.

4. Ibid., 2.

5. Ibid.

6. Ibid.

7. Lovemore, "The Use of DDT," 5–6.

8. Ibid., 8.

9. Ibid., 6–7.

10. Ibid., 4.

11. Cockbill, "Tsetse and Trypanosomiasis Report," 1.

12. Interview, Mavasa.

13. Farrell, "Insecticide Spraying Operations," 5.

14. Ibid.

15. Report 1960, 245.

16. Ibid.

17. Interview, Mavasa.

18. A *comet assay* is a sensitive and rapid technique for quantifying and analyzing DNA damage in individual cells. It is used often in cancer tests.

14 Chemoprophylactics

1. R. W. Bryson, "The Position To-day in Chemo-therapy and Chemo-prophylaxis of Animal Trypanosomiasis" [n.d., likely 1958], 2. SACEMA/TA.

2. Ibid.

3. R. Bryson, "The Place of Chemicals in the Control of Bovine Trypanosomiasis" [no date, but post 1955], 1. SACEMA/TA.

4. D. A. Lawrence, "Chemotherapy in Animal Trypanosomiasis in Southern Rhodesia," 1956, 1. SACEMA/TA.

5. Ibid.

6. Ibid.

7. Ibid., 7–8.

8. Ibid., 7–8.

9. Bryson, "The Position To-day," 2.

10. Ibid.

11. Ibid., 5.

12. Bryson, "The Place of Chemicals," 6.

13. Bryson, "Recent Advances," 5–6.

14. Ibid., 5.

15. Ibid.

16. Annual Report 1967, 30.

17. Report 1964, 19.

18. Report 1965, 18.

19. Ibid., 20.

20. Report 1966, 14.

21. Ibid.

22. Ibid., 15.

23. Annual Report 1967, 29–31.

24. Report 1964, 15–16.

25. Report 1965, 20.

26. M. A. Bolton, "Report to the Government of Southern Rhodesia on Investigations into the Immunological Response of Cattle and Laboratory Animals to Trypanosomiasis," 1965. KALRO/TA.

27. Annual Report 1967, 30.

28. Ibid., 31.

29. Ibid., 32.

30. Lawrence, "Chemotherapy," 2.

31. Bryson, "The Place of Chemicals," 6.

32. Lawrence, "Chemotherapy," 1, 4–5.

33. Lawrence, "Chemotherapy," 1, 4–5.

34. Bryson, "The Place of Chemicals," 3.

35. Report 1959, 196–210.

36. Bryson, "The Place of Chemicals," 3.

37. Report 1961, 13.

38. Report 1965, 13.

39. Report 1960, 242.

40. Report 1961.

41. Interview, Solomon Bvekenya and Vhumbululu Ndhlovu.

42. Report 1965, 13.

43. W. P. Boyt to Director of Veterinary Services, 1966, 2. SACEMA/TA.

44. Ibid.

45. Ibid., 3.

46. Annual Report 1967, 16.

47. Annual Report 1968, 16.

48. Ibid., 17.

49. Annual Report 1969, 19.

50. Report 1956, 147, 149.

51. Bryson, "The Place of Chemicals," 2.

52. Ibid., 3.

53. Ibid.

54. Bryson, "The Position To-day," 1.

55. Bryson, "The Place of Chemicals," 6.

56. Ibid., 6.

57. Annual Report 1973, 3.

58. Ibid., 2.

59. Ibid., 9.

60. "Trypanosomiasis: The Need to Modify the Approach to the Control of Nagana in the Face of Changed Political and Security Situations" [n.d., likely 1975–1976], 1. SACEMA/TA.

61. Ibid.

62. Ibid.

Unleashed: *Mhesvi* in a Time of War

1. "Trypanosomiasis: The Need to Modify the Approach to the Control of Nagana in the Face of Changed Political and Security Situations," 1972, 2. SACEMA/TA.

2. Ibid.

3. Ibid., 3.

4. Ibid.

5. Ibid.

6. Ibid.

7. Lovemore, "A Note on Tsetse," 2.

8. Interview, Mavasa.

9. Ibid.

10. AGR 9/6/78 P. G. Dix (District Commissioner, Gokwe), to the Senior Tsetse Field Officer, Gokwe, "Livestock in Trypanosomiasis Areas," June 7, 1978. SACEMA/TA.

11. G2/135 W. P. Boyt (Chief Veterinary Officer, Trypanosomiasis), to the Director of Veterinary Services, "Trypanosomiasis: Livestock Policy," June 23, 1978, 1. SACEMA/TA.

12. Ibid., 1–2.

13. "Cordon Sanitaire: Defoliation Efforts," July 26, 2014, https://fatfox9.wordpress.com/tag/special-forces/.

14. Ibid.

15. Ibid.

16. Ibid.

17. Ibid.

References

Interviews and Fieldwork Archives

The oral and visual archives for this monograph are excerpted from two projects I initiated in northern and southern Zimbabwe between 2001 and 2017. The Black Bvekenyas (hence BBP) project (2001–present) is a digital documentation of four generations of a hunting family, the descendants of the early twentieth-century ivory poacher, Cecil Bvekenya Barnard. The author had extensive initial interviews with one of Bvekenya's grandsons, Solomon Bvekenya, who then conducted the rest of the interviews. The videotapes are marked as BBP/MOV …, the "MOV …" denoting the specific tape's ID. The Zambezi Valley Interviews/Gokwe/Nembudziya (hence ZVI/G/N) project (2016–2017) was a short-term research that sought local perspectives of antitsetse operations for purposes of this book. The goal was to create a digital archive that would be deposited with the local communities for future use. Asa Mudzimu, who comes from Nembudziya, conducted the set of interviews in that area. The videos are marked ZVI/G/N, with a number denoting the specific tape. The author lived at Chiroti Secondary School, close to two former tsetse gates, Deve and Chiroti, just a few miles from Sanyati River bridge, west of which lies Gokwe. This entire area, from Magunje in the east, through Zvipani, into Gokwe to the west, was deep inside *mhesvi* territory until the 1990s.

The Black Bvekenyas Project (BBP), Chibwedziva Area, Southeastern Zimbabwe

Interview with Elias Chauke, Malipati Secondary School (MSS), Sengwe. August 7, 2001. BBP/MSS.

Interview with Gazeni Komundela, Chibwedziva. December 11, 2010. BBP/MOV0B4.

Interview with Hanyani Chauke, Bhaule. January 10, 2011. BBP/MOV0A6.

Interview with Mugocha Julius Mavasa, Chibwedziva. January 1, 2011. BBP/MOVOF3.

Interview with Pastor Staff Masungwini, Mashamba Village, Chibwedziva. December 6, 2010. BBP/MOV098.

Interview with Piet Barnard, Mapikule Village, Chibwedziva. December 6, 2010. BBP/MOV09F.

Interview with Solomon Bvekenya and Vhumbululu Ndhlovu, Malipati Business Center (MBC). July 11, 2001. BBP/MBC.

Interview with Solomon Bvekenya, Nthlaveni Block H Hill, Makuleke, South Africa. November 4, 2010. BBP/MOV04C.

Interview with Solomon Bvekenya, Makuleke, South Africa. November 1, 2010. BBP/MOV04A.

Interview with Timothy Sumbani, Chibwedziva. December 14, 2010. BBP/MOV0BF.

Interview with Watson Machiukele, Chibwedziva. December 6, 2016. BBP/MOV09C.

Panning: Chipinda Pools Clips. December 6, 2010. BBP/MOV0EA.

Panning: Chipinda Pools Clips. December 6, 2010. BBP/MOV0E9.

Zambezi Valley Interviews/Gokwe/Nembudziya (ZVI/G/N), Northwestern Zimbabwe

Interview with Willias Chabata, Nembudziya, Gokwe. May 18, 2016. ZVI/G/N/1.

Interview with Evelyn Musengi, Nembudziya, Gokwe. March 14, 2016. ZVI/G/N/2.

Interview with Reuben Mavenge, Nembudziya. March 30, 2016. ZVI/G/N/3.

Interview with Ambuya Misi, Nembudziya. March 16, 2016. ZVI/G/N/4.

Interview with Raymond Muzanenhamo, Nembudziya. May 18, 2016. ZVI/G/N/5.

Interview with Fema Ngonda, Nembudziya. March 15, 2016. ZVI/G/N/6.

Interview with man who requested anonymity, Nembudziya. March 14, 2016. ZVI/G/N/7.

Interview with Cleto Zharare, Nembudziya. March 25, 2016. ZVI/N/8.

Published Primary Sources

ACIL Tasman. 2009. *The Economic Impact of the Bushranger Project*. Melbourne: ACIL Tasman.

Aliu, Y. O., M. Mamman, and A. S. Peregrine. 1993. "Pharmacokinetics of Diminazene in Female Boran (*Bos indicus*) Cattle." *Journal of Veterinary Pharmacology and Therapeutics* 16 (3): 291–300.

Allen, J. R., W. A. Hargraves, M. T. Hsia, and F. S. Lin. 1979a. "Comparative Toxicology of Chlorinated Compounds on Mammalian Species." *Pharmacology & Therapeutics* 7 (3): 513–547.

Allen, R. H., M. Gottlieb, E. Clute, M. J. Pongsiri, J. Sherman, and G. I. Obrams. 1979b. "Breast Cancer and Pesticides in Hawaii: The Need for Further Study." *Environmental Health Perspectives* 105:679–683.

Allsopp, R. 1990. *A Practical Guide to Aerial Spraying for Tsetse Control: Aerial Spraying Research and Development Project Final Report: Volume 2*. Harare, Zimbabwe: EEC Delegation.

Allsopp, R. 1991. *Aerial Spraying Research and Development Project: Final Report 1: Technical Report and Accounts*. Chatham: UK Natural Resources Institute.

Anderson, C. J. 1856. *Lake Ngami, Or Explorations and Discoveries During Four Years' Wanderings in the Wilds of Southwestern Africa*. New York: Harper & Brothers Publishers.

Allsopp, R., and B. S. Hursey. 1986. *Integrated Chemical Control of Tsetse Flies (*Glossina *spp.) in Western Zimbabwe 1984–1985*. London: Tropical Development and Research Institute Report.

Ashton, E. H. 1937. "Notes on the Political and Judicial Organization of the Tawana." *Bantu Studies* 11:67–83.

Austen, E. E. 1903. *A Monograph of the Tsetse Flies*. London: British Museum of Natural History.

Austen, E. E. 1908. "The Dependence or Non-Dependence of Tsetse Flies upon Big Game." *Journal of the Society for the Preservation of the Wild Fauna of the Empire* 4:11–25.

Ayotte, P., S. Giroux, E. Dewailly, M. Hernández Avila, P. Farias, R. Danis, and C. Villanueva Díaz. 2001. "DDT Spraying for Malaria Control and Reproductive Function in Mexican Men." *Epidemiology* 12:366–367.

Bailey, N. M. 1968. "Oral Berenil in the Treatment and Prophylaxis of Human Trypanosomiasis." *Transactions of the Royal Society of Tropical Medicine and Hygiene* 62:122.

Baines, T. 1864. *Exploration in South-West Africa*. London: Longman.

Baines, T. 1877. *The Gold Regions of South Eastern Africa*. London: Edward Stanford.

Baldry, D. A. T. 1969. "A Note on the Parasitism of the Pupa of *Glossina palpalis* (R.-D.) by Anastatus Motsch. Sp. (Hym., Eupelmidae)." *Bulletin of Entomological Research* 59 (3): 529–531.

Baldry, D. A. T. 1971. "The Control of *Glossina pallidipes* in East Africa by the Aerial Application of Dieldrin Invert Emulsion Sprays." *OAU/STRC Publication* no. 105: 311–320.

Baldwin, W. F., and G. D. Chant. 1971. "Use of Nitrogen during Irradiation to Improve Competitiveness in Sterile Males of *Rhodnius prolixus*." In *Proceedings, Symposium: Sterility Principle for Insect Control or Eradication*, 469–475.Vienna: IAEA.

Barnes, J. A. 1951. *Marriage in a Changing Society: A Study in Structural Change among the Fort Jameson Ngoni*. London: Oxford University Press.

Barrass, R. 1960. "The Settling of Tsetse Flies *Glossina morsitans* Westwood (Diptera, Muscidae) on Cloth Screens." *Entomologia Experimentalis et Applicata* 3 (1): 59–67.

Baumhover, A. H., A. J. Graham, B. A. Bitter, D. E. Hopkins, W. D. New, F. H. Dudley, and R. C. Bushland. 1955. "Screw-worm Control through Release of Steriled Flies." *Journal of Economic Entomology* 48:462–466.

Bernhard, F. O., ed. and trans. 1971. *Karl Mauch, African Explorer*. Cape Town: Struik.

Bent, T. 1892. *The Ruined Cities of Mashonaland: Being a Record of Excavation and Exploration in 1891*. London: Longman's Green & Co.

Bergman, A., J. J. Heindel, T. Kasten, K. A. Kidd, S. Jobling, M. Neira, R. T. Zoeller, G. Becher, P. Bjerregaard, R. Bornman, I. Brandt, A. Kortenkamp, D. Muir, M. N. Drisse, R. Ochieng, N. E. Skakkebaek, A. S. Byléhn, T. Iguchi, J. Toppari, and T. J. Woodruff. 2013. "The Impact of Endocrine Disruption: A Consensus Statement on the State of the Science." *Environmental Health Perspectives* 121:A104–A106.

Biesheuvel, S. 1952. "The Study of African Ability Part I: The Intellectual Potentialities of Africans." *African Studies* 11 (2): 45–57.

Blair, D. M. 1939. "Human Trypanosomiasis in Southern Rhodesia, 1911–1938." *Transactions of the Royal Society of Tropical Medicine and Hygiene* 32 (6): 729–742.

Blair, W. B., and H. D. Catling. 1974. "Outbreaks of African Armyworm, *Spodoptera exempta* (Walker) (Lepideptera: Noctuidae), in Rhodesia, South Africa, Botswana and South-west Africa from February to April 1972." *Rhodesia Journal of Agricultural Research* 12:57–67.

Boryslawsky, M., A. C. Garrod, J. T. Pearson, and D. Woodhead. 1985. "Processes Involved in the Uptake and Elimination of Organic Micropollutants in Bivalve Molluscs." *Marine Environmental Research* 17: 310.

Bold, J. D. 1949. "Fanagalo Is Becoming the Lingua Franca of Southern Africa." *The Outspan*, May 13.

Bold, J. D. 1952. *Fanagalo Dictionary*. Cape Town: Central News Agency.

Bradshaw, B. F. 1877–1878. "The Tsetse Fly." *Transactions of the South African Philosophical Society* 1 (2): 51–55.

BTTC (Branch of Tsetse and Trypanosomiasis Control). 1968. "The Sengwa Research Project: A Study of Game-Tsetse Fly Relations." *Rhodesia Science News* 2 (11): 236.

Bulpin, T. V. 1954. *The Ivory Trail.* Cape Town: T. V. Bulpin.

Burnett, G. F. 1954. "The Effect of Poison Bait Cattle on Populations of *Glossina morsitans* and *G. swynnertoni.*" *Bulletin of Entomological Research* 45:411–421.

Burnett, G. F., J. Robinson, and J. G. Leroux. 1957. "Comparative Trials of DDT and Dieldrin for the Control of the Riverside Tsetse *Glossina palpalis fuscipes.*" *East Africa Agricultural Journal* 22 (3): 142–146.

Bursell, E. 1959. "Determination of the Age of Tsetse Puparia by Dissection." *Proceedings of the Royal Entomological Society* 34 (1–3): 23–24.

Bursell, E. 1967. "Tsetse Physiology in Relation to Tsetse Control." *Proceedings and Transactions of the Rhodesia Scientific Association* 52 (1): 33–35.

Burtt, E. 1946. "Incubation of Tsetse Pupae: Increased Transmission-rate of Trypanosoma rhodesiense in *Glossina morsitans.*" *Annals of Tropical Medicine and Parasitology* 40:18–28.

Bush, B., A. H. Bennett, and J. T. Snow. 1986. "Polychlorobiphenyl congeners, p,p'-DDE and Sperm Function in Humans." *Archives of Environmental Contamination and Toxicology* 15:33–41.

Buxton, P. A. 1955. "The Natural History of Tsetse Flies." *Memoirs of the London School of Hygiene and Tropical Medicine* 10. London: Lewis.

Cabral, A. 1974. *Revolution in Guinea.* London: Stage 1.

Callaway, H. 1870. *The Religious System of the Amazulu.* Natal: Springvale.

"Call to Decentralize Cancer Treatment Services." 2012. *Chronicle,* July 26.

Carpenter, G. D. H. 1912. "Progress Report on Investigations into the Bionomics of *Glossina palpalis,* July 27, 1910, to August 5, 1911." *Report of the Sleeping Sickness Commission of the Royal Society* 12:79–108.

Castilhos, J. A. A., N. Fenzl, S. M. Guillen, and F. S. Nascimento. 2000. "Organochlorine and Organophosphorus Pesticide Residues in the Atoya River Basin, Chinandega, Nicaragua." *Environmental Pollution* 110 (3): 523–533.

CDC (Centers for Disease Control and Prevention). 2016. "Biomonitoring Summary: 2,4,5-Trichlorophenoxyacetic Acid." CDC. Last updated December 23. https://www.cdc.gov/biomonitoring/2,4,5-TrichlorophenoxyaceticAcid_BiomonitoringSummary.html.

CEC (Commission for Environmental Cooperation). 2006. *North American Regional Action Plan (NARAP) on Lindane and Other Hexachlorocyclohexane (HCH) Isomers.* http://www3.cec.org/islandora/en/item/11602-north-american-regional-action-plan-narap-lindane-and-other-en.pdf

Césaire, A. [1955] 2000. *Discourse on Colonialism.* Translated by J. Pinkham. New York: Monthly Review Press.

Chadwick, P. R., J. S. S. Beesley, P. J. White, and H. T. Matechi. 1964. "An Experiment on the Eradication of *G. swynnertoni* by Insecticidal Treatment of its Resting Sites." *Bulletin of Entomological Research* 55:411–419.

Chandler, R. L. 1952. "Comparative Tolerance of West African N'Dama Cattle to Trypanosomiasis." *Annals of Tropical Medicine and Parasitology* 46:127–134.

Chandler, R. L. 1958. "Studies on the Tolerance of N'Dama Cattle to Trypanosomiasis. *Journal of Comparative Pathology* 68:253–260.

Chapman, J. 1868. *Travels in the Interior of South Africa.* 2 vols. London: Bell & Daldy.

Chapman, N. G. 1976. "Aerial Spraying of Tsetse Flies (*Glossina* spp.) in Rhodesia with Ultra Low Volumes of Endosulfan." *Transactions of the Rhodesia Scientific Association* 57:12–21.

Chikuhwa, J. 2006. *Zimbabwe: The Rise to Nationhood.* Bloomington, IA: AuthorHouse.

Chokunonga, E., M. Z. Borok, Z. M. Chirenje, A. M. Nyakabau, and R. Makunike-Mutasa. 2013. *Cancer Incidence in Harare: Triennal Report 2010–2012.* Harare: National Cancer Registry Zimbabwe.

Chorley, J. K. 1933. "Traps for Tsetse Flies of the 'Crinoline' and 'Ventilator' Forms." *Bulletin of Entomological Research* 24:315–317.

Chorley, J. K. 1938. "A Short Historical Account of Tsetse in Southern Rhodesia." *Proceedings of the Rhodesia Science Association* 36:41–51.

"The Civilizing Influence of Roads: Intercommunication Developments an Urge to Advancement." 1929–1930. *British South Africa Annual*: 144.

Clark, B. 1857. "The Tzetze of Africa Identified with Cestrus Bovis." *Zoologist* 15:5720–5721.

Cockbill, G. F. 1961. "Tests on the Relative Toxicity of Certain Commercial Insecticides to Tsetse Fly." Unpublished report.

Cockbill, G. F. 1967. "The History and Significance of Trypanosomiasis Problems in Rhodesia." *Proceedings and Transactions of the Rhodesia Scientific Association* 52 (1): 7–15.

Cockbill, G. F. 1968. "Recent Trends in Tsetse Control and Research." *The Rhodesia Science News* 2 (11): 232.

Cole, D. T. 1953. "Fanagalo and the Bantu Languages in South Africa." *African Studies* 12:1–9.

Curtis, C. F. 1968. "Radiation Sterilization and Effect of Multiple Mating of Female in *Glossina austeni*." *Journal of Insect Physiology* 14:1365–1380.

Curtis, C. F. 1969. "The Production of Partially Sterile Mutants in *Glossina austeni*." *Genetic Research* 13:289–301.

Curtis, C. F. 1970. "The Stimulation of Emergence from *Glossina* pupae by Gamma Rays." *Acta Tropica* 27:89–93.

Curtis, C. F. 1971. "Induced Sterility in Insects." *Advances in Reproductive Physiology* 5:120–165.

Curtis, C. F. 1972. "Sterility from Crosses between Sub-species of the Tsetse Fly *Glossina morsitans*." *Acta Tropica* 29:250–268.

Curtis, C. F., J. Curtis, and H. J. Hamann. 1973. "Experiments on Radio-Activation Labelling of the Tsetse Fly *Glossina morsitans*." *International Journal of Applied Radiation & Isotopes* 24:535–543.

Curtis, C. F., and A. M. Jordan. 1970. "Calculations of the Productivity of *Glossina austeni* Newst. Maintained on Goats and on Lop-eared Rabbits." *Bulletin of Entomological Research* 59:651–658.

Curtis, C. F., and P. A. Langley. 1972. "Use of Nitrogen and Chilling in the Production of Radiation-Induced Sterility in the Tsetse Fly *Glossina morsitans*." *Entologia Experimentalis et Applicata* 15:360–376.

Dame, D. A., and H. R. Ford. 1966. "Effect of the Chemosterilant Tepa on *Glossina morsitans* Westw." *Bulletin of Entomological Research* 56:649–658.

Dame, D. A., and H. R. Ford. 1968. "Multiple Mating of *Glossina morsitans* Westw. and Its Potential Effect on the Sterile Male Technique." *Bulletin of Entomological Research* 58:213–219.

Dame, D. A., and P. K. I. Mackenzie. 1968. "Transmission of *Trypanosoma congolense* by Chemosterilised male *Glossina morsitans*." *Annals of Tropical Medicine and Parasitology* 62:372–374.

Dame, D. A., and C. H. Schmidt. 1970. "The Sterile Male Technique against Tsetse Flies, *Glossina* spp." *Bulletin of the Entomological Society of America* 16:24–30.

Davies, H., and P. Blasdale. 1960. "The Eradication of *Glossina morsitans submorsitans* Newst. and *G. tachinoides* Westw. in Part of a River Flood Plain in Northern Nigeria by Chemical Means: Part III." *Bulletin of Entomological Research* 51:265–270.

Dean, G. J. W. 1969. "Observations on Sex Attraction and Mating Behaviour of the Tsetse Fly *Glossina morsitans morsitans* Vanderplank." *Bulletin of Entomological Research* 59:355–365.

Dean, G. J. W., and S. A. Clements. 1969. "Effect of Gamma Radiation on the *Glossina pallidipes* Aust." *Bulletin of Entomological Research* 58 (4): 755–762.

Dean, G. J. W., S. A. Clements, and J. Paget. 1969. "Observations on Some Possible Attractants of Tsetse Flies (*Glossina morsitans* Westw. and *G. pallidipes* Aust.)." *Bulletin of Entomological Research* 59 (3): 423–434.

Dean, G. J. W., D. A. Dame, and D. R. Birkenmeyer. 1969. "Field Cage Evaluation of the Competitiveness of Male *Glossina morsitans orientalis* Vanderplank Sterilized with Tepa or Gamma Iradiation." *Bulletin of Entomological Research* 59 (2): 339–344.

Dean, G. J. W., R. J. Phelps, and B. Williamson. 1968. "Sterilization with Gamma-rays and Field Investigations into the Breeding of *Glossina morsitans*." In *Control of Livestock Insect Pests by the Sterile-Male Technique*, 31–36. Rome: FAO/IAEA.

Dean, G. J. W., and S. M. Wortham. 1969. "Effect of Gamma Radiation on Tsetse Fly *Glossina morsitans* Westw." *Bulletin of Entomological Research* 58 (3): 505–519.

Deichmann, W. B., and W. E. MacDonald. 1977. "Organochlorine Pesticides and Liver Cancer Deaths in the United States, 1930–1972." *Ecotoxicology and Environmental Safety* 1 (1): 89–110.

Derbyshire, D. 2010. "Why Elephants Are NOT Afraid of Mice (But Are Terrified by Ants)." *Daily Mail*, September 2. http://www.dailymail.co.uk/sciencetech/article-1308415/Elephants-NOT-afraid-mice-terrified-ants.html#ixzz4A3IkpZiw.

de Zulueta, J. 1990. "Forty Years of Malaria Eradication in Sardinia: A New Appraisal of a Great Enterprise." *Parassitologia* 32 (2): 231–236.

Dicke, B. H. 1932. "The Tsetse-Fly's Influence on South African History." *South African Journal of Science* 29:792–796.

Doke, C. M. 1947. "Bantu Wisdom-Lore." *African Studies* 6:101–120.

Dolan, R. B. 1990. "Homidium Bromide as a Chemoprophylactic for Cattle Trypanosomiasis in Kenya." *Acta Tropica* 47:137–144.

Dowler, M. E., D. Schillinger, and R. J. Connor. 1989. "Notes on the Routine Intravenous Use of Isometamidium in the Control of Bovine Trypanosomiasis on the Kenya Coast." *Tropical Animal Health and Production* 21 (1): 4–10.

Driberg, J. H. 1928. "Primitive Law in East Africa." *Africa* 1:63–72.

Duerden, J. E. 1925. "Genetics and Eugenics in South Africa: Heredity and Environment." *South African Journal of Science* 22:59–72.

du Toit, R. 1954. "Trypanosomiasis in Zululand and the Control of Tsetse Flies by Chemical Means." *Onderstepoort Journal of Veterinary Research* 26:317–387.

du Toit, R. 1959. "The Eradication of the Tsetse Fly (*Glossina pallidipes*) from Zululand, Union of South Africa." In *Advances in Veterinary Science*, vol. 5, edited by C. A. Brandly, 227–240. Chicago: University of Chicago Press.

du Toit, T., and E. B. Kluge. 1949. "The Application of Insecticides by Aircraft for the Control of Tsetse Flies in South Africa." *British Science News* 2 (20): 246–250.

EATTRRO [East African Tsetse and Trypanosomiasis Research and Reclamation Organization]. 1953. *Notes on the Field Studies of Tsetse Flies in East Africa*. Nairobi: EATTTRO.

Ellioít, R., and R. A. Fitz-John. 1953. "Trials of a Portable Fog Generator for Mosquito Control in West Africa." *Monthly Bulletin of the Ministry of Health and the Public Health Laboratory Service* 12:178–186.

Encyclopaedia Britannica. 1888. *Encyclopaedia Britannica*, 9th ed., s.v. "Tsetse-fly (*Glossina morsitans*)."

Epstein, S. 1994. "Environmental and Occupational Pollutions Are Avoidable Causes of Breast Cancer." *International Journal of Health Services* 24: 145–150.

Eskenazi, B., J. Chevrier, L. G. Rosas, H. A. Anderson, M. S. Bornman, H. Bouwman, A. Chen, B. A. Cohn, C. de Jager, D. S. Henshel, F. Leipzig, J. S. Leipzig, E. C. Lorenz, S. M. Snedeker, and D. Stapleton. 2009. "The Pine River Statement: Human Health Consequences of DDT Use." *Environmental Health Perspectives* 117 (9): 1359–1367.

Espir, M. L., J. W. Hall, J. G. Shirreffs, and D. L. Stevens. 1973. "Impotence in Farm Workers Using Toxic Chemicals." *British Medical Journal* 1:423–425.

Essig, E. O. 1949. "Man's Six-Legged Competitors." *Scientific Monthly* 69 (1): 15–19.

Fairbairn, H., and H. J. Watson. 1955. "The Transmission of *Trypanosome vivax* by *Glossina palpalis*." *Annals of Tropical Medicine and Parasitology* 49:250–259.

Fanon, F. [1952] 1967. *Black Skin, White Masks*. Translated by C. L. Markmann. New York: Grove Press.

Fantham, H. B. 1925. "Some Factors in Eugenics." *South African Journal of Science* 22:400–424.

Fantham, H. B. 1927. "Some Thoughts on Biology and the Race." *South African Journal of Science* 24:1–20.

FAO [Food and Agriculture Organization]. 1977. *Insecticides and Application Equipment for Tsetse Control*. Rome: FAO. http://www.fao.org/docrep/004/x6502e/X6502E01.htm.

Fenske, R. A. 1997. "Pesticide Exposure Assessment of Workers and their Families." *Occupational Medicine* 60:221–237.

Ferrière, C. 1935. "Les Hymenopteres Parasites des Mouches Tsetse." *Mittelungen der Schweizerischen Gesellschaft* 16:328–340.

Field. 1907. October 5, 620.

"Fighting the Cancer Scourge: Together It Is Possible!" 2012. *Herald*, February 14.

Finaughty, W. 1916. *The Recollections of William Finaughty: Elephant Hunter, 1864–1875*. Philadelphia: J. B. Lippincott Company.

Fleschner, C. A. 1959. "Biological Control of Insect Pests." *Science* 129 (3348): 537–544.

Fleming, A. 1910. *Report on the Public Health, Southern Rhodesia, 1909*. Salisbury: Government Printer.

Fleming, A. 1913. "Trypanosomiasis in Southern Rhodesia." *Transactions of the Society of Tropical Medicine and Hygiene* 6:298–310.

Ford, J. 1971. *The Role of the Trypanosomiases in African Ecology: A Study of the Tsetse Fly Problem*. Oxford: Clarendon Press.

Ford, J., J. P. Glasgow, D. L. Johns, and J. R. Welch. 1959. "Transect Fly-rounds in Field Studies of *Glossina*." *Bulletin of Entomological Research* 50 (2): 275–285.

Frere, B. 1881. "On Temperate South Africa." *Journal of the Royal Geographical Society* 3 (1): 1–19.

Frost, S. W. 1937. "The Insect Motif in Art." *Scientific Monthly* 44 (1): 77–83.

Fuller, C. 1923. *Tsetse in the Transvaal and Surrounding Territories: An Historical Review*. Pretoria: Division of Entomology, Union of South Africa Government.

Gatehouse, A. G. 1972. "Host-finding Behaviour of Tsetse Flies." *Behavioural Aspects of Parasite Transmission* 51:83–95.

Gelfand, M. 1964. *Witch Doctor: Traditional Medicine Man of Rhodesia*. London: Harvill Press.

Gil, Y., and C. Sinfort. 2005. "Emission of Pesticides to the Air during Sprayer Application: A Bibliographic Review." *Atmospheric Environment* 39:5183–5193.

Gladen, B. C., D. P. Sandler, S. H. Zahm, F. Kamel, A. S. Rowland, and M. C. Alavanja. 1998. "Exposure Opportunities of Families of Farmer Pesticide Applicators." *American Journal of Industrial Medicine* 34:581–587.

Glasgow, J. P. 1960. "Ecological Effects of Tsetse Fly Control in Particular as a Consequence of Bush Clearing." *IUCN Symposium-Warszawa* 7:85–92.

Gomes, J., O. L. Lloyd, and D. M. Revitt. 1999. "Influence of Personnel Protection, Environmental Hygiene and Exposure to pesticides on the Health of Immigrant Farm Workers in a Desert Country." *International Archives of Occupational Environmental Health* 24:40–45.

Goodier, R., and J. B. Phipps. 1961. "A Revised Check List of the Vascular Plants of the Chimanimani Mts." *Kirkia* 1:44–66.

Gordon, J. A. 1973. "Clinical Features of Kaposi's sarcoma amongst Rhodesian Africans." *Central African Journal of Medicine* 19:1–6.

Gordon-Cumming, R. 1850. *Five Years of a Hunter's Life in the Far Interior of South Africa 2*. London: John Murray.

Government of Southern Rhodesia. 1924. *Handbook for the Use of Prospective Settlers on the Land*. London: Southern Rhodesia Department of Agriculture and Lands.

GOZ (Government of Zimbabwe). 2013. *The National Cancer Prevention and Control Strategy for Zimbabwe 2013–2017*. Harare: Government Printer.

Gray, E. 1939. "Some Riddles of the Nyanja People." *Bantu Studies* 8:251–291.

Gray, E. 1944. "Some Proverbs of the Nyanja People." *African Studies* 3:101–128.

Gunn, D. L., J. F. Graham, E. C. Jaques, F. C. Perry, W. G. Seymour, T. M. Telford, J. Ward, E. N. Wright, and D. Yeo. 1948a. "Aircraft Spraying against the Desert Locust in Kenya, 1945." *Anti-Locust Bulletin* 4:1–121.

Gunn, D. L., H. A. F. Lea, D. H. Botha, S. Callaway, J. R. Clackson, A. Immelman, J. J. Taljaard, and J. Ward. 1948b. *Locust Control by Aircraft in Tanganyika in 1947*. London and Pretoria: Anti-Locust Research Centre and Locust Control and Research Section, Department of Agriculture.

Hadaway, A. B., and F. Barlow. 1965. "Studies on the Deposition of Oil Drops." *Annals of Applied Biology* 55:267–274.

Hall, R. N. 1905. *Great Zimbabwe, Mashonaland, Rhodesia*. London: Methuen & Co.

Hamner, C. L., and H. B. Tukey. 1944. "Herbicidal Action of 2,4-Dichlorophenoxyacetic Acid and Trichlorophenoxyacetic Acid on Bindweed." *Science* 100:154–155.

Hamutyinei, M. A. 1992. *Chiparurangoma chaMudyanadzo*. Gweru: Mambo Press.

Harries, L. 1942a. "Makua Song-Riddles from the Initiation Rites." *African Studies* 1:27–46.

Harries, L. 1942b. "Some Riddles of the Makua People." *African Studies* 1:275–291.

Harries, L. 1947. "Some Riddles of the Mwera People." *African Studies* 6:21–34.

Harris, R. T. H. P. 1938. "The Control and Possible Extermination of the Tsetse Fly by Trapping." *Acta Conventus Tertii de Tropicis atque Malariae Morbis* 1:663–677.

Harris, W. C. 1839. *The Wild Sports of Southern Africa*. London: John Murray.

Heaversedge, R. C. 1968. "Variation in the Size of Insect Parasites of Puparia of *Glossina* spp." *Bulletin of Entomological Research* 58 (1): 153–159.

Hill, E. F., W. E. Dale, and J. W. Miles. 1971. "DDT Intoxication in Birds: Subchronic Effects and Brain Residues." *Toxicology and Applied Pharmacology* 20:502–514.

Hocking, K. S., C. W. Lee, J. S. S. Beesley, and H. T. Matechi. 1966. "Aircraft Applications of Insecticides in East Africa: XVI—Airspray Experiment with Endosulfan against *Glossina morsitans* Westw., *G. swynnertoni* Aust. and *G. pallidipes* Aust." *Bulletin of Entomological Research* 56 (4): 737–744.

Hoernlé, A. W. 1933. "New Aims and Methods in Social Anthropology." *South African Journal of Science* 30:74–92.

Holub, E. 1880. "Journey through Central Africa, from the Diamond Fields to the Zambesi." *Proceedings of the Royal Geographical Society and Monthly Record of Geography* 2 (3): 166–182.

Holub, E. 1881. "On the Central South African Tribes from the South Coast to the Zambesi." *Journal of the Anthropological Institute of Great Britain and Ireland* 10:2–20.

Homeida, A. M., E. A. El Amin, S. E. Adam, and M. M. Mahmoud. 1981. "Toxicity of Diminazene Aceturate (Berenil) to Camels." *Journal of Comparative Pathology* 91:355–360.

Hopkin-Jenkins, K. 1948. *Basic Bantu*. Pietermaritzburg: Shuter & Shooter.

Howard, L. O. 1919. "Entomology and the War." *Scientific Monthly* 8 (2): 109–117.

Huang, H. T., and Pei Yang. 1987. "The Ancient Cultured Citrus Ant." *Bioscience* 37 (9): 665–671.

Hudson Manufacturing Company. 1963. "Reach More Pest Control Markets with These Hudson Power Sprayers." *Weeds and Turf*, December.

Hursey, B. S., and R. Allsopp. 1984. *The Eradication of Tsetse Flies (*Glossina *spp.) from Western Zimbabwe by Integrated Aerial and Ground Spraying*. Harare, Zimbabwe: Tsetse and Trypanosomiasis Control Branch.

"Insects Chemical Warfare." 1962. *The Science News-Letter* 82 (21): 335.

Isherwood, F., B. J. Duffy, J. P. Glasgow, F. Lee-Jones, and B. Weitz. 1961. "Further Studies of the Food of Tsetse Flies." *Journal of Animal Ecology* 30 (2): 259–266.

Itard, J. 1968. "Stérilisation des mâles de *Glossina techinoides* Westw, par Irradiation aux Rayon Gamma." *Revue d'élevage et de Médecine Vétérinaire des pays Tropicaux* 21:479–491.

Itard, J. 1970. "Stérilisation des males de *Glossina orsitans* et *Glossina techinoides* Westw, par Irradiation aux Rayon Gamma." In *Proceedings of the Symposium on Tsetse Breeding under Laboratory Conditions and Its Practical Application*, 369–384. Lisbon.

Jack, R. W. 1927. "Some Environmental Factors Relating to the Distribution of *Glossina Morsitans* Westw. in Southern Rhodesia." *South African Journal of Science* 24:457–475.

Jack, R. W. 1930. "Tsetse Fly: Traffic Control." *Rhodesia Agricultural Journal* 5:493–501.

Jack, R. W. 1933. "Experiments with Tsetse Fly Traps against Glossina Morsitans in Southern Rhodesia." *Rhodesia Agricultural Journal* 30 (5): 393–398.

Jack, R. W. 1934. "Tsetse Fly and Game." *Rhodesia Agricultural Journal* 31:259–294.

Jack, R. W. 1939. *Studies in the Physiology and Behaviour of Glossina morsitans Westw.: Southern Rhodesia Memoirs of the Department of Agriculture 1*. Salisbury: Government Printer.

Jack, R. W. 1941. "Notes on the Behaviour of *Glossina pallidipes* and *G. brevipalpis*, and Some Comparisons with *G. morsitans*." *Bulletin of Entomological Research* 31 (4): 407–430.

Jackson, C. H. N. 1949. "The Biology of Tsetse Flies." *Biological Reviews of the Cambridge Philosophical Society* 24:174–199.

"Jane Ndlovu: Unsung Hero of Struggle." 2015. *Sunday Mail*, April 18.

Jemal, A., F. Bray, D. Forman, M. O'Brien, J. Ferlay, M. Center, and D. M. Parkin. 2012. "Cancer Burden in Africa and Opportunities for Prevention." *Cancer* 118 (18): 4372–4384.

Johnstone, D. R. 1985. "Monitoring the Coarse Aerosols Used for Non-residual Sprays in Tsetse Control." *OAU/STRC Publication* 113:384–392.

Johnstone, D. R., R. Allsopp, J. F. Cooper, and H. M. Dobson. 1988. "Predicted and Observed Droplet Deposition on Tsetse Flies (*Glossina morsitans*) Following Aerosol Application from Aircraft." *Pesticide Science* 22:107–121.

Johnstone, D. R., J. F. Cooper, and H. M. Dobson. 1987. "The Availability and Fall-Out of an Insecticidal Aerosol Dispersed from Aircraft during Operations for Control of Tsetse Fly in Zimbabwe." *Atmospheric Environment* 21:2311–2321.

JSPWFE. 1907. *Journal of the Society for the Preservation of the Wild Fauna of the Empire* 3:24–26, 43–44.

Junod, H. A. 1918. "Native Customs in Relation to Smallpox among the Ba-Ronga." *South African Journal of Science* 16:694–702.

Junod, H. A. 1920. "The Magic Conception of Nature amongst Bantus." *South African Journal of Science* 17:76–85.

Kachala, R. 2010. "Systematic Review: Epidemiology of Oesophageal Cancer in Sub-Saharan Africa." *Malawi Medical Journal* 22 (3): 65–70.

Kadzatsa, W., and E. Chokunonga. 2016. "The Status and Challenges of Cancer Care in Zimbabwe." *ASCO Daily News*, May 26.

Kamel, F., and J. A. Hoppin. 2004. "Association of Pesticide Exposure with Neurologic Dysfunction and Disease." *Environmental Health Perspectives* 112 (9): 950–958.

Kannan, K., S. Tanabe, J. P. Giesy, and R. Tatsukawa. 1997. "Organochlorine Pesticides and Polychlorinated Bipheny F ls in Foodstuffs from Asian and Oceanic Countries." *Review of Environmental Contamination and Toxicology* 152:1–55.

Kendrick, J. A., and N. Allsop. 1974. "Aerial Spraying with Endosulfan against *Glossina morsitans morsitans* in the Okavango Delta Area of Botswana." *PANS* 20:392–399.

Kerr, W. M. 1886a. *A Journey from Cape Town Overland to Lake Nyassa*. London: Royal Geographical Society.

Kerr, W. M. 1886b. *A Journey from Cape Town Overland to Lake Nyassa*. London: Royal Geographical Society.

Kinabo, L. D. B., and J. A. Bogan. 1988. "Solid-Phase Extraction and Ion-Pair Reverse-Phase HPLC of Isometamedium in Bovine Serum and Tissues." *Acta Tropica* 45:165–170.

Kirk, J. 1865. "On the 'Tsetse' Fly of Tropical Africa." *Journal of the Linnean Society* 8:149–156.

Kirkby, W. W., and P. Blasdale. 1960. "The Eradication of *Glossina morsitans submorsitans* Newst. and *Glossina tachinoides* Westw. in Part of a River Flood Plain in Northern Nigeria by Chemical Means." *Bulletin of Entomological Research* 51 (2): 253–264.

Knipling, E. F. 1963. "Potential Role of the Sterility Principle for Tsetse Fly Eradication." *WHO/Vector Control* 27:1–17.

Koeman, J. H., W. M. J. D. Boer, A. F. Feith, H. H. de Iongh, P. C. Spliethoff, B. K. Na'isa, and U. Spielberger. 1978. "Three Years' Observation on Side Effects of Helicopter Applications of Insecticides Used to Exterminate Glossina Species in Nigeria." *Environmental Pollution* 15 (1): 31–59.

Kumar, M., and A. Kumar. 2007. "Application and Health Effects of Pesticides Commonly Used in India." ECO Services International, May. http://www.eco-web.com/edi/070526.html.

LaChance, L. E., C. H. Schmidt, and R. C. Bushland. 1967. "Radiation-Induced Sterilization." In *Pest Control: Biological, Physical and Selected Chemical Methods*, edited by W. W. Kilgore and R. L. Doutt, 147–196. New York: Academic Press.

Langley, P. A. 1970. "Post General Development of Thoracic Flight Vasculature in Tsetse Flies, *Glossina austeni* and *Glossina morsitans*." *Entomologia Experimentalis et Applicata* 13:133–140.

Langley, P. A. 1972. "The Role of Physical and Chemical Stimuli in the Development of *in vitro* Feeding Techniques for Tsetse Flies *Glossina* spp. (Dipt., Glossinidae)." *Bulletin of Entomological Research* 62 (2): 215–228.

Langley, P. A., and H. Maly. 1971. "Control of Mediterranean Fruit Fly (Ceratitus capitate [Wied] Using Sterile Males: Effects of Nitrogen and Chilling during Gamma Irradiation of Puparia." *Entomologia Experimentalis et Applicata* 14:137–146.

Laws, E. R. Jr., A. Curley, and F. J. Biros. 1967. "Men with Intensive Occupational Exposure to DDT." *Archives of Environmental Health* 15:766–775.

Lee, C. W. 1969. "Aerial Applications of Insecticides for Tsetse Fly Control in East Africa." *Bulletin of the World Health Organization* 41:261–268.

Lee, C. W. 1977. "New Developments in Tsetse Fly Control Using Aircraft." *Agricultural Aviation* 18 (3): 6–17.

Lee, C. W., and A. W. D. Miller. 1966. "Trials with Devices for Atomising Insecticides by Exhaust Gases from an Aircraft Engine." *Agricultural Aviation* 8:17–22.

Lee, C. W., P. O. Park, and J. A. Gledhill. 1967. "A Trial to Control Tsetse Flies by Aerial Application of Aerosols in Barotseland, Zambia." Tropical Pesticide Research Institute, miscellaneous report no. 635.

Lee, C. W., G. G. Pope, and G. Bowles. 1977. "Aerosol Studies Using a Cessna 310 Aircraft Fitted with Micronair Equipment for Tsetse Fly Control in Botswana, 1975." Centre for Overseas Pest Research, miscellaneous report no. 29.

Lee, C. W., G. G. Pope, J. A. Kendrick, G. Bowles, and H. G. Wiggett. 1975. "Aerosol Studies Using an Aztec Aircraft Fitted with Micronair Equipment for Tsetse Control in Botswana." Centre for Overseas Pest Research, miscellaneous report no. 18.

Lee, E. W., J. T. Johnson, and C. D. Garner. 1989. "Inhibitory Effects of Benzene Metabolites on Nuclear DNA Synthesis in Bone Marrow Cells." *Journal of Toxicology & Environmental Health* 26:277–291.

Lener, W. 1963. "Go to the Ant." *American Biology Teacher* 25 (1): 25–28.

Le Roux, J. G., and D. C. Platt. 1968. "Application of a Dieldrin Invert Emulsion by Helicopter for Tsetse Control." *OAU/ISTRC Publication* 102:219–229.

Lévy-Bruhl, L. 1923. *Primitive Mentality*. London: Allen & Unwin.

Lewillon, R. 1945. "Le piégeage expérimental de Glossiizapalpalis à la mission médicale du Kwango." *Rec. Trav. Sci. nzéd. Congo belge* 4:45–57.

Livingstone, C. 1861–1862. "On the Batoka Country." *Proceedings of the Royal Geographical Society of London* 6 (1): 32–36.

Livingstone, D. 1851. "Second Visit to the South African Lake, Ngami: Extract of a Letter from the Rev. Dr. Livingston, under Date Kolobeng, 24th August, 1850." *Journal of the Royal Geographical Society of London* 21:18–24.

Livingstone, D. 1854. "Explorations into the Interior of Africa." *Journal of the Royal Geographical Society of London* 24:291–302.

Livingstone, D. [1857] 2001. *Missionary Travels and Researches in South Africa.* London: John Murray.

Livingstone, D., and C. Livingstone. 1865. *Narrative of an Expedition to the Zambesi and Its Tributaries.* London: John Murray.

Lloyd, B. G. 1950. *Kitchen-Kafir Grammar and Vocabulary.* Cape Town: Central News Agency.

Lloyd, H. M. 1935. "Notes on the Bionomics of *Glossina swynnertoni* Austen." *Bulletin of Entomological Research* 26 (2): 439–468.

Lloyd, L. 1912. "A New Nematode Parasite of *Glossina morsitans*." *Journal of the London School of Tropical Medicine* 2:41.

Lukwa, N. 1994. "Do Traditional Mosquito Repellents Work as Mosquito Larvicides?" *Central African Journal of Medicine* 40 (11): 306–309.

Manirakiza, P., A. Covaci, L. Nizigiymana, G. Ntakimazi, and P. Schepens. 2002. "Persistent Chlorinated Pesticides and Polychlorinated Biphenyls in Selected Fish Species from Lake Tanganyika, Burundi, Africa." *Environmental Pollution* 117 (3): 447–455.

Marshall, G. A. K. 1938. "Swynnerton's Obituary Notice in Nature." *Nature* 142:198–199.

Martin, S., S. D. Harlow, M. F. Sowers, M. P. Longnecker, D. Garabrant, D. L Shore, and D. P. Sandler. 2002. "DDT Metabolite and Androgens in African-American Farmers." *Epidemiology* 13:454–458.

Matibe, P. 2009. *Madhinga Bucket Boy: From Boyhood in Colonial Rhodesia to Manhood in Zimbabwe, A Native Son's Story.* Dallas: Mbedzi Publishing.

Maund, E. A. 1890. "Zambezia, the New British Possession in Central South Africa." *Proceedings of the Royal Geographical Society* 12 (1): 649–655.

Maund, E. A. 1891. "On Matabele and Mashona Lands." *Proceedings of the Royal Geographical Society* 13 (1): 1–21.

Maxie, M. G., and G. J. Losos. 1977. "Release of Trypanosoma Congolense from the Micro-circulation of Cattle by Berenil." *Veterinary Parasitology* 3:277–281.

McLaughlin, D. 1998. *"Fooling with Nature: Silent Spring Revisited." Frontline,* June 2.

Menzie, C. M. 1972. "Fate of Pesticides in the Environment." *Annual Review of Entomology* 17: 199–222.

Mhombera, W. 2012. "Hail Our Freedom Fighters!" *Zimbabwe Broadcasting Corporation.* http://www.zbc.co.zw archives.

Millais, J. G. 1895. *A Breath from the Veldt.* London: Henry Sotheran.

Mitchell, B. L. 1963. "A New Aspect of the Effect of Bush Fires in Connection with the Tsetse *Glossina morsitans.*" *Kirkia* 3:26–28.

Mitchell, B. L., and B. Steele. 1956. *A Report on the Distribution of Tsetse Flies in Nyasaland.* Zomba: Government Printer.

Moffat, R. 1846. *Missionary Labours and Scenes of Southern Africa.* London: John Snow.

Moffat, R. 1856. "Visit to Moselekatse, King of the Matebele." *Journal of the Royal Geographical Society of London* 26:84–109.

Mohr, E. 1876. *To the Victoria Falls of the Zambesi.* Translated by N. D'Anvers. London: Sampson Low.

Moloo, S. K. 1972. "Mermithid Parasite of *Glossina brevipalpis* Newstead." *Annals of Tropical Medicine and Parasitology* 66 (1): 159.

Moloo, S. K. 1973. "A New Trap for *Glossina pallidipes* Aust. and *G. fuscipes* Newst. (Dipt., Glossinidae)." *Bulletin of Entomological Research* 63 (2): 231–236.

Moloo, S. K., and S. B. Kutuza. 1990. "Expression of Resistance to Isometamidium and Diminazene in *Trypanosoma congolense* in Boran Cattle Infected by *Glossina morsitans centralis.*" *Acta Tropica* 47:79–89.

Monclaro, F. [1571] 1975. "Narrative (Copy) by Father Francisco de Monclaro of the Society of Jesus, of the Expedition to Monomotapa, Led by Franscisco Barreto." In *Documents on the Portuguese in Mozambique and Central Africa 8,* edited by National Archives of Zimbabwe, 325–429. Salisbury: National Archives of Rhodesia.

"More Motor-Cars in Southern Rhodesia: Preponderance of American Models." 1938. *The African World* (April 18): 21.

Morris, K. R. S. 1960. "Trapping as a Means of Studying the Game Tsetse, *Glossina pallidipes* Aust." *Bulletin of Entomological Research* 51 (3): 533–557.

Morris, K. R. S. 1961. "Effectiveness of Traps in Tsetse Surveys in the Liberian Rain Forest." *American Journal of Tropical Medicine and Hygiene* 10 (6): 905–913.

Morris, K. R. S., and M. G. Morris. 1949. "The Use of Traps against Tsetse in West Africa." *Bulletin of Entomological Research* 39 (4): 491–523.

Morris, M. G. 1950. "The Persistence of Toxicity in DDT-Impregnated Hessian and Its Use on Tsetse Traps." *Bulletin of Entomological Research* 41 (2): 259–288.

Morrow, M. G. 1952. "Plants That Kill Insects." *Science News Letter* 61 (21): 330–332.

Moses, M., E. S. Johnson, W. K. Anger, V. W. Burse, S. W. Horstman, R. J. Jackson, R. G. Lewis, K. T. Maddy, R. McConnell, and W. J. Meggs. 1993. "Environmental Equity and Pesticide Exposure." *Toxicology and Industrial Health* 9:913–959.

Mugabe, R. G. 1978. "The Patriotic Front is a Creation of Zimbabwe: Interview with Robert Mugabe, ZANU." In *Zimbabwe: The Final Advance, Documents on the Zimbabwe Liberation Movement,* edited by the Liberation Support Movement (LSM), 28–31. Oakland, CA: LSM Press.

Munjanja, A. 1972. *Zvirahwe.* Gweru: Mambo Press.

Mwevura, H., O. A. Amir, M. Kishimba, P. Berggren, and H. Kylin. 2010. "Organohalogen Compounds in Blubber of Indo-pacific Bottlenose Dlphin (Tursiops aduncus) and Spinner Dolphin (Stenella longirostris) from Zanzibar, Tanzania." *Environmental Pollution* 158 (6): 2200–2207.

Nash, R. G., and E. A. Woolson. 1967. "Persistence of Chlorinated Hydrocarbon Insecticides in Soils." *Science* 157:924–927.

Nash, T. A. M. 1930. "A Contribution to Our Knowledge of the Bionomics of *Glossina morsitans.*" *Bulletin of Entomological Research* 21 (2): 201–256.

"The New Pioneers: A Tribute to the South Africans in General Motors South African Limited." 1929–1930. *British South Africa Annual*: 102.

Nobbs, E. A. 1927. "The Native Cattle of Southern Rhodesia." *South African Journal of Science* 24:328–342.

NPIC (National Pesticide Information Center). 2015. "2,4-D Technical Fact Sheet." Corvallis: NPIC.

Oba, G. 2009. "Harnessing Pastoralists' Indigenous Range Management Knowledge for Drought-Resilient Livelihood Systems in the Horn of Africa." WISP, IUCN, and FAO, September. http://cmsdata.iucn.org/downloads/harnessing_pastoralists_indigenous_knowledge_review_2009_1.pdf.

Oswell, W. C. 1894. *William Cotton Oswell, Hunter and Explorer.* London: William Heinemann.

Pan, I., J. L. Daniels, B. D. Goldman, A. H. Herring, A. M. Siega-Riz, and W. J. Rogan. 2009. "Lactational Exposure to Polychlorinated Biphenyls, Dichlorodiphenyltrichloroethane, and Dichlorodiphenyldichloroethylene and Infant Neurodevelopment: An Analysis of the Pregnancy, Infection, and Nutrition Babies Study." *Environmental Health Perspectives* 117 (3): 488–494.

Park, P. O., J. A. Gledhill, N. Alsop, and C. W. Lee. 1972. "A Large-scale Scheme for the Eradication of *Glossina morsitans* Newst. in the Northern Guinea Vegetation Zone in Relation to Control Using Insecticides." *Entomologist* 105:143–152.

Phelps, R. J. 1964. "Parasites of Tsetse Puparia." *Report of the Agricultural Research Council of Central Africa*: 41–42.

Phelps, R. J. 1966. "The Sterile-Male Technique in Relation to Tsetse Control." *Proceedings and Transactions of the Rhodesia Scientific Association* 52 (1): 29–32.

Phelps, R. J. 1968. "A Falling Cage for Sampling Tsetse Flies (*Glossina* Diptera)." *Rhodesia Journal of Agricultural Research* 6:47–53.

Phillips, J. 1929. "The Application of Ecological Research Methods to the Tsetse (*Glossina* spp.) Problem in Tanganyika Territory: A Preliminary Account." *South African Journal of Science* 26:438.

Pilson, R. D., and P. K. I. MacKenzie. 1972. "A Preliminary Report on the Relative Attractiveness of Certain Domestic and Wild Animals to Tsetse Flies." *Rhodesia Science News* 6 (1): 21–24.

Pilson, R. D., and B. M. Pilson. 1967. "Behaviour Studies of *Glossina morsitans* Westw. in the Field." *Bulletin of Entomological Research* 57 (2): 227–257.

Pinto, A. A. 1879. "Major Serpa Pinto's Journey across Africa." *Proceedings of the Royal Geographical Society and Monthly Record of Geography* 1 (8): 481–489.

Posselt, F. W. T. 1935. *Fact and Fiction: A Short Account of the Natives of Southern Rhodesia*. Bulawayo: Government of Southern Rhodesia.

Potts, W. H. 1930. "A Contribution to the Study of Numbers of Tsetse Fly (*Glossina morsitans* Westw.) by Quarantine Methods." *South African Journal of Science* 27:491–497.

Potts, W. H. 1933. "Observations on *Glossina morsitans* Westw. in East Africa." *Bulletin of Entomological Research* 24 (2): 293–300.

Potts, W. H. 1958. "Sterilisation of Tsetse-Flies (*Glossina*) by Gamma Radiation." *Annals of Tropical Medicine and Parasitology* 52:484–499.

Ramesh, A., S. Tanabe, H. Iwata, R. Tatsukawa, A. N. Subramanian, D. Mohan, and V. K. Venugopalan. 1990. "Seasonal Variation of Persistent Organochlorine Insecticide Residues in Velar River Waters in Tamil Nadu, South India." *Environmental Pollution* 67 (4): 289–304.

Rendle, A. B., E. G. Baker, S. Moore, and A. Gepp. 1911. "A Contribution to Our Knowledge of the Flora of Gazaland: Being an Account of the Collections Made by C. F. M. Swynnerton." *Journal of the Linnaean Society, Botany* 40:1–245.

"Report of the Conference of the Parties of the Stockholm Convention on Persistent Organic Pollutants on the Work of Its Fourth Meeting." 2009. Paper presented at the fourth meeting of the Convention on Persistent Organic Pollutants, Geneva, Switzerland, May 4–8.

"Report of the Expert Group on the Assessment of the Production and Use of DDT and Its Alternatives for Disease Vector Control." 2010. Paper presented at the third Expert Group Meeting, Stockholm, Sweden.

Rheinallt-Jones, J. D. 1926. "The Need of a Scientific Basis for South African Native Policy." *South African Journal of Science* 23:79–91.

Rivero-Rodriguez, L., V. H. Borja-Aburto, C. Santos-Burgoa, S. Waliszewskiy, C. Rios, and V. Cruz. 1997. "Exposure Assessment for Workers Applying DDT to Control Malaria in Veracruz, Mexico." *Environmental Health Perspectives* 105 (1): 98–101.

Robertson, A. G., and E. B. Kluge. 1968. "The Use of Insecticide in Arresting an Advance of *Glossina morsitans* Westwood in the South-east Lowveld of Rhodesia." *Proceedings and Transactions of the Rhodesia Scientific Association* 53 (1): 17–33.

Rodhain, J., C. Pons, F. van den Branden, and J. Bequaert. 1913. *Rapport sur les travaux de la mission scientifique du Katanga*. Bruxelles: Hayez.

Rogers, D. J., and D. T. Smith. 1977. "A New Electric Trap for Tsetse Flies." *Bulletin of Entomological Research* 67 (1): 153–159.

Rowan, D. L., and J. B. Rasmussen. 1992. "Why Don't Great Lakes Fish Reflect Environmental Concentrations of Organic Contaminants?—An Analysis of Between-Lake Variability in the Ecological Partition of PCBs and DDT." *Journal of Great Lakes Research* 18:724–741.

Salazar-Garcia, F., E. Gallardo-Díaz, P. Cerón-Mireles, D. Loomis, and V. H. Borja-Aburto. 2004. "Reproductive Effects of Occupational DDT Exposure among Male Malaria Control Workers." *Environmental Health Perspectives* 112 (5): 542–547.

Selous, F. C. 1881. *A Hunter's Wanderings in Africa*. London: Richard Bentley.

Selous, F. C. 1893. "Twenty Years in Zambesia." *Geographical Journal* 1 (4): 289–322.

Selous, F. C. 1907. [Untitled note]. *Field*, November 9, 835.

Selous, F. C. 1908. *African Nature Notes and Reminiscences*. London: Macmillan.

Semenza, J. C., P. E. Tolbert, C. H. Rubin, L. J. Guillette Jr., and R. J. Jackson. 1997. "Reproductive Toxins and Alligator Abnormalities at Lake Apopka, Florida." *Environmental Health Perspectives* 105 (10): 1030–1032.

Simpson, H. R. 1958. "The Effect of Sterilised Males on a Natural Tsetse Fly Population." *Biometrics* 14:159–173.

Smith, D., D. O'Connell, I. Dey, J. Ashton, M. Goodman, T. Lyons, and W. Kay. 2008. "Falling into the Abyss." *The Times* (London), July 6.

Southwood, T. R. E. 1977. "Entomology and Mankind: Insects over the Ages Have Greatly Affected Man's Health and Food Supply and Have Played an Important Role as Religious and Cultural Symbols." *American Scientist* 65 (1): 30–39.

Spielberger, U., and U. Abdurrahim. 1971. "Pilot Trial of Discriminative Aerial Application of Persistent Alienation Deposits to Eradicate *Glossina morsitans submorsitans* in the Anchau and Ikara Forest Reserves, Nigeria." OAU/ISTRC publication no. 105: 271–281.

SRG (Southern Rhodesia Government). 1955. *What the Native Land Husbandry Act Means to the Rural African and to Southern Rhodesia*. Salisbury: Government Printer.

Stafford, J. 1944. "Insect War May Backfire." *Science News Letter* 46 (6): 90–92.

Stephen, L. E. 1958. "Suramin Complexes: IV: Ethidium Bromide Complex: A Large-scale Laboratory Trial of Its Prophylactic Activity in Cattle." *Annals of Tropical Medicine and Parasitology* 52:417–426.

Stephen, L. E., and J. Williamson. 1958. "Suramin Complexes: V: Ethidium Complex: Attempts to Overcome the Injection-site Reaction in Cattle." *Annals of Tropical Medicine and Parasitology* 52:427–442.

Stephens, J. W., and H. B. Fantham. 1910. "On the Peculiar Morphology of a Trypanosome from a Case of Sleeping Sickness and the Possibility of Its Being a New Species (*T. rhodesiense*)." *Annals of Tropical Medicine and Parasitology* 4:343–350.

"Summer and Rain in Rhodesia: Native Providence, Arms and Murderers Coming In." 1897. *African Review*, February 27.

Sutherland, I. A., S. K. Moloo, P. H. Holmes, and A. S. Peregrine. 1991. "Therapeutic and Prophylactic Activity of Isometamidium Chloride against a Tsetse-Transmitted Drug-Resistant Clone of *Trypanosoma congolense* in Boran Cattle." *Acta Tropica* 49 (1): 57–64.

Swynnerton, C. F. M. 1908a. "Further Notes on the Birds of Gazaland." *Ibis* 9 (2): 1–107.

Swynnerton, C. F. M. 1908b. "A List of Mammals Collected by Mr. C. F. M. Swynnerton in Northern Gazaland (Portuguese East Africa) and the Melsetter District of Rhodesia." *Annals & Magazine of Natural History* 1 (4): 303–307.

Swynnerton, C. F. M. 1921a. "An Examination of the Tsetse Problem in North Mossurize." *Bulletin of Entomological Research* 11 (4): 315–385.

Swynnerton, C. F. M. [1921b] 1960. "Game Circular No. 734/96 to Supplement Secretariat Circular No. 74 of 1921: Tsetse Control." *East African Agricultural and Forestry Journal*, July, 31–34.

Swynnerton, C. F. M. 1933. "Some Traps for Tsetse Flies." *Bulletin of Entomological Research* 24 (1): 69–106.

Swynnerton, C. F. M. 1936. *The Tsetse Flies of East Africa*. London: Royal Entomological Society.

Symes, C. B., A. B. Hadaway, F. Barlow, and W. Galley. 1948. "Field Experiments with DDT and Benzene Hexachloride against Tsetse (*Glossina palpalis*)." *Bulletin of Entomological Research* 38 (4): 591–612.

Sydenham, E. W., P. G. Thiel, W. F. O. Marasas, G. S. Shephard, D. J. van Schalkwyk, and K. R. Koch. 1990. "National Occurrence of Some Fusarium Mycotoxins in Corn from Low- and High-Oesophageal-Cancer-Prevalence Areas of the Transkei." *South African Journal Agriculture and Food Chemistry* 38:1900–1903.

Taberer, W. S. 1905. "Mashonaland Natives." *Journal of the Royal African Society* 4 (15): 311–336.

Tangley, L. 1985. "Releasing Engineered Organisms in the Environment." *Bioscience* 35 (8): 470–473.

Tapson, R. R. 1944. "Some Notes on the Mrozwi Occupation of the Sebungwe District." *NADA* 21:29.

Tarry, D. W. 1967. "Control of *Glossina morsitans submorsitans* Newst. in Northern Nigeria by Riverine Spraying with Motorized Sprayers." *Annals of Tropical Medicine and Parasitology* 61:465–471.

Theal, G. C. 1901. *Records of South-Eastern Africa*. Vol. 7. Salisbury: National Archives of Rhodesia.

Thomson, R. 2001. *The Adventures of Shadrek: Southern Africa's Most Infamous Elephant Poacher*. Long Beach: Safari Press.

Thomson, W. E. F. 1947. "Nematodes in Tsetse." *Annals of Tropical Medicine and Parasitology* 41 (1): 164.

Thone, F. 1927. "Nature Ramblings: Parasitic Allies." *Science News-Letter* 12 (339): 233.

Thone, F. 1937. "Without Benefit of Insects. " *Science News-Letter* 32 (858): 186–188.

Thoreau, H. D. 1854. *Walden or, Life in the Woods*. Boston: Ticknor and Fields.

Thorsteinson, A. J., G. K. Bracken, and W. Hanec. 1964. "The Manitoba Horse Fly Trap." *Canadian Entomologist* 96 (1–2): 166.

Time. 1971. "Environment: Sue the Bastards." October 18.

Tompa, A., J. Major, and M. G. Jakab. 1994. "Monitoring of Benzene-Exposed Workers for Genotoxic Effects of Benzene: Improved-Working-Condition-Related Decrease in the Frequencies of Chromosomal Aberrations in Peripheral Blood Lymphocites." *Mutation Research* 304:159–165.

Trypanosomiasis Committee of Southern Rhodesia. 1946. "The Scientific Basis of the Control of *Glossina morsitans* by Game Destruction." *Society for the Preservation of the Fauna of the Empire* 54:10–16.

"Tsetse Fly the Winner in Terror War." 1977. *Rhodesia Herald*, March 3.

Tucker, M. R., and D. E. Pedgley. 1983. "Rainfall and Outbreaks of the African Armyworm, *Spodoptera exempta* (Walker) (Lepidoptera: Noctuidae)." *Bulletin of Entomological Research* 73 (2): 195–199.

Turner, D. A. 1972. "Laboratory Investigations of Host-Seeking Behaviour in Tsetse Flies." *Rhodesia Science News* 6 (1): 25–27.

UNEP (United Nations Environment Programme). 2001. *Stockholm Convention on Persistant Organic Pollutants*. Stockholm: United Nations.

USDHHS (United States Department of Health and Human Services). 2002. *Toxicological Profile for DDT, DDE, and DDD*. Atlanta, GA: Agency for Toxic Substances and Disease Registry.

Vale, G. A. 1968. "Residual Insecticides for Use against Tsetse Flies in Rhodesia." *PANS* 15:213–218.

Vale, G. A. 1969. "Mobile Attractants for Tsetse Flies." *Arnoldia* 4 (33): 1–7.

Vale, G. A. 1971. "Artificial Refuges for Tsetse (*Glossina* spp.)." *Bulletin of Entomological Research* 61 (2): 331–350.

Vale, G. A. 1972. "Some Aspects of the Field Study of Tsetse Flies." *Rhodesia Science News* 6 (1): 10–13.

Vale, G. A. 1973. "The Responses of Tsetse to Their Host Animals." PhD thesis, University of London.

Vale, G. A. 1974a. "New Field Methods for Studying the Responses of Tsetse Flies (Diptera, Glossinidae) to Hosts." *Bulletin of Entomological Research* 64 (2): 199–208.

Vale, G. A. 1974b. "The Responses of Tsetse Flies (Diptera, Glossinidae) to Mobile and Stationary Baits." *Bulletin of Entomological Research* 64 (4): 545–588.

Vale, G. A., and J. W. Hargrove. 1975. "Field Attraction of Tsetse Flies (Diptera: Glossinidae) to Ox Odour: The Effects of Dose." *Transactions of the Rhodesia Scientific Association* 56 (3): 46–50.

Vale, G. A., and R. J. Phelps. 1974. "Notes on the Host-finding Behaviour of Tabanidae (Diptera)." *Arnoldia* 36 (6): 1–6.

Vale, G. A., J. W. Hargrove, G. F. Cockbill, and R. J. Phelps. 1986. "Field Trials of Baits to Control Populations of *Glossina morsitans morsitans* Westwood and *G. pallidipes* Austen (Diptera: Glossinidae)." *Bulletin of Entomological Research* 76 (2): 179–193.

Vale, G. A., D. R. Hall, and A. J. E. Gough. 1988. "The Olfactory Responses of Tsetse Flies, *Glossina* spp. (Diptera: Glossinidae), to Phenols and Urine in the Field." *Bulletin of Entomological Research* 78 (2): 293–300.

Vanderplank, F. L. 1944. "Studies of the Behaviour of the Tsetse-Fly (*Glossina pallidipes*) in the Field: The Attractiveness of Various Baits." *Journal of Animal Ecology* 13 (1): 39–48.

Vanderplank, F. L. 1947. "Experiments in the Hybridisation of Tsetse-Flies (*Glossina*, Diptera) and the Possibility of a New Method of Control." *Transactions of the Royal Entomological Society of London* 98:1–18.

van Hoof, L. 1928. "Sleeping Sickness in the Semliki Valley (Belgian Congo)." In *Final Report of the League of Nations International Committee on Human Trypanosomiasis*, 329–345. Geneva: League of Nations, Health Organization.

van Zyl, H. J. 1941. "Praises in Northern Sotho." *Bantu Studies* 15:119–156.

Videla, L. A., S. B. Barros, and V. B. Junqueira. 1990. "Lindane-Induced Liver Oxidative Stress." *Free Radical Biology and Medicine* 9 (2): 169–179.

Wallis, J. P. R. 1976a. *The Matabele Journals of Robert Moffat 1829–1860. Vol. 1.* Salisbury: National Archives of Rhodesia.

Wallis, J. P. R. 1976b. *The Matabele Journals of Robert Moffat 1829–1860. Vol. 2.* Salisbury: National Archives of Rhodesia.

"War against Insects." 1964. *Science News* 86 (5): 74–75.

Waterston, J. 1915. "Notes on African Chalcidoidea—II." *Bulletin of Entomological Research* 5 (4): 343–372.

Waterston, J. 1916. "Chalcidoidea Bred from *Glossina morsitans* in Nyasaland." *Bulletin of Entomological Research* 6 (4): 381–393.

Waterston, J. 1917. "Chalcidoidea Bred from *Glossina* in the Northern Territories, Gold Coast." *Bulletin of Entomological Research* 8 (2): 178–179.

Weitz, B., and C. H. N. Jackson. 1955. "Host Animals of *Glossina morsitans* at Daga Iloi." *Bulletin of Entomological Research* 46 (3): 531–538.

Wells, L. H. 1929. "Variations in the Muscles of the Bantu Foot." *South African Journal of Science* 26:787–798.

Wells, L. H. 1931. "The Foot of the South African Native." *American Journal of Physical Anthropology* 15:185–289.

Wells, L. H. 1938. "The Status of Bushman as Revealed by a Study of Endocranial Casts." *South African Journal of Science* 34:365–398.

Wells, L. H. 1952. "Physical Measurements of Northern Bushmen." *Man* 52:75–98.

Whitelaw, D. D., E. A. Gault, P. H. Homles, I. A. Sutherland, F. J. Rowell, A. Phillips, and G. M. Urquhart. 1991. "Development of an Enzyme-Linked Immunosorbent Assay for the Detection and Measurement of the Trypanocidal Drug Isometamidium Chloride (Samorin) in Cattle." *Research in Veterinary Science* 50 (2): 185–189.

Whiteside, E. F. 1949. "An Experiment in Control of Tsetse with DDT-Treated Oxen." *Bulletin of Entomological Research* 40 (1): 123–134.

Whiteside, E. F. 1962. "The Control of Cattle Trypanosomiasis with Drugs in Kenya: Methods and Costs." *East African Agricultural and Forestry Journal* 28 (2): 67–73.

WHO (World Health Organization). 1979. *DDT and its Derivatives.* Geneva: World Health Organization.

Wilkinson, P. R. 1948. "Tests of Insecticides on Cattle." *East African Agricultural Journal* 14 (1): 12–17.

Williamson, J., and R. S. Desowitz. 1956. "Prophylactic Activity of Suramin Complexes in Animal Trypanosomiasis." *Nature* 177:1074–1075.

Williamson, J., and L. E. Stephen. 1960. "A Test for Drug-Resistant Trypanosomes in Experimental Tsetse-fly Challenge of Cattle." *Annals of Tropical Medicine and Parasitology* 54 (3): 366–370.

Winter, J. A. 1914. "The Mental and Moral Capabilities of the Natives, Especially of Sekukuniland (Eastern Transvaal)." *South African Journal of Science* 1:371–383.

Wolfe, H. R., and J. F. Armstrong. 1971. "Exposure of Formulating Plant Workers ot DDT." *Archives of Environmental Health* 33:169–176.

Wolff, M. S., and P. G. Toniolo. 1995. "Environmental Organochlorine Exposure as a Potential Eyiologic Factor in Breast Cancer." *Environmental Health Perspectives* 103 (7): 141–145.

Woodcock, K. E. 1949. *Peripheral Vegetation Spraying on Ziribanje Island, Lake Victoria.* Brazzaville, Bureau Permanent Interafricain de Tsetse et Trypanosomiase, no. 79.

Yáñez, L. I., V. H. Borja-Aburto, E. Rojas, H. de la Fuente, R. González-Amaro, H. Gómez, A. A. Jongitud, and F. Díaz-Barriga. 2004. "DDT Induces DNA Damage in

Blood Cells: Studies in vitro and in Women Chronically Exposed to This Insecticide." *Environmental Research* 94 (1): 18–24.

Zhou, R., L. Zhu, and Q. Kong. 2007. "Persistent Chlorinated Pesticides in Fish Species from Qiantang River in East China." *Chemosphere* 68 (5): 838–847.

Zhou, R., L. Zhu, Y. Chen, and Q. Kong. 2008. "Concentrations and Characteristics of Organochlorine Pesticides in Aquatic Biota from Qiantang River in China." *Environmental Pollution* 151 (1): 190–199.

Secondary Sources

Akintoye, S. A. 1969. "The Ondo Road Eastwards of Lagos, c. 1870–95." *Journal of African History* 10 (4): 581–598.

Akurang-Parry, K. O. 2000. "Colonial Forced Labor Strategies for Road-Building in the Gold Coast (Southern Ghana) and International Anti- Forced Labor Pressures, 1900–1940." *African Economic History* 28:1–25.

Akurang-Parry, K. O. 2002. "'The Loads Are Heavier than Usual': Forced Labor by Women and Children in the Central Province, Gold Coast (Colonial Ghana), ca. 1900–1940." *African Economic History* 30:31–51.

Alexander, J., J. McGregor, and T. O. Ranger. 2000. *Violence and Memory: One Hundred Years in the "Dark Forests" of Matabeleland.* Oxford: James Currey Ltd.

Alexander, J., and T. Ranger. 1998. "Competition and Integration in the Religious History of North-Western Zimbabwe." *Journal of Religion in Africa* 28 (1): 3–31.

Amato, J. 2004. *On Foot: A History of Walking.* New York: NYU Press.

Anderson, J. L. 2001. "Plant Hormones in War and Peace: Science, Industry, and Government in the Development of Herbicides in 1940s America." *Isis* 92:291–316.

Anderson, J. L. 2005. "War on Weeds: Iowa Farmers and Growth-Regulator Herbicides." *Technology and Culture* 46 (4): 719–744.

Apffel-Marglin, F. 1990. *Dominating Knowledge: Development, Culture, and Resistance.* Oxford: Clarendon Press.

Armah, A. K. 1968. *The Beautyful Ones Are Not Yet Born.* New York: Houghton Mifflin.

Armah, A. K. 2010. *Remembering the Dismembered Continent.* Popenguine: Per Ankh.

Azarya, V. 2001. "The Nomadic Factor in Africa: Dominance or Marginality." In *Nomads in the Sedentary World*, edited by Anatoly M. Khazanov and Andre Wink, 250–284. London: Curzon Press.

Bailey, R. 2002. "Silent Spring at 40." Reason, June 12.

Bain, G. 1992. *De Havilland: A Pictorial Tribute*. London: AirLife.

Barfield, T. J. 1993. *The Nomadic Alternative*. Englewood Cliffs, NJ: Prentice-Hall.

Bassett, M. T., L. Levy, E. Chokunonga, B. Mauchaza, J. Ferlay, and D. M. Parkin. 1995. "Cancer in the European Population of Harare, Zimbabwe, 1990–1992." *International Journal of Cancer* 63:24–28.

Beisel, U., and C. Boëte. 2013. "The Flying Public Health Tool: Genetically Modified Mosquitoes and Malaria Control." *Science as Culture* 22 (1): 38–60.

Beisel, U., A. Kelly, and N. Tousignant. 2013. "Knowing Insects: Hosts, Vectors and Companions of Science." *Science as Culture* 22 (1): 1–15.

Bekele, S. 1982. "The Railway Trade and Politics: A Historical Survey 1896–1935." Master's thesis, Addis Ababa University.

Bijker, W. 1997. *Of Bicycles, Bakelites, and Bulbs: Towards a Theory of Sociotechnical Change*. Cambridge, MA: MIT Press.

Bohme, S. 2015. *Toxic Injustice: A Transnational History of Exposure and Struggle*. Berkeley: University of California Press.

Bourdillon, M. F. C. 1978. "The Cults of Dzivaguru and Karuva amongst the North-Eastern Shona Peoples." In *Guardians of the Land*, edited by. J. M. Schoffeleers, 44–53. Gwelo: Mambo Press.

Bowman, J. L. 1987. "'Legitimate Commerce' and Peanut Production in Portuguese Guinea, 1840s–1880s." *Journal of African History* 28 (1): 87–106.

Bradshaw, E., and M. Schonland. 1974. "Smoking, Drinking and Oesophageal Cancer in African Males in Johannesburg, South Africa." *British Journal of Cancer* 30:157–163.

Bransom, A. 2005. *The Tiger Moth Story*. 5th ed. Manchester: Crécy Publishing Ltd.

Brooks, G. E. 1975. "Peanuts and Colonialism: Consequences of the Commercialization of Peanuts in West Africa, 1830–70." *Journal of African History* 16 (1): 29–54.

Brown, K., and D. Gilfoyle. 2010. *Healing the Herds: Disease, Livestock Economies, and the Globalization of Veterinary Medicine*. Athens, OH: Ohio University Press.

Bulliet, R. 1990. *The Camel and the Wheel*. New York: Columbia University Press.

Bwititi, K. 2015. "The Seed Planted at Sikombela." *Sunday Mail*, July 12.

Callon, M. 1986. "Some Elements of a Sociology of Translation: Domestication of the Scallops and the Fishermen of St. Brieuc Bay." In *Power, Action and Belief*, edited by J. Law, 196–233. London: Routledge.

Carney, J. 2001. *Black Rice: The African Origins of Rice Cultivation in the Americas*. Cambridge, MA: Harvard University Press.

Carney, J., and R. N. Rosomoff. 2009. *In the Shadow of Slavery: Africa's Botanical Legacy in the Atlantic World*. Berkeley: University of California Press.

Carruthers, J. 1995. *The Kruger National Park: A Social and Political History*. Durban: Natal University Press.

Carson, R. 1962. *Silent Spring*. Boston: Houghton Mifflin.

CEC (Commission for Environmental Cooperation. 2006. *North American Regional Action Plan (NARAP) on Lindane and Other Hexachlorocyclohexane (HCH) Isomers*. CEC, November 30. http://www.cec.org/files/PDF/POLLUTANTS/LindaneNARAP-Nov06_en.pdf.

Chapin, G., and R. Wasserstrom. 1981. "Agricultural Production and Malaria Resurgence in Central America and India." *Nature* 293:181–185.

Chen, H., G. J. D. Smith, S. Y. Zhang, K. Qin, J. Wang, K. S. Li, R. G. Webster1, J. S. M. Peiris, and Y. Guan 2005. "H5N1 Virus Outbreak in Migratory Waterfowl." *Nature* 436:191–192.

Chikowero, M. 2015. *African Music, Power, and Being in Colonial Zimbabwe*. Bloomington: Indiana University Press.

Chilundo, A. 1995. "The Economic and Social Impact of the Rail and Road Transportation Systems in the Colonial District of Moçambique (1900–1961)." PhD thesis, University of Minnesota.

Chirenje, Z. M., S. Rusakaniko, T. Chipato, W. Mpanju-Shumbusho, E. Ngwalle, L. W. Kirumbi, P. Makuta-Tlebere, and S. Kaggwa. 1998. "Situation Analysis for Cervical Cancer Diagnosis and Treatment in Zimbabwe." *Central African Journal of Medicine* 44 (12): 307–310.

Chirenje, Z. M., T. Chipato, J. Kasule, and S. Rusakaniko. 1999. "Visual Inspection of the Cervix as a Primary Means of Cervical Cancer Screening: Results of a Pilot Study." *Central African Journal of Medicine* 45 (2): 30–33.

Chiteji, F. 1979. "The Development and Socio-economic Impact of Transportation in Tanzania, 1884–Present." PhD thesis, Michigan State University.

Clarence-Smith, W. G. 1994. "African and European Cocoa Producers on Fernando Póo, 1880s to 1910s." *Journal of African History* 35 (2): 179–199.

Clark, J. F. M. 2001. "Bugs in the System: Insects, Agricultural Science, and Professional Aspirations in Britain, 1890–1920." *Agricultural History* 75 (1): 83–114.

Clark, N. 2013. "Mobile Life: Biosecurity Practices and Insect Globalization." *Science as Culture* 22 (1): 16–37.

Collins, H. M., and S. Yearley. 1992. "Epistemological Chicken." In *Science as Practice and Culture*, edited by A. Pickering, 301–326. Chicago: University of Chicago Press.

Costa-Neto, E. M. 2002. "The Use of Insects in Folk Medicine in the State of Bahia, Northeastern Brazil, with Notes on Insects Reported Elsewhere in Brazilian Folk Medicine." *Human Ecology* 30 (2): 245–263.

Coulston, F. 1985. "Reconsideration of the Dilemma of DDT for the Establishment of an Acceptable Daily Intake." *Regulatory Toxicology and Pharmacology* 5:332–382.

Creswell, T. 2006. *On the Move: Mobility in the Modern Western World*. London: Routledge.

Crosby, A. 1986. *Ecological Imperialism: The Biological Expansion of Europe, 900–1900*. Cambridge: Cambridge University Press.

Cwerner, S., S. Kesslring, and J. Urry, eds. 2009. *Aeromobilities*. London: Routledge.

Daneel, M. L. 1970. *The God of the Matopo Hills*. The Hague: Mouton.

Daniel, P. 2002. *Lost Revolutions: The South in the 1950s*. Chapel Hill: University of North Carolina Press.

Davis, F. R. 2014. *Banned: A History of Pesticides and the Science of Toxicology*. New Haven, CT: Yale University Press.

Darwin, C. 1859. *On the Origin of Species by Means of Natural Selection, or the Preservation of Favoured Races in the Struggle for Life*. London: John Murray.

Dingle, H. 1996. *Migration: The Biology of Life on the Move*. New York: Oxford University.

Diouf, M. 2000. "The Senegalese Murid Diaspora and the Making of a Vernacular Cosmopolitanism." *Public Culture* 12 (3): 679–702.

Dlamini, N. 2001. "Race Relations in Swaziland: The Case of Havelock Asbestos Mine, 1939–1964." MA diss., University of Swaziland.

Douglas, M. [1966] 2002. *Purity and Danger: An Analysis of the Concept of Pollution and Taboo*. London: Routledge.

Douglas, M., and A. Wildavsky. 1982. *Risk and Culture: An Essay on the Selection of Technological and Environmental Dangers*. Berkeley: University of California Press.

Draulans, D., and E. Van Krunkelsven. 2002. "The Impact of War on Forest Areas in Democratic Republic of Congo." *Oryx* 36 (1): 35–40.

Dube, F. 2009. "Colonialism, Cross-border Movements, and Epidemiology: A History of Public Health in the Manica Region of Central Mozambique and Eastern Zimbabwe and the African Response, 1890–1980." PhD thesis, University of Iowa.

Dube, S., N. R. Dlamini, A. Mafunga, M. Mukai, and Z. Dhlamini. 2013. "A Survey of Entomophagy Prevalence in Zimbabwe." *African Journal of Food, Agriculture, Nutrition and Development* 13 (1): 7242–7253.

Dubois, C. 1997. *Djibouti 1888–1967: Feritage ou Frustration?* Paris: L'Harmattan.

Dumett, R. E. 1983. "African Merchants of the Gold Coast, 1860–1905: Dynamics of Indigenous Entrepreneurship." *Comparative Studies in Society and History* 25 (4): 661–693.

Dunlap, T. R. 1981. *DDT: Scientists, Citizens and Public Policy*. Princeton, NJ: Princeton University Press.

Dunlap, T. R., ed. 2008. *DDT, Silent Spring, and the Rise of Environmentalism: Classic Texts*. Seattle: University of Washington Press.

Edgerton, E. 2007. *Shock of the Old: Technology and Global History since 1900*. New York: Oxford University.

Elliott, J. E. 1994. "Chlorinated Hydrocarbons and Shell Thinning in Eggs of (Accipiter) Hawks in Ontario, 1986–1989." *Environmental Pollution* 86 (2): 189–200.

Enserink, M. 2007. "Welcome to Ethiopia's Fly Factory." *Science* 317 (5836): 310–313.

EPA (Environmental Protection Agency). 1975. *DDT Regulatory History: A Brief Survey (to 1975)*. Washington, DC: Environmental Protection Agency: History Office.

Faber, R. A., R. W. Risebrough, and H. M. Pratt. 1972. "Organochlorines and Mercury in Common Egrets and Great Blue Herons." *Environmental Pollution* 3 (2): 111–112.

Falola, T. 2008. *The Power of African Cultures*. Rochester, NY: Rochester University Press.

Feierman, S. 1985. "Struggles for Control: The Social Roots of Health and Healing in Modern Africa." *African Studies Review* 28 (2–3): 73–147.

Feierman, S., and J. M. Janzen, eds. 1992. *The Social basis of Health and Healing in Africa*. Berkeley: University of California Press.

Fitzgerald, D. 2003. *Every Farm a Factory: The Industrial Ideal in American Agriculture*. New Haven, CT: Yale University Press.

Fitzgerald, D. 1990. *The Business of Breeding: Hybrid Corn in Illinois, 1890–1940*. Ithaca, NY: Cornell University Press.

Flannery, M. C. 2002. "Insects: Beautiful and Useful." *American Biology Teacher* 64 (5): 376–381.

Flint, L. S. 2003. "State-Building in Central Southern Africa: Citizenship and Subjectivity in Barotseland and Caprivi." *International Journal of African Historical Studies* 36 (2): 393–428.

Flower, K. B., J. A. Hoppin, C. F. Lynch, A. Blair, C. Knott, D. L. Shore, and D. P. Sandler. 2004. "Cancer Risk and Parental Pesticide Application in Children of Agricultural Health Study Participants." *Environmental Health Perspectives* 112 (5): 631–635.

Foster, V., W. Butterfield, C. Chen, and N. Pushak. 2008. *Building Bridges: China's Growing Role as Infrastructure Financier for Sub-Saharan Africa. Trends and Policy Options*, no. 5. Washington, DC: World Bank. http://documents.worldbank.org/curated/en/936991468023953753/Building-bridges-Chinas-growing-role-as-infrastructure-financier-for-Sub-Saharan-Africa.

French, R. 1975. *Antivivisection and Medical Science in Victorian Society*. Princeton, NJ: Princeton University Press.

Frøslie, A., G. Holt, and G. Norheim. 1986. "Mercury and Persistent Chlorinated Hydro-carbons in Owls Strigiformes and Birds of Prey Falconiformes Collected in Norway during the Period 1965–1983." *Environmental Pollution* 11:91–108.

Garlake, P. S. 1978. "Pastoralism and Zimbabwe." *Journal of African History* 19 (4): 479–493.

Georgino, G. N. 1975. *Autos: Encyclopédie complète*. Paris: Courtille.

Georgino, G. N. 2001. *P–Z*. Vol. 3 of *The Beaulieu Encyclopedia of the Automobile*. Chicago: Fitzroy Dearborn Publishers.

Gewald, J.-B., S. Luning, and K. van Walraven, eds. 2009. *The Speed of Change: Motor Vehicles and People in Africa, 1890–2000*. Leiden: Brill.

Giblin, J. 1990. "Trypanosomiasis Control in African History: An Evaded Issue?" *Journal of African History* 31:59–80.

Ginsburg, S. 2000. *Securing Human Mobility in the Age of Risk: New Challenges for Travel, Migration, and Borders*. Washington, DC: Migration Policy Institute.

Gladwell, M. 2001. "The Mosquito Killer." *New Yorker*, July 2.

Glynn, T., J. R. Seffrin, O. W. Brawley, N. Grey, and H. Ross. 2010. "The Globalization of Tobacco Use: 21 Challenges for the 21st Century." *CA: A Cancer Journal for Clinicians* 60:50–61.

Goucher, C. 1981. "'Iron Is Iron 'Till It Is Rust': Trade and Ecology in the Decline of West African Iron Smelting." *Journal of African History* 22:179–189.

Goucher, C. 2004. "The Memory of Iron: African Technology in the Americas." Paper presented at the National Park Service, Places of Cultural Memory Conference, Atlanta, GA.

Green, W. 1964. *Aircraft Handbook*. London: Macdonald & Co. (Publishers) Ltd.

Greenberg, D. S. 1963. "Pesticides: White House Advisory Body Issues Report Recommending Steps to Reduce Hazard to Public." *Science* 140 (3569): 878–879.

Green-Simms, L. 2009. "Postcolonial Automobility: West Africa and the Road to Globalization." PhD thesis, University of Minnesota.

Gudhlanga, E. S., and G. Makaudze. 2012. "Indigenous Knowledge Systems: Confirming a Legacy of Civilisation and Culture on the African Continent." *Prime Journal of Social Science* 1 (4): 72–77.

Gyekye, K. 1997. *Tradition and Modernity: Philosophical Reflections on the African Experience*. New York: Oxford University Press.

Harrison, J. L. 2011. *Pesticide Drift and the Pursuit of Environmental Justice*. Cambridge, MA: MIT Press.

Hart, J. 2016. *Ghana on the Go. African Mobility in the Age of Motor Transportation*. Bloomington: Indiana University Press.

Headrick, D. R. 1981. *The Tools of Empire: Technology and European Imperialism in the Nineteenth Century*. New York: Oxford University Press.

Headrick, D. R. 2010. *Power over Peoples: Technology, Environments, and Western Imperialism, 1400 to the Present*. Princeton, NJ: Princeton University Press.

Heap, S. 1990. "The Development of Motor Transport in the Gold Coast, 1900–39." *Journal of Transport History* 11 (1): 19–37.

Heap, S. 2000. "Transport and Liquor in Colonial Nigeria." *Journal of Transport History* 21 (1): 28–53.

Hecht, G. 2012. *Being Nuclear: Africans and the Global Uranium Trade*. Cambridge, MA: MIT Press.

Helmreich, S. 2009. *Alien Ocean: Anthropological Voyages in Microbial Seas*. Berkeley: University of California Press.

Henige, D. 1977. "John Kabes of Komenda: An Early African Entrepreneur and State Builder." *Journal of African History* 18 (1): 1–19.

Herbert, E. 1975. "Smallpox Inoculation in Africa." *Journal of African History* 16 (4): 539–559.

Herman, S. G., and J. B. Bulger. 1979. "Effects of a Forest Application of DDT on Nontarget Organisms." *Wildlife Monographs* 69:3–62.

Herskovits, M. J. 1940. *Economic Anthropology: The Economic Life of Primitive Peoples*. New York: Knopf.

Hoffman, B., J. M. Taw, and D. Arnold. 1991. *Lessons for Contemporary Counterinsurgencies: The Rhodesian Experience*. Santa Monica, CA: RAND.

Holmes, H. 2004. *Avro: The History of an Aircraft Company*. Wiltshire, UK: Crowood Press.

Hoppe, K. 2003. *Lords of the Fly: Sleeping Sickness Control in British East Africa, 1900–1960*. London: Praeger.

Hotson, F. W. 1983. *The de Havilland Canada Story*. Toronto: CANAV Books.

Ingold, T. 2011. *Being Alive: Essays on Movement, Knowledge and Description*. New York: Routledge.

Irele, A. 1983. "In Praise of Alienation." An inaugural lecture, University of Ibadan, Nigeria.

Jackson, A. J. 1965. *Avro Aircraft since 1908*. London: Putnam.

Jacobsen, A. 2014. *Operation Paperclip: The Secret Intelligence Program to Bring Nazi Scientists to America*. New York: Little, Brown and Company.

Janzen, J. M. 1978. *The Quest for Therapy: Medical Pluralism in Lower Zaire*. Berkeley: University of California Press.

Janzen, J. M. 1982. *Lemba, 1650–1930: A Drum of Affliction in Africa and the New World*. New York: Garland Publishers.

Janzen, J. M. 1992. *Ngoma: Discourses of Healing in Central and Southern Africa*. Berkeley: University of California Press.

Jasanoff, S. 1995. *Science at the Bar*. Cambridge, MA: Harvard University Press.

Jeater, D. 2005. "Imagining Africans: Scholarship, Fantasy, and Science in Colonial Administration, 1920s Southern Rhodesia." *International Journal of African Historical Studies* 38 (1): 1–26.

Kaba, L. 1981. "Archers, Musketeers, and Mosquitoes: The Moroccan Invasion of the Sudan and the Songhay Resistance (1591–1612)." *Journal of African History* 22:457–475.

Karp, I., and D. A. Masolo, eds. 2000. *African Philosophy as Critical Inquiry*. Bloomington: Indiana University Press.

Kazembe, T. C., and S. Nkomo. 2010. "Lake Ngami, Or Explorations and Discoveries during Four Years' Wanderings in the Wilds of Southwestern Africa." *Indian Journal of Traditional Knowledge* 9 (2): 394–397.

Keifer, M. C., and R. K. Mahurin. 1997. "Chronic Neurologic Effects of Pesticide Overexposure." *Occupational Medicine* 12 (2): 291–304.

Kinkela, D. 2011. *DDT and the American Century: Global Health, Environmental Politics, and the Pesticide That Changed the World*. Chapel Hill: University of North Carolina Press.

Kirksey, S. E., and S. Helmreich. 2010. "The Emergence of Multispecies Ethnography." *Cultural Anthropology* 25 (4): 545–576.

Kjekshus, H. [1977] 1996. *Ecology Control and Economic Development in East African History*. 2nd ed. London: James Currey.

Kline, R. 2000. *Consumers in the Country: Technology and Social Change in Rural America*. Baltimore: Johns Hopkins University Press.

Kopytoff, I. 1986. "The Cultural Biography of Things: Commoditization as Process." In *The Social Life of Things: Commodities in Cultural Perspective*, edited by A. Appadurai, 64–91. Cambridge: Cambridge University Press.

Kosmin, B. A. 1971. "On the Imperial Frontier: The Pioneer Community of Salisbury in November 1897." *Rhodesian History* 2:25–37.

Kuhn, T. S. 1962. *The Structure of Scientific Revolutions*. Chicago: University of Chicago Press.

Kumar, P. 2012. *Indigo Plantations and Science in Colonial India, 1670–1920*. Cambridge: Cambridge University Press.

Lan, D. 1985. *Guns and Rain*. London: James Currey.

Latour, B. 1993. *We Have Never Been Modern*, translated by Catherine Porter. Cambridge, MA: Harvard University Press.

Latour, B., and S. Woolgar. [1979] 1986. *Laboratory Life: The Construction of Scientific Facts*. 2nd ed. Princeton, NJ: Princeton University Press.

Law, R. 1980. *The Horse in West African History*. Oxford: Oxford University Press.

Law, R. 1989. "'Between the Sea and the Lagoons': The Interaction of Maritime and Inland Navigation on the Precolonial Slave Coast." *Cahiers d'Etudes Africaines* 114:209–237.

Lawrence, J. A., C. M. Foggin, and R. A. Norval. 1980. "The Effects of War on the Control of Diseases of Livestock in Rhodesia (Zimbabwe)." *Veterinary Record* 107:82–85.

Lear, L. 2007. *Rachel Carson: Witness for Nature*. New York: Henry Holt & Co.

Le Baron, B. 1966. "Négritude: A Pan-African Ideal?" *Ethics* 76 (4): 267–276.

Le Billon, P. 2001. "The Political Ecology of War: Natural Resources Armed Conflicts." *Political Geography* 20 (5): 561–584.

Le Couteur, D. G., A. J. McLean, M. C. Taylor, B. L. Woodham, and P. G. Board. 1999. "Pesticides and Parkinson's Disease." *Biomedicine & Pharmacotherapy* 53 (3): 122–130.

Lester, C. 1996. *Protection of Light Skinned Vehicles against Landmines—A Review*. Melbourne: DSTO Aeronautical and Maritime Research Laboratory.

Levy, S. 2007. "Mosquito Modifications: New Approaches to Controlling Malaria." *Bioscience* 57 (10): 816–821.

Lezra, E. 2014. *The Colonial Art of Demonizing Others: A Global Perspective*. London: Routledge.

Likaka, O. 2009. *Naming Colonialism and Rural Society and Cotton in the Belgian Congo*. Madison: University of Wisconsin Press.

Livingston, J. 2012. *Improvising Medicine: An African Oncology Ward in an Emerging Cancer Epidemic*. Durham, NC: Duke University Press.

Lyons, M. 1992. *The Colonial Disease: A Social History of Sleeping Sickness in Northern Zaire, 1900–1940*. Cambridge: Cambridge University Press.

Machin, I. 2002. *Antbears and Targets for Zulu Assegais: The Levying of Forced Labour and Military Service by the Colonial State of Natal*. Howick, South Africa: Brevitas.

Machingaidze, V. E. M. 1980. "The Development of Settler Capitalist Agriculture in Southern Rhodesia, with Particular Reference to the Role of the State, 1908–39." PhD diss., University of London.

Mack, P. 1990. *Viewing the Earth: The Social Construction of the Landsat Satellite System*. Cambridge, MA: MIT Press.

Mamdani, M. 2001. *When Victims Become Killers: Colonialism, Nativism, and the Genocide in Rwanda*. Princeton, NJ: Princeton University Press.

Mangoma, J. F., M. Z. Chirenje, M. J. Chimbari, and S. K. Chandiwana. 2006. "An Assessment of Rural Women's Knowledge, Constraints and Perception on Cervical Cancer Screening: The Case of Two Districts in Zimbabwe." *African Journal of Reproductive Health* 10 (1): 91–103.

Mapara, J., and G. Makaudze. 2016. "The Interface between Toponyms, Hydronyms and Geography: The Case of Selected Shona Names from Three Provinces in Zimbabwe." *South African Journal of African Languages* 36 (2): 243–249.

Marasas, W. F. O., S. J. van Rensburg, and C. J. Mirocha. 1979. "Incidence of Fusarium Species and the Mycotoxins Deoxynivalenol and Zearalenone in Corn Produced in Esophageal-Cancer Areas in Transkei." *Journal of Agriculture and Food Chemistry* 27:1108–1112.

Marx, L. 1964. *The Machine in the Garden: Technology and the Pastoral Ideal in America*. New York: Oxford University Press.

Mavhunga, C. C. 2007. "Big Game Hunters, Bacteriologists, and Tsetse Fly Entomology in Colonial Southeast Africa: The Selous-Austen Debate Revisited, 1905–1940s." *Icon* 12:75–117.

Mavhunga, C. C. 2011. "Vermin Beings: On Pestiferous Animals and Human Game." *Social Text* 29 (1): 151–176.

Mavhunga, C. C. 2013. "Cidades Esfumaçadas: Energy and the Rural-Urban Connection in Mozambique." *Public Culture* 25 (2): 261–271.

Mavhunga, C. C. 2014. *Transient Workspaces: Technologies of Everyday Innovation in Zimbabwe*. Cambridge, MA: MIT Press.

Mavhunga, C. C., and M. Spierenburg. 2007. "A Finger on the Pulse of the Fly: Hidden Voices of Colonial Anti-Tsetse Science on the Rhodesia and Mozambique Borderlands, 1945–56." *South African Historical Journal* 58:117–141.

Maviya, J., and D. Gumbo. 2005. "Incorporating Traditional Natural Resource Management Techniques in Conventional Natural Resources Management Strategies: A Case of Mopane Worms (*Amacimbi*) Management and Harvesting in the Bulalimamangwe District, Zimbabwe." *Journal of Sustainable Development in Africa* 7 (2): 95–107.

McCulloch, J. 2002. *Asbestos Blues: Labour, Capital, Physicians and the State in South Africa*. Oxford: James Currey.

McCulloch, J. 2005. "Dust, Disease and Labour at Havelock Asbestos Mine, Swaziland." *Journal of Southern African Studies* 31 (2): 251–266.

McGregor, J. A., and T. O. Ranger. 2000. "Displacement and Disease: Epidemics and Ideas about Malaria in Matabeleland, Żimbabwe 1945–1996." *Past & Present* 167 (1): 203–237.

McShane, C. 1994. *Down the Asphalt Path: The Automobile and the American City*. New York: Columbia University Press.

Meeran, R. 2003. "Cape Pic: South African Mineworkers' Quest for Justice." *International Journal of Occupational and Environmental Health* 9 (3): 218–227.

Meinwald, J. 1990. "The Insect as Organic Chemist." *Bulletin: American Academy of Arts and Sciences* 43 (7): 27–34.

Merton, R. 1973. *The Sociology of Science: Theoretical and Empirical Investigations*. Chicago: University of Chicago Press.

Mignolo, W. 2011. *The Darker Side of Western Modernity: Global Futures, Decolonial Options*. Durham, NC: Duke University Press.

Mindell, D. 2008. *Digital Apollo: Human and Machine in Spaceflight*. Cambridge, MA: MIT Press.

Mitchell, T. 2002. *Rule of Experts: Egypt, Techno-Politics, Modernity*. Berkeley: The University of California Press.

Mlambo, A. 1996. "The Cold Storage Commission: A Colonial Parastatal 1938–1963." *Zambezia* 23 (1): 53–71.

Mom, G. 2004. *The Electric Vehicle*. Baltimore: Johns Hopkins University Press.

Mom, G., G. Clarsen, N. Kim, C. Seiler, K. Möser, D. Müller, N. Terranova, and R. Volti. 2011. "Hop on the Bus, Gus." *Transfers* 1 (1): 1–13.

Monson, J. 2009. *Africa's Freedom Railway: How a Chinese Development Project Changed Lives and Livelihoods in Tanzania*. Bloomington: Indiana University Press.

Montgomery, M. R., and G. Foster. 1992. *A Field Guide to Airplanes*. 2nd ed. Boston: Houghton Mifflin Company.

Morgan, D. P., and C. C. Roan. 1971. "Absorption, Storage, and Metabolic Conversion of Ingested DDT and DDT Metabolites in Man." *Archives of Environmental Health* 22 (3): 301–315.

Morgan, D. P., and C. C. Roan. 1972. "Loss of DDT from Storage in Human Body Fat." *Nature* 238 (5361): 221–223.

Moyo, J. 2012. "Cancer: Zimbabwe's New Headache." *Herald*, October 3.

Mudenge, S. I. G. 1988. *A Political History of Munhumutapa*. Harare: Zimbabwe Publishing House.

Mudimbe, V. Y. 1988. *The Invention of Africa*. Bloomington: Indiana University Press.

Mullin, L. J. 1994. "Chirinda Forest in Legend and History." In *Chirinda Forest: A Visitors Guide*, edited by J. R. Timberlake and P. Shaw, 12–28. Harare: Zimbabwe Forestry Commission.

Murombedzi, J. 1994. "Dynamics of Conflict in Environmental Management Policy in the Context of the Communal Areas Management Programme for Indigenous Resources." PhD diss., University of Zimbabwe.

Mutasa, D. E., S. Nyota, and J. Mapara. 2008. "*Ngano*: Teaching Environmental Education Using the Shona Folktale." *Journal of Pan African Studies* 2 (3): 33–54.

Mutimukuru-Maravanyika, T. 2010. "Can We Learn Our Way to Sustainable Management? Adaptive Collaborative Management in Mafungautsi State Forest, Zimbabwe." PhD diss., Wageningen University.

Naimark, N. 1995. *The Russians in Germany*. Cambridge, MA: Harvard University Press.

Nash, L. 2007. *Central Valley, Inescapable Ecologies: A History of Environmental Disease and Knowledge*. Berkeley: University of California Press.

Nass, M. 1992. "Anthrax Epizootic in Zimbabwe, 1978–1980: Due to Deliberate Spread?" *PSR Quarterly* 2:198–209.

Nass, M. 1992–1993. "Zimbabwe's Anthrax Epizootic." *Covert Action* 43:12–18.

Ngulube, P. 2017. *Handbook of Research on Theoretical Perspectives on Indigenous Knowledge Systems in Developing Countries*. Hershey, PA: IGI Global.

Nyambara, P. 1999. "A History of Land Acquisition in Gokwe, Northwestern Zimbabwe, 1945–1997." PhD diss., Northwestern University.

Nyambara, P. 2001. "The Closing Frontier: Agrarian Change, Immigrants and the 'Squatter Menace' in Gokwe, 1980–1990s." *Journal of Agrarian Change* 1 (4): 534–549.

Nyambara, P. 2002. "Madheruka and Shangwe: Ethnic Identities and the Culture of Modernity in Gokwe, Northwestern Zimbabwe, 1963–79." *Journal of African History* 43 (2): 287–306.

Oreskes, N., and E. M. Conway. 2010. "Denial Rides Again: The Revisionist Attach on Rachel Carson." In *Merchants of Doubt: How a Handful of Scientists Obscured the Truth on Issues from Tobacco Smoke to Global Warming*, edited by N. Oreskes and E. M. Conway, 216–239. New York: Bloomsbury.

Osseo-Asare, A. 2014. *Bitter Roots: The Search for Healing Plants in Africa*. Chicago: University of Chicago Press.

Packard, R. 1989. *White Plague, Black Labor: Tuberculosis and the Political Economy of Health and Disease in South Africa*. Berkeley: University of California Press.

Packard, R. 2007. *The Making of a Tropical Disease: A Short History of Malaria*. Baltimore: Johns Hopkins Press.

Palmer, R. 1977. *Land and Racial Discrimination in Rhodesia*. London: Heinemann.

Peperell, R., and C. Smith. 1987. *Piper Aircraft and their Forerunner*. Tonbridge: Air-Britain.

Perkins, J. 2003. *Insects, Experts and the Insecticide Crisis: The Quest for New Pest Management Strategies*. New York: Plenum.

Phillips, E. H. 1986. *Wings of Cessna, Model 120 to the Citation III*. Eagin, MN: Flying Books.

Phimister, I. 1978. "Meat and Monopolies: Beef Cattle in Southern Rhodesia, 1890–1938." *Journal of African History* 19 (3): 391–414.

Phimister, I. 1994. *Wangi Kolia: Coal, Capital and Labour in Colonial Zimbabwe 1894–1954*. Harare, Zimbabwe: Baobab Books.

Pikirayi, I. 2001. *The Zimbabwe Culture: Origins and Decline of Southern Zambezian States*. Walnut Creek, CA: AltaMira Press.

Pirie, G. 1993. "Slaughter by Steam: Railway Subjugation of Ox-wagon Transport in the Eastern Cape and Transkei, 1886–1910." *International Journal of African Historical Studies* 26 (2): 319–343.

Pirie, G. 1997. "Brutish Bombelas: Trains for Migrant Gold Miners in South Africa, c. 1900–1925." *Journal of Transport History* 18:31–44.

Pirie, G. 2009. *Air Empire: British Civil Aviation, 1919–39*. Manchester: Manchester University Press.

Pirie, G. 2011. "Non-urban Motoring in Colonial Africa in the 1920s and 1930s." *South African Historical Journal* 63 (1): 59–81.

Plane and Pilot. 1978. *Aircraft Directory*. Santa Monica, CA: Werner & Werner Corp.

Quiggin, J., and T. Lambert. 2008. "Rehabilitating Carson." *Prospect*, May 24.

Raffa, K. F. 1989. "Genetic Engineering of Trees to Enhance Resistance to Insects." *Bioscience* 39 (8): 524–534.

Raffles, H. 2010. *Insectopedia*. New York: Vintage Books.

Ranger, T. 1999. *Voices from the Rocks: Nature, Culture, and History in the Matopos Hills of Zimbabwe*. Indianapolis: Indiana University Press.

Rennie, J. K. 1978. "Four Transformations of the Musikavanhu Territorial Cult in Rhodesia." In *Guardians of the Land*, edited by J. M. Schoffeleers, 44–53. Gwelo: Mambo Press.

Reverby, S. 2009. *Examining Tuskegee: The Infamous Syphilis Study and its Legacy*. Chapel Hill: University of North Carolina Press.

Reverby, S. 2011. "'Normal Exposure' and Inoculation Syphilis: A PHS 'Tuskegee' Doctor in Guatemala, 1946–1948." *Journal of Policy History* 23 (1): 6–28.

Richards, P. 1996. *Fighting for the Rain Forest: War, Youth and Resources in Leone*. Oxford: James Currey.

Ritvo, H. 1987. *The Animal Estate: The English and Other Creatures in the Victorian Age*. Cambridge, MA: Harvard University Press.

Robinson, R. E. 1991. "Introduction: Railway Imperialism." In *Railway Imperialism*, edited by R. E. Robinson, C.B. Davis and K.E. Wilburn, 1–6. London: Greenwood Press.

Rome, A. 2001. *The Bulldozer in the Countryside: Suburban Sprawl and the Rise of American Environmentalism*. New York: Cambridge University Press.

Russell, E. 1999. "The Strange Career of DDT: Experts, Federal Capacity, and Environmentalism in World War II." *Technology and Culture* 40:776–781.

Russell, E. 2001. *War and Nature: Fighting Humans and Insects with Chemicals from World War I to Silent Spring*. Cambridge: Cambridge University Press.

Ryan, J. R., and J. F. Glarum. 2008. *Biosecurity & Bioterrorism: Containing and Preventing Biological Threats*. Amsterdam: Elsevier.

Sadasivaiah, S., Y. Tozan, and J. G. Breman. 2007. "Dichlorodiphenyltrichloroethane (DDT) for Indoor Residual Spraying in Africa: How Can It Be Used for Malaria Control?" *American Journal of Tropical Medicine and Hygiene* 77 (6): 249–263.

Samasuwo, N. 2003. "Food Production and War Supplies: Rhodesia's Beef Industry during the Second World War, 1939–1945." *Journal of Southern African Studies* 29 (2): 487–502.

Schivelbusch, W. 1977. *The Railway Journey*. Berkeley: University of California Press.

Schroeter, D. J. 1988. *Merchants of Essaouira: Urban Society and Imperialism in Southwestern Morocco, 1844–1886*. Cambridge: Cambridge University Press.

Scott, P. 1954. "Migrant Labor in Southern Rhodesia." *Geographical Review* 44 (1): 29–48.

Segal, I., S. G. Reinach, and M. de Beer. 1988. "Factors Associated with Oesophageal Cancer in Soweto, South Africa." *British Journal of Cancer* 58 (5): 681–686.

Seiler, C. 2008. *Republic of Drivers: A Cultural History of Automobility in America*. Chicago: University of Chicago Press.

Senghor, L. 1964. *On African Socialism*. London and New York: Praeger.

Serres, M. 1982. *The Parasite*. Translated by L. R. Schehr. Baltimore: Johns Hopkins University Press.

Shafey, O., M. Eriksen, and J. Mackay. 2009. *The Tobacco Atlas*. 3rd ed. Atlanta: American Cancer Society.

Sheldon, K. 2002. *Pounders of Grain: A History of Women, Work, and Politics in Mozambique*. Los Angeles: University of California Press.

Sheller, M., and J. Urry, eds. 2004. *Tourism Mobilities: Places to Play, Places in Play*. London: Routledge.

Sheller, M., and J. Urry, eds. 2006. *Mobile Technologies of the City*. London: Routledge.

Sheriff, A. 2010. *Dhow Culture of the Indian Ocean: Cosmopolitanism, Commerce and Islam*. New York: Columbia University Press.

Simpson, R. 2005. *The General Aviation Handbook*. Leicester: Midland Publishing.

Sodhy, P. 1991. *The US-Malaysian Nexus: Themes in Superpower-Small State Relations*. Kuala Lumpur: Institute of Strategic and International Studies.

Soto-Laveaga, G. 2009. *Jungle Laboratories: Mexican Peasants, National Projects, and the Making of the Pill*. Durham, NC: Duke University Press.

Spinage, C. A. 2003. *Cattle Plague: A History*. New York: Kluwer.

Stapleton, D. 2005. "A Lost Chapter in the Early History of DDT: The Development of Anti-Typhus Technologies by the Rockefeller Foundation's Louse Laboratory, 1942–1944." *Technology and Culture* 46 (3): 513–540.

Stephens, J. L. 1994. "The Political Economy of Transport in Mozambique: Implications for Regional Development." PhD thesis, University of Sussex.

Sterne, M. 1967. "Distribution and Economic Importance of Anthrax." *Federation Proceedings* 26:1493–1495.

Stigger, P. 1971. "Volunteers and the Profit Motive in the Anglo-Ndebele War, 1893." *Rhodesian History* 2:11–23.

Stine, J. K., and J. Tarr. 1998. "At the Intersection of Histories: Technology and the Environment." *Technology and Culture* 39 (4): 601–640.

Sunseri, T. 2002. *Vilimani: Labor Migration and Rural Change in Early Colonial Tanzania*. Portsmouth, NH: Heinemann.

Swart, S. 2010. *Riding High: Horses, Humans and History in South Africa*. Johannesburg: Wits University Press.

Tannock, J., W. W. Howells, and R. J. Phelps. 1983. "Chlorinated Hydrocarbon Pesticide Residues in Eggs of Some Birds in Zimbabwe." *Environmental Pollution* 5 (2): 147–155.

Tarr, J. 1996. *The Search for the Ultimate Sink: Urban Pollution in Historical Perspective*. Akron, OH: University of Akron Press.

Taverne, D. 2005. "The Harm that Pressure Groups Can Do." In *Panic Nation*, edited by S. Feldman and M. Vincent, 29–42. London: John Black.

Thornton, J. 1990–1991. "Precolonial African Industry and the Atlantic Trade, 1500–1800." *African Economic History* 19:1–19.

Tilley, H. 2011. *Africa as Living Laboratory: Empire, Development and the Problem of Scientific Knowledge, 1870–1950*. Chicago: University of Chicago Press.

Tousignant, N. 2013. "Insects-as-Infrastructure: Indicating, Project Locustox and the Sahelization of Ecotoxicology." *Science as Culture* 22 (1): 108–131.

Towa, M. 1971. *Léopold Sédar Senghor: Négritude ou servitude*. Yaoundé: Clé.

Towa, M. 1979. *L'idée d'une philosophie africaine*. Yaoundé: Clé.

Turner, J. 1980. *Reckoning with the Beast: Animals, Pain, and Humanity in the Victorian Mind*. Baltimore: Johns Hopkins University Press.

van Onselen, C. 1974. "The 1912 Wankie Colliery Strike." *Journal of African History* 12 (2): 275–289.

van Onselen, C. 1982a. *Studies in the Social and Economic History of the Witwatersrand 1886–1914: Volume I: New Babylon*. Johannesburg: Ravan Press.

van Onselen, C. 1982b. *Studies in the Social and Economic History of the Witwatersrand, 1886–1914: Volume II: New Nineveh*. Johannesburg: Ravan Press.

van Rensburg, S. J. 1981. "Epidemiologic and Dietary Evidence for a Specific Nutritional Predisposition to Esophageal Cancer." *Journal of the National Cancer Institute* 67:243–251.

van Rensburg, S. J., E. S. Bradshaw, D. Bradshaw, and E. F. Rose. 1985. "Oesophageal Cancer in Zulu Men, South African: A Case-Control Study." *British Journal of Cancer* 51:399–405.

Waldman, L. 2007. "When Social Movements Bypass the Poor: Asbestos Pollution, International Litigation and Griqua Cultural Identity." *Journal of Southern African Studies* 3 (3): 577–600.

Walker, B. L. 2010. *Toxic Archipelago: A History of Industrial Disease in Japan*. Seattle: University of Washington Press.

Wallcave, L., S. Bronczyk, and R. Gingell. 1974. "Excreted Metabolites of 1,1,1-trichloro-2,2-bis(p-chlorophenyl)ethane in the Mouse and Hamster." *Journal of Agricultural and Food Chemistry* 22 (5): 904–908.

wa Thiong'o, N. 1985. "The Commitment of the Intellectual." *Review of African Political Economy* 32:18–24.

Wenger, A., and R. Wollenmann, eds. 2007. *Bioterrorism: Confronting a Complex Threat*. Boulder, CO: Lynne Rienner Publishers.

Werbner, R. 1989. *"Regional Cult of God Above" in his Ritual Passage: Sacred Journey*. Manchester: Manchester University Press.

White, L. 1993. *White, Bridging the Zambesi: A Colonial Folly*. London: Macmillan.

White, L. 1995. "Tsetse Visions: Narratives of Blood and Bugs in Colonial Northern Rhodesia, 1931–9." *Journal of African History* 36 (2): 219–245.

Williams, R. 1987. *Fordsony Farmall and Poppin' Johnny: A History of the Farm Tractor and Its Impact on America*. Urbana: University of Illinois Press.

Wood, J. R. T. 2005. "The Pookie: A History of the World's First Successful Landmine Detector Carrier." www.jrtwood.com/article_pookie.asp.

Worby, E. 2002. "Remaking Labour, Reshaping Identity: Cotton, Commoditization and the Culture of Modernity in Northwestern Zimbabwe." PhD diss., McGill University.

Wynne, B. 1996. "May the Sheep Safely Graze? A Reflexive View of the Expert-Lay Knowledge Divide." In *Risk, Environment and Modernity: Towards a New Ecology*, edited by S. Lash, B. Szerszynski, and B. Wynne, 44–83. London: Sage.

Yang, R. Z., and C. S. Tang. 1988. "Plants Used for Pest Control in China: A Literature Review." *Economic Botany* 42 (3): 376–406.

Zegeye, A., and R. Muponde, eds. 2012. "Social Lives of Mobile Telephony." Special issue, *African Identities* 10 (2).

Zhao, Z. 1994. "Paving a Way to Eldorado: Roads, Railways and Political Economy in Matabeleland, 1888–1914." PhD thesis, Yale University.

Index